AF388963

Aquatic Emerging Contaminants and Their Ecotoxicological Consequences

Aquatic Emerging Contaminants and Their Ecotoxicological Consequences

Editors

François Gagné
Stefano Magni
Valerio Matozzo

MDPI • Basel • Beijing • Wuhan • Barcelona • Belgrade • Manchester • Tokyo • Cluj • Tianjin

Editors
François Gagné
Environment and Climate Change Canada
Canada

Stefano Magni
University of Milan
Italy

Valerio Matozzo
University of Padova
Italy

Editorial Office
MDPI
St. Alban-Anlage 66
4052 Basel, Switzerland

This is a reprint of articles from the Topic published online in the open access journals *Toxics* (ISSN 2305-6304), *Journal of Marine Science and Engineering* (ISSN 2077-1312), *Journal of Xenobiotics* (ISSN 2039-4713), and *Water* (ISSN 2073-4441) (available at: https://www.mdpi.com/topics/Aquatic_Contaminants_Ecotoxicological).

For citation purposes, cite each article independently as indicated on the article page online and as indicated below:

LastName, A.A.; LastName, B.B.; LastName, C.C. Article Title. *Journal Name* **Year**, *Volume Number*, Page Range.

ISBN 978-3-0365-6533-0 (Hbk)
ISBN 978-3-0365-6534-7 (PDF)

Cover image courtesy of Stefano Magni

Contents

About the Editors

François Gagné

François Gagné, Ph.D., obtained his bachelor and master degrees in Biochemistry (toxicology) followed by a Ph.D. from the Environmental Sciences Department of Metz University in France in 1996. Since then, he has become a research scientist at Environment and Climate Canada in biochemical ecotoxicology. His research interests reside in the interaction of xenobiotics at the molecular level in living systems and non-linear responses. He was involved in various research programs with non-linear methods in ecotoxicology (fractal behavior of biochemical processes, rhythmic responses to toxics), toxicogenomic analysis for cumulative effects of climate changes, municipal/industrial effluent ecotoxicity, nanotechnology, pharmaceuticals, oil sands and critical elements for technology. François is also involved in various national and international ecotoxicology networks and scientific committees as an advisor and international reviewer. He has published over 250 research papers (Google H index = 59 with over 12,750 citations) and 8 book chapters, given 80 workshops presentations, and recently wrote a book on biochemical methods in ecotoxicology (*Biochemical Ecotoxicology*, Elsevier Inc., 2014). His research work is listed in the top 2% of the most cited scientist in his field.

Stefano Magni

Stefano Magni, Ph.D., is a Researcher of Ecology at the Department of Biosciences of the University of Milan. His research interest concerns the ecotoxicology of emerging aquatic contaminants. After receiving his master's degree in Biology (Biodiversity and Evolutionary Biology), he earned a Ph.D. in Animal Biology at the University of Milan concerning the evaluation of adverse effects of Pharmaceuticals and Personal Care Products (PPCPs) and illicit drugs in freshwater invertebrates. From his post-doc to today, his research concerns the monitoring and the toxicity evaluation of (micro)plastics in freshwater environments and Wastewater Treatment Plants (WWTPs) using different ecotoxicological tools represented by biomarkers of cellular stress, oxidative damage, neurotoxicity and cyto-genotoxicity and "Omics" techniques. In 2017, he was a visiting researcher at the Laboratory of Prof. François Gagné at Environment and Climate Change Canada, Montréal, Canada, regarding the evaluation of the neurotoxic effects of (micro)plastics. He is on the editorial board of the *Journal of Xenobiotics* and has authored and co-authored more than 40 publications in international scientific, peer-reviewed journals and books.

Valerio Matozzo

Valerio Matozzo, Ph.D., is an Associate Professor of Ecology at the University of Padova, where he teaches "Ecology", "Applied Ecology" and "Marine Ecotoxicology". He oversees the Laboratory of Ecophysiology and Ecotoxicology of Marine Invertebrates. His main scientific interest concerns the use of marine invertebrates as bioindicators in both laboratory and field conditions. Particular attention is paid to the evaluation of the effects of emerging contaminants (e.g., nanoparticles, pharmaceuticals, herbicides) on marine species. He is a faculty member of the PhD program in Biosciences (Evolution, Ecology and Conservation curriculum) at the University of Padova. He has participated in several national and international research projects. Prof. Matozzo is a member of the Editorial Boards of the *Journal of Xenobiotics*, the *Journal of Toxicology, ISJ—Invertebrate Survival Journal* and the *Journal of Marine Science and Engineering* and is an Associate Editor of *Frontiers in Physiology* and *Frontiers in Marine Science*. In 2022, he was featured in the list of the top 2% most widely cited scientists—the World's Top 2% Scientists—released by Stanford University. He serves as a reviewer for various national and international research proposals and for more than 50 international scientific journals. He has authored and co-authored more than 110 publications in international scientific, peer-reviewed journals.

Preface to "Aquatic Emerging Contaminants and Their Ecotoxicological Consequences"

This book concerns the collection of 15 scientific articles (13 research papers and 2 reviews), published in *Toxics*, *Waters*, the *Journal of Xenobiotics*, and the *Journal of Marine Science and Engineering* regarding the Topic "Aquatic Emerging Contaminants and Their Ecotoxicological Consequences".

We proposed the above-mentioned topic because of the importance of filling the knowledge gaps, in terms of presence and toxicity, associated with the impact of emerging contaminants in aquatic ecosystems. In support of the Environmental Risk Assessment (ERA), needed for the protection of ecosystems, which concerns the characterization of both the exposure to and the effects of contaminants, the following issues were considered: (i) the research and monitoring of emerging contaminants in both seawaters and freshwaters and (ii) adverse effect evaluation of emerging contaminants at different levels of biological organization and trophic levels, including their uptake measurement in biological organisms.

In this context, emerging contaminants concern a wide plethora of both chemical and physical pollutants not yet considered in regular monitoring programs. This class includes pharmaceuticals and personal care products (PPCPs), illicit drugs, nanomaterials, and micro(nano)plastic polymers. Some evidence related to PPCP monitoring and their effects (specifically referred to Ceftriaxone and Levonorgestrel), as well as the phenomenon of antibiotic resistance in aquatic ecosystems, were considered in this issue. We also included studies investigating the occurrence and effects of nano- and microplastics with their additives (Bisphenol A) and plastic's interaction with other emerging pollutants such as antibiotics (Enrofloxacin). In the plastic context, it is very important to also highlight the potential impact of the *non*-conventional plastic materials, represented, e.g., by tire particles, which massively contribute to the total plastic pollution. For this reason, a paper on the effect of end-of-life tires (ELTs) in the aquatic ecosystem was reported in this book. On the other hand, a new category of emerging pollutants, represented by the so-called ionic liquids, applied in a wide plethora of uses was considered and published in this book with a study on the potential effects of 1-butyl-3-methylimidazolium chloride ([BMIM]Cl).

We also included some studies with existing/legacy contaminants such as heavy metals, polycyclic aromatic hydrocarbons (PAHs), polybromodiphenyl ethers (PBDE), and pesticides, conferring to this volume a more holistic view of (aquatic) pollution.

François Gagné, Stefano Magni, and Valerio Matozzo

Editors

Journal of
Xenobiotics

Review

A Review of Methods for Removal of Ceftriaxone from Wastewater

Petro Karungamye [1,2,*], Anita Rugaika [2], Kelvin Mtei [2] and Revocatus Machunda [2]

[1] Department of Chemistry, The University of Dodoma (UDOM), Dodoma P.O. Box 338, Tanzania
[2] School of Materials Energy Water and Environmental Sciences, The Nelson Mandela African Institution of Science and Technology, Arusha P.O. Box 447, Tanzania; anita.rugaika@nm-aist.ac.tz (A.R.); kelvin.mtei@nm-aist.ac.tz (K.M.); revocatus.machunda@nm-aist.ac.tz (R.M.)
* Correspondence: petrokarungamye@gmail.com or karungamyep@nm-aist.ac.tz; Tel.: +255-763750792

Abstract: The presence of pharmaceuticals in surface water and wastewater poses a threat to public health and has significant effects on the ecosystem. Since most wastewater treatment plants are ineffective at removing molecules efficiently, some pharmaceuticals enter aquatic ecosystems, thus creating issues such as antibiotic resistance and toxicity. This review summarizes the methods used for the removal of ceftriaxone antibiotics from aquatic environments. Ceftriaxone is one of the most commonly prescribed antibiotics in many countries, including Tanzania. Ceftriaxone has been reported to be less or not degraded in traditional wastewater treatment of domestic sewage. This has piqued the interest of researchers in the monitoring and removal of ceftriaxone from wastewater. Its removal from aqueous systems has been studied using a variety of methods which include physical, biological, and chemical processes. As a result, information about ceftriaxone has been gathered from many sources with the searched themes being ceftriaxone in wastewater, ceftriaxone analysis, and ceftriaxone removal or degradation. The methods studied have been highlighted and the opportunities for future research have been described.

Keywords: antibiotics; ceftriaxone; wastewater treatment; degradation; removal

Citation: Karungamye, P.; Rugaika, A.; Mtei, K.; Machunda, R. A Review of Methods for Removal of Ceftriaxone from Wastewater. *J. Xenobiot.* **2022**, *12*, 223–235. https://doi.org/10.3390/jox12030017

Academic Editors: François Gagné, Stefano Magni and Valerio Matozzo

Received: 3 June 2022
Accepted: 27 July 2022
Published: 2 August 2022

Publisher's Note: MDPI stays neutral with regard to jurisdictional claims in published maps and institutional affiliations.

1. Introduction

Pharmaceutical traces, and their metabolites and degradation products have been found in both surface and ground water across the globe [1]. Antibiotics contribute a higher proportion to this in pharmaceutical wastewater [2] due to their significant use [3]. They are used to treat different diseases and bacterial infections in human beings and other animals [4–6]. Wastewater containing such complex components becomes difficult to treat [7,8]. Their high solubility in aqueous systems, longer half-life [9], and low biodegradability [10] makes them accumulate over time. Ceftriaxone (refer Figure 1) is a type of antibiotic used to treat a variety of bacterial illnesses. It is a 3rd generation cephalosporin that inhibits the formation of mucopeptide in bacterial cell walls [11]. Its systematic chemical name is [6R-[6a,7b,(Z)]]-5-thia-1-azabicclo-[4.2.0]-oct-2-ene-2-carboxylicacid,7-[[(2-amino-4-thiazolyl)(methoxyimino)-acetyl]amino]-8-oxo-3-[[(1,2,5,6-tetrahydro-2-methyl-5,6-dioxo-1-2,4–triazin-3-yl)-thio]methyl]]-, disodium salt [12]. It is widely used in clinical settings due to its strong antibacterial effect, good lactamase tolerance, good clinical effect, low toxicity, and low allergic reaction [13].

Like other 3rd generation cephalosporins, this antibiotic is less effective against Gram-positive bacteria compared to first-generation medicines, but it has a far larger spectrum of activity against Gram-negative bacteria [14]. Ceftriaxone has been useful for the treatment of infections caused by susceptible organisms in the lower respiratory tract, abdomen, skin and soft tissue, pelvic area, bone and joint, meninges, and urinary tract [15]. Based on intramuscular injections, ceftriaxone is 100% bioavailable and it is removed by biliary and renal excretion [16].

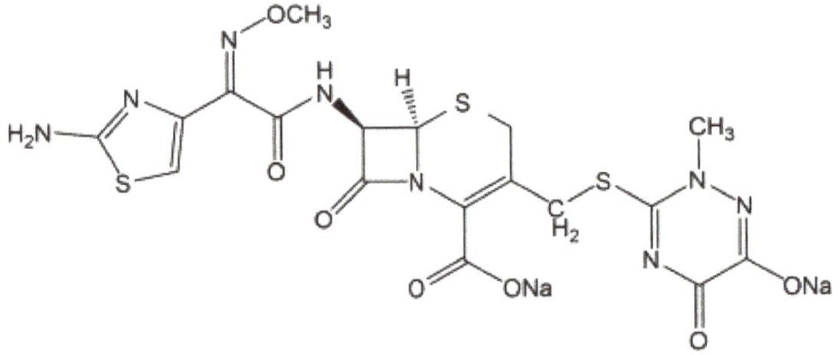

Figure 1. Chemical structure of ceftriaxone [17].

The antibiotics used for animals and humans' treatment enter the environment via urine and feces, thus optimizing and/or limiting antibiotic use, which is essential to minimize contamination of the environment [18]. It is believed that around 40–90% of the prescribed antibiotic dose (depending on the class of pharmaceutical) is excreted as a parent compound in the active form in the feces and urine, and when it finally reaches the environment it causes soil, water, and plant contamination [19–21]. The use of excessive doses of antibiotics in livestock farming can pollute agro-ecosystems through either the application of contaminated manure as fertilizer in agriculture, or the irrigation of farms with wastewater [22,23]. Another source of concern comprises the improper disposal of leftover, expired, or unused pharmaceuticals which are released into sewage systems [24]. Due to the incomplete removal of pharmaceutical compounds and their metabolites by conventional treatment technologies, several pharmaceutical residues have been detected in wastewater effluents. This makes antibiotics present in wastewater treatment plants sludge and, finally, effluent [25–27]. Hospital effluents also comprise a significant source of antibiotics and antibiotic-resistant microorganisms in the environment [28].

Although antibiotic residue quantities in aquatic environments range from ng/L to μg/L, their continual discharge and persistence may have unexpected consequences for non-target aquatic organisms [11]. Antibiotics in water resources can generate a wide range of issues, including toxicity on aquatic organisms such as bacteria, algae, crabs, and fish, and increased antibiotic resistance in bacteria [5,29]. According to the WHO [23], antibiotic resistance is one of the three biggest dangers to human health. As a result, developing effective and environmentally friendly methods to break down those antibiotics in the aquatic environment is critical [2,4,30]. Being an antibiotic, ceftriaxone has similar effects. Due to its widespread usage in medicine and veterinary medicine, ceftriaxone contributes significantly to environmental pollution [9].

Several studies show that ceftriaxone aqueous solution is unstable, with a stability that is pH and temperature-dependent. The ideal pH for ceftriaxone stability in aqueous solution is 7.5, and when maintaining this pH for more than 6 h at 37 °C, only around 10% of ceftriaxone can be degraded. However, degradation occurs more quickly at lower or higher pH levels. The aqueous solution of ceftriaxone is stable for 4 days at room temperature in the presence and absence of light, and that ceftriaxone is stable for a longer amount of time at lower temperatures, but it decomposes after a specific period of time [31]. More characteristics of ceftriaxone are presented in Table 1.

Table 1. Characteristics of ceftriaxone sodium [16].

Characteristics	Value
Physical properties	Crystalline white powder
Solubility	Soluble in water (app. 40 g/100 mL at 25 °C)
Ionization constants (pKa)	4.1 (enolic OH), 3.2 (NH_3^+) and 3 (COOH)
Route of elimination	By glomerular filtration, ceftriaxone is eliminated unaltered in the urine. Bile excretes around 35–45% of a given dosage of ceftriaxone.

Ceftriaxone in wastewater has been reported by various researchers. For instance, research was conducted in India to examine the effluent of selected health care establishments and municipal wastewater treatment plants. The study findings indicated that the results for ceftriaxone ranged from 1.25–29.15 µg/mL [32]. The antibiotics have been proven in several publications as emergent contaminants in the aquatic environments around the world. However, the majority of the findings are from outside Africa [33]. Therefore, the purpose of this literature review was to analyze the information available in relation to the techniques for the removal of ceftriaxone from wastewater systems. The reviewed literature employed electronic databases, manual searches of reference lists from chosen electronic publications, and internet search engines to find relevant literature on the occurrence, concentrations, and techniques used to examine ceftriaxone in wastewater. The expressions ceftriaxone in wastewater, ceftriaxone analysis, and ceftriaxone removal or degradation were searched in Google Scholar, PubMed, Science Direct, Scopus, Taylor & Francis online, Web of Science, and Wiley Online Library. The search was limited to articles written in the English language.

2. Methods Used to Analyze Antibiotics

Various methods have been developed to detect and quantify antibiotics in various types of samples. The referred methods include chromatographic, spectrophotometric, and electrochemical methods [34]. High performance liquid chromatography (HPLC) is, by far, the most extensively utilized instrumental method in pharmaceuticals analysis [35].

2.1. Chromatographic Methods

Pharmaceuticals and their metabolites have been analyzed using a variety of chromatographic methods. Such methods can be used alone or hyphenated with mass spectrometry. Mass spectrometry-based approaches, particularly liquid chromatography, coupled with tandem mass spectrometry (LC/MS/MS) can reach extraordinarily high degrees of specificity compared with immunoassay or even chromatographic detection utilizing detectors such as UV or fluorescence. The specificity and sensitivity of a chromatographic method are controlled by chromatographic conditions such as choice of mobile phase and analytical column, detector, and sample preparation [36]. For antibiotic analysis, the analytical method is selected based on the characteristics of the analyzed antibiotic, which includes solubility in water and organic solvents or acid-base properties [37].

Thin-layer chromatography (TLC) is one of the most important analytical methods used to determine the qualitative and semiquantitative levels of pharmaceuticals in various types of samples [34]. TLC is usually applied as the quick, easy, and straightforward procedure. The effective separation is determined by the sample's properties as well as the properties of the stationary and mobile phases [38]. TLC can successfully be used for preliminary screening of the pharmaceutical compounds. It is commonly employed in contemporary analysis as a separation method to determine the presence or absence of antibiotics over a predetermined concentration level [38,39]. It can also be used to evaluate and categorize pure and impure antibiotic preparations as well as assay antibiotics quantitatively in bulk or pharmaceutical preparations [40,41].

Gas chromatography (GC) is a commonly used analytical technique that combines separation chromatographic stage with measurement capacity. GC employs the gas as the mobile phase and coating inside the long capillary column or, less typically, the tiny particles of a solid material packed in a column as the stationary phase. The sample in GC should be able to evaporate so that it flows with the gaseous mobile phase. The temperature gradient to which the chromatographic column is subjected is frequently utilized to speed up the elution of less volatile substances that would otherwise take a long time to elute. The detector signals for the sample's eluting components are used for quantitative and qualitative analysis [42]. GC is a useful technique for evaluating pharmaceutically relevant substances [43] and impurities [44]. Many pharmaceutical chemicals, however, cannot be gas chromatographed in their natural state and must be transformed into stable and volatile derivatives in order to accomplish successful GC elution and separation. The derivatives are sometimes created in order to attain the appropriate sensitivity, selectivity, or specificity for a given separation [44,45].

High-performance liquid chromatography (HPLC) is a chromatographic technique that can separate a mixture of substances, and it is used in biochemistry and analytical chemistry to identify, quantify, and purify different components of the mixture [46]. HPLC employs various types of the stationary phases, and the pump that drives the mobile phase(s) and analyte through the column and detector to provide a characteristic retention time for the analyte. The retention period of an analyte varies according to the strength of its interactions with the stationary phase, solvent(s) ratio/composition utilized, and flow rate of the mobile phase [47]. HPLC has a number of advantages, including low organic solvent utilization, minimal sample volume, quick analysis, and high chromatographic resolution [48]. Apart from conventional HPLC, other sophisticated HPLC-based techniques have been widely applied for the determination of pharmaceuticals, including antibiotics in various samples. The referred methods include liquid chromatography—mass spectrometry (LC-MS) [49,50], ultra-high performance liquid chromatography-MS/MS (UHPLC-MS/MS) [51–53], and liquid chromatography linked to tandem mass spectrometry (LC-MS/MS) [54,55].

2.2. Spectrophotometric Methods

Spectrophotometric methods are based on the creation of the complex between the pharmaceutical and the reagent [12]. The intensity of the color is used to calculate pharmaceutical concentration. The complex generated by the pharmaceutical and reagent can either be charge transfer or ion-pair in nature. The charge transfer complex, also known as the electron donor-acceptor complex, transfers a fraction of electrical charge between molecules. Coulomb attraction holds oppositely charged ions together in solution in the ion-pair complex [56]. Some antibiotics have been analyzed using spectrophotometric methods including amoxicillin [40], azithromycin [41,57], tetracycline, doxycycline [58], and cefixime trihydrate [59]. They have also been used to analyze gentamicin sulfate [60], cefadroxil, ceftazidime, cefazolin sodium, cefoperazone sodium, cefaclor, cephaprin sodium, cefotaxime sodium, and cefuroxime sodium [61].

2.3. Electrochemical Methods

The measurement of the current, charge, and potential is utilized in electrochemical techniques to characterize an analyte's chemical reactivity and detect the concentration. The basic electrochemical signals that serve as analytical signals constitute current, charge, and potential [62]. These techniques include cyclic voltammetry, chronoamperometry, electro-chemical impedance spectroscopy, and potentiometry [63]. In comparison to separation and spectral methods, electrochemical methods offer practical advantages such as operation simplicity, satisfactory sensitivity, a wide linear concentration range, low instrument cost, miniaturization capability, suitability for real-time detection, and less sensitivity to matrix effects [34,64]. Due to advances in electronics and computer sciences, the electroanalysis of pharmaceutically active substances is actively involved in new study fields of various methodologies. Due to their great sensitivity and selectivity, many innovative electroanalytical techniques have been effectively employed for trace analyses of essential pharmaceutically active substances [65]. The electrochemical analysis of active pharmaceuticals is based on redox processes that occur via electron transfer channels [66]. Electrochemical methods have been used for the analysis of antibiotics such as clarithromycin and azithromycin [67], diclofenac [68], and cefixime [69].

2.4. Methods Studied for Analysis of Ceftriaxone in Aquatic and Biological Samples

Ceftriaxone levels have been estimated using a variety of techniques including HPLC, high performance thin layer chromatography, capillary electrophoresis, and spectrophotometry [37,70]. Literature shows a higher proportion of the usage of HPLC in the analysis of ceftriaxone in the aqueous and biological samples [14]. The methods studied for the analysis of ceftriaxone include high-performance liquid chromatography coupled with mass spectrometry detection (HPLC-MS) [14,71,72], high-performance liquid chromatography with detection by ultraviolet (HPLC-UV) [14,73–76], and high-performance liquid chromatography coupled with sequential mass spectrometry (HPLC-MS/MS) [14,77–81]. The referred methods studied for the analysis of ceftriaxone also include ultra-performance liquid chromatography with detection by ultraviolet (UPLC-UV) [14,82] and ultra-performance liquid chromatography coupled with sequential mass spectrometry (UPLC-MS/MS) [49,83,84]. The linear range, limit of detection, and recovery of these methods are given in Table 2.

Table 2. Comparison of chromatographic methods used for determination of ceftriaxone [85].

Type of Technique	Sample Used	Limit of Detection (μgL^{-1})	Range of Linearity (μgL^{-1})	% Recovery
HPLC-UV	Hospital wastewater	2.0	5.0–600	152.38
HPLC-MS/MS	Human plasma		3.0–300	87.35
HPLC	Sterile powder for injection		20–150	99.42
HPLC	Human urine	0.05	0.24–250	97.73–100.7
RP-HPLC	Pharmaceutical formulation	0.51–1.54	2.5–25	>98.1

Absorption spectroscopy methods such as ultraviolet (UV) [14,86–90], infrared spectroscopy [14,88,90–92], spectrofluorimetry [14,93], microbiological methods [14,94,95], and capillary zone electrophoresis [96] have also been used for the analysis of ceftriaxone. When used as an identification technique, UV has limited selectivity because multiple compounds may have the same or similar spectra. As a result, this technique is typically supplemented with additional spectroscopic techniques such as IR for positive analyte confirmation [35].

3. Methods Used for Removal of Antibiotics from Wastewater

The selection of the method for wastewater treatment depends on the characteristics of the wastewater and features such as costs, feasibility, efficiency, practicability, dependability, impact on the environment, sludge production, difficulty in operation, pretreatment demands, and the formation of potentially dangerous by-products which characterize the relevant method [97]. The potential of various techniques to remove antibiotics from wastewater systems has been investigated. Among those techniques are constructed wetlands, biological treatment, advanced oxidation processes (AOPs), and membrane technology [23].

3.1. Constructed Wetland

A constructed wetland (CW) wastewater treatment system utilizes the combined influence of microbes, plants, and soil to remove the pollutants from wastewater. The wastewater is treated through microbial decomposition, adsorption, plant uptake, ion exchange, co-precipitation, and filtration [98]. The suitability of CWs for the elimination of some pharmaceuticals and personal care products (PPCPs) has recently been studied [26].

Diclofenac, ibuprofen, naproxen, ketoprofen, salicylic acid, triclosan, sulfamethoxazole, carbamazepine, clofibric acid, atenolol, and caffeine are some of the pharmaceuticals that have been investigated in constructed wetlands [99,100]. The average removing efficiencies of constructed wetlands are 93% (monensin), 89% (ofloxacin), 87% (oxytetracycline), 83% (sulfapyridine), 80% (caffeine), 79% (salicylic acid), 72% (atenolol), 72% (furosemide), 69% (doxycycline), 68% (codeine), 67% (diltiazem), 64% (acetaminophen), 62% (naproxen), 57% (ibuprofen), 56% (metoprolol), and 51% (sulfadiazine) to some studied pharmaceuticals [101]. Several studies have shown that physico-chemical decomposition, photodegradation, adsorption by wetland soil and plants, and biodegradation (microbial activity) comprise the mechanisms used to remove antibiotics from wastewater in CWs [67,68]. Antibiotics can accumulate in plants by water transport and passive absorption and high quantities of antibiotics in water or soil can be harmful to plant development and metabolic activity [102]. Since there are very few informative publications on the decontamination of antibiotics using CWs, this area of research could benefit from combined support from other disciplines, primarily soil science, botany, environmental chemistry, and chemical engineering [103].

3.2. Biological Treatment

The microorganisms utilize organic compounds and nutrients to gain energy and build the blocks for their growth in biological treatment methods. Despite the presence of high density and diverse consortium of microorganisms in activated sludge, antibiotics cannot be completely removed in biological treatment methods [104]. Some reasons for the incomplete removal of antibiotics in biological methods include relatively low concentration of antibiotics in the wastewater, which leads to a lack of enzymes responsible for antibiotic biodegradation and inhibitory or toxic properties of antibiotics that can stop the microorganism activity responsible for antibiotic biodegradation, antibiotic properties, and operation conditions [18]. Different biological treatment methods have been investigated in relation to the removal of antibiotics from wastewater. For instance, using a biological aerated filter system (BAF), 89–91% of nine antibiotics were removed from swine wastewater. Those antibiotics include oxytetracycline, leucomycin, lincomycin, ofloxacin, trimethoprim, norfloxacin, sulfamonomethoxine, sulfamethazine, and sulfachloropyridazine [105]. Using anaerobic digestion, 65% tetracyclines and 85% of quinolones were removed from swine wastewater after 16d hydraulic retention time (HRT) [106].

Another study indicated that the lab-scale intermittently aerated sequencing batch reactor (IASBR) was applied to treat anaerobically digested swine wastewater. The results from the referred study show that 87.9% tetracyclines were removed, and 96.2% sulfonamides were removed at about 3–5 d HRT [107]. The elimination of antibiotics using the sequencing-batch membrane bioreactor (SMBR) was investigated for the treatment of swine wastewater. Nine antibiotics, which were divided into sulfonamides, tetracyclines, and fluoroquinolones, and three categories of frequently used veterinary antibiotics were investigated. The results demonstrated that SMBR effectively removed sulfonamides and tetracyclines (90%), whereas fluoroquinolones were removed less effectively (70%) [108]. Many antibiotics have been identified in the literature as being resistant to biodegradation. While some antibiotics can be partially decomposed, the majority of antibiotics including ciprofloxacin, metronidazole, ceftriaxone, ofloxacin, and trimethoprim are not biodegradable [73,74]. More research is needed to understand the factors affecting the process and possibility of improving the degradation of pharmaceuticals.

3.3. Advanced Oxidation Processes (AOP)

AOPs comprise water and wastewater treatment technologies that use powerful oxidizing agents such as hydroxyl radical (OH•), ozone (O_3), chloride (Cl^-), and superoxide radical (O_2^-) [109]. The generated species react with the medium's organic molecules [110] to start a series of oxidation reactions until all of the components have been mineralized to CO_2 and H_2O [111]. AOP methods can be divided according to the source of OH• production with UV–hydrogen peroxide processes, with Fenton and photo-Fenton, ozone-based processes, photocatalysis, and sonolysis being the most common [112]. Such methods have proven to be effective at removing a wide range of contaminants in general and antibiotics in particular [110]. Electrochemical oxidation was used to study the removal of tetracycline (TC) antibiotics from the livestock wastewater. The electrochemical treatment of the TC in aqueous solutions for 6 h with a Ti/IrO_2 anode and Na_2SO_4 electrolyte resulted in concentrations decreasing from $100\ mgL^{-1}$ to less than $0.6\ mgL^{-1}$ [113].

With sinusoidal alternating electro-Fenton (SAEF), the removal efficiency and the mechanism of TC degradation were studied. According to the findings, the removal rates of TC were 94.87% in optimal conditions [114]. A study was done to examine the efficacy of three AOPs for removing antibiotics from wastewater: ozonation, photo-Fenton process, and heterogeneous photocatalytic process with a TiO_2 semiconductor. The ozonation process was discovered to be effective at removing all types of antibiotics [115]. The majority of the literature to date, however, has been devoted to bench- or pilot-scale experiments.

The use of AOPs on a large scale is still a work in progress. The high operational cost of AOPs, especially when compared to the conventional methods that are routinely used today, is likely to be the greatest challenge for the development of AOPs on an industrial scale [110]. Further research is needed to address the challenges associated with AOPs in attempt to make the processes affordable and useful in the real wastewater treatments.

3.4. Membrane Technology

A membrane is described as a thin layer, film, or sheet that serves as a specific barrier between two phases which may be vapor, gas, or liquid. To put it in another way, a membrane is the boundary between two adjacent phases that function as a selective barrier to control the movement of species between the two compartments. Membrane technology includes the associated engineering and scientific techniques for transporting or excluding the parts, species, or substances from membranes [116]. Ultrafiltration (UF), electrodialysis (ED), membrane distillation (MD), microfiltration (MF), nanofiltration (NF), particle filtration (PF), pervaporation (PV), reverse osmosis (RO), and membrane bioreactor (MBR) are just a few of the membrane-based technologies that have been developed based on the impurities that need to be removed and the method of application [117,118].

Various membrane technologies have been evaluated for pharmaceutical removal at both the pilot and full-scale levels [119]. The membrane technology is preferred due to significant reductions in equipment size, energy requirements, and low capital costs. It has the potential to close the economic and sustainability gap with low or no chemical usage, environmental friendliness, and ease of access for many [120]. A few studies have investigated the removal of antibiotics from wastewater using membrane technology. For instance, one study on wastewater treatment indicate that the rate of antibiotic removal was 87% when UV/ozone and nanofiltration were used [121]. The combination of nanofiltration and reverse osmosis technologies was utilized to treat swine wastewater and efficiently removed various antibiotic resistant genes [122]. As a conclusion, additional research on the use of membrane technology to remove antibiotics from wastewater should be done.

4. Methods Studied for Removal of Ceftriaxone from Water and Wastewater

The techniques studied regarding the removal of ceftriaxone from aqueous systems include photochemical degradation, ion ex-change, chemical oxidation, biological treatment, and adsorption [123]. Table 3 summarizes some of the studies on the methods for the removal of ceftriaxone from wastewater.

Table 3. Methods for removal of ceftriaxone from aqueous solution.

Method	Results	Reference
Chemical oxidation	Degradation occurs through Type I and Type II mechanisms.	[124]
UVC/H_2O_2 and UVC	At a solution pH of 5 and an H_2O_2 concentration of 10 mg/L, the most ceftriaxone degradation was observed. Pseudo-first- and second-order kinetics models with reaction rate constants of 0.0165 and 0.0012 min^{-1}, respectively, better represent UVC/H_2O_2 and UVC processes.	[11]
O_3/UV/Fe_3O_4@TiO_2	Maximum ceftriaxone removal 92.40% Organic carbon reduction 72.5% Optimal conditions, time: 30 min, photocatalyst dosage: 2 g/L, pH: 9, initial ceftriaxone concentration: 10 mg/L, and ozone dosage: 0.2 g/h)	[125]
Immobilized TiO_2 and ZnO	Results revealed that photodegradation using UV/TiO_2 process was more effective than photodegradation using the UV/ZnO process. Ceftriaxone photodegradation followed pseudo-first-order kinetics in both systems.	[126]
Electrochemical in aqueous solutions containing sodium halides	Ceftriaxone gradually decomposes, but not fully, in the presence of fluoride ions in about 60 min without yielding a reaction product. The electro (degradation/transformation) of ceftriaxone is practically complete in 10 and 5 min with completion of the electro-transformation reaction, which take 60 and 30 min, respectively. Ceftriaxone and the iodide ions formed instantaneous interactions.	[127]
Heterogeneous catalytic AOP γ-Fe_2O_3 encapsulated NaY zeolites solid adsorbent	initial concentration of 20 mg/L, catalyst 1.17 g/L, H_2O_2 30 mM, and UV light, ceftriaxone may be effectively removed within 90 min at pH 4.0. The adsorption mechanism was investigated using the kinetic and isotherm model, and the results demonstrate that the model and data are in good agreement.	[128]

5. Conclusions

Despite the fact that ceftriaxone is one of the most commonly prescribed antibiotics in health facilities, this review demonstrates that there is little information on its occurrence in the environmental samples. Due to potential consequences of their presence in the environment, it is necessary to examine and monitor their presence. The majority of studies on the strategies for the degradation or removal of ceftriaxone from various samples are based on AOPs. The most significant disadvantage of the AOP methods is their expense, which comprise the operating and maintenance costs associated with the system's needs for energy and chemical reagents. Despite the evidence that some approaches such as biological procedures cannot remove ceftriaxone, further research is needed to study the possibilities of other alternatives such as constructed wetland systems. The majority of the reviewed studies were conducted on a small scale in the laboratory under controlled environments. Alternative research is required to determine the feasibility and effectiveness of the techniques for degrading ceftriaxone in wastewater by involving the complex mixtures of contaminants and variations in weather conditions.

Author Contributions: Conceptualization, P.K., A.R. and K.M.; methodology, R.M.; resources, K.M. and R.M.; writing—original draft preparation, P.K.; writing—review and editing, P.K., A.R., K.M. and R.M.; supervision, A.R., K.M. and R.M. All authors have read and agreed to the published version of the manuscript.

Funding: This research received no external funding.

Institutional Review Board Statement: Not applicable.

Informed Consent Statement: Not applicable.

Data Availability Statement: Not applicable.

Conflicts of Interest: The authors declare no conflict of interest.

References

1. Shi, X.; Karachi, A.; Hosseini, M.; Yazd, M.S.; Kamyab, H.; Ebrahimi, M.; Parsaee, Z. Ultrasound wave assisted removal of Ceftriaxone sodium in aqueous media with novel nano composite g-C3N4/MWCNT/Bi2WO6 based on CCD-RSM model. *Ultrason. Sonochemistry* **2019**, *68*, 104460. [CrossRef] [PubMed]
2. Zhao, Y.; Liang, X.; Shi, H.; Wang, Y.; Ren, Y.; Liu, E.; Zhang, X.; Fan, J.; Hu, X. Photocatalytic activity enhanced by synergistic effects of nano-silver and ZnSe quantum dots co-loaded with bulk g-C3N4 for Ceftriaxone sodium degradation in aquatic environment. *Chem. Eng. J.* **2018**, *353*, 56–68. [CrossRef]
3. Kordestani, B.; Yengejeh, R.J.; Takdastan, A.; Neisi, A. A new study on photocatalytic degradation of meropenem and ceftriaxone antibiotics based on sulfate radicals: Influential factors, biodegradability, mineralization approach. *Microchem. J.* **2019**, *146*, 286–292. [CrossRef]
4. Amiri, S.; Sohrabi, M.R.; Motiee, F. Optimization Removal of the Ceftriaxone Drug from Aqueous Media with Novel Zero-Valent Iron Supported on Doped Strontium Hexaferrite Nanoparticles by Response Surface Methodology. *ChemistrySelect* **2020**, *5*, 5831–5840. [CrossRef]
5. Kordestani, B.; Takdastan, A.; Yengejeh, R.J.; Neisi, A. Photo-Fenton oxidative of pharmaceutical wastewater containing meropenem and ceftriaxone antibiotics: Influential factors, feasibility, and biodegradability studies. *Toxin Rev.* **2018**, *39*, 292–302. [CrossRef]
6. Kaur, B.; Kuntus, L.; Tikker, P.; Kattel, E.; Trapido, M.; Dulova, N. Photo-induced oxidation of ceftriaxone by persulfate in the presence of iron oxides. *Sci. Total Environ.* **2019**, *676*, 165–175. [CrossRef]
7. Thalji, M.R. Nanotechnologies for Removal of Pharmaceuticals from Wastewater. *Med. Pharm. Sci.* **2021**, *1*, 25–28.
8. Puddoo, H.; Nithyanandam, R.; Nguyenhuynh, T.; Taylor's University Malaysia. Degradation of the Antibiotic Ceftriaxone by Fenton Oxidation Process and Compound Analysis. *J. Phys. Sci.* **2017**, *28*, 95–114. [CrossRef]
9. AttariKhasraghi, N.; Zare, K.; Mehrizad, A.; Modirshahla, N.; Behnajady, M.A. Achieving the Enhanced Photocatalytic Degradation of Ceftriaxone Sodium Using CdS-g-C3N4 Nanocomposite under Visible Light Irradiation: RSM Modeling and Optimization. *J. Inorg. Organomet. Polym. Mater.* **2021**, *31*, 3164–3174. [CrossRef]
10. Mahmoud, M.E.; El-Ghanam, A.M.; Mohamed, R.H.A.; Saad, S.R. Enhanced adsorption of Levofloxacin and Ceftriaxone antibiotics from water by assembled composite of nanotitanium oxide/chitosan/nano-bentonite. *Mater. Sci. Eng. C* **2019**, *108*, 110199. [CrossRef]
11. Khorsandi, H.; Teymori, M.; Aghapour, A.A.; Jafari, S.J.; Taghipour, S.; Bargeshadi, R. Photodegradation of ceftriaxone in aqueous solution by using UVC and UVC/H2O2 oxidation processes. *Appl. Water Sci.* **2019**, *9*, 81. [CrossRef]

12. Owens, H.M.; Dash, A.K. Ceftriaxone Sodium: Comprehensive Profile. *Profiles Drug Subst. Excip. Relat. Methodol.* **2003**, *30*, 21–57. [CrossRef]

13. Zhang, P.O.; Zhang, L.; Ma, C. Degradation of Ceftriaxone Sodium in Pharmaceutical Wastewater by Photocatalytic Oxidation. In Proceedings of the 2020 2nd World Congress on Chemistry, Biotechnology and Medicine (WCCBM 2020), Zurich, Switzerland, 6–8 April 2020; pp. 217–221. [CrossRef]

14. Da Trindade, M.T.; Salgado, H.R.N. A Critical Review of Analytical Methods for Determination of Ceftriaxone Sodium. *Crit. Rev. Anal. Chem.* **2018**, *48*, 95–101. [CrossRef] [PubMed]

15. Rohimmahtunnissa, A.; Alfan, D.A.; Firdayani, A.S. The Influence Study of the Mole Ratio Reactant in Ceftriaxone Sodium Synthesis Against The Yield of The Production. *Int. J. Innov. Eng. Sci. Res.* **2018**, *2*, 6. [CrossRef]

16. Scholar, E. Ceftriaxone. In *xPharm: The Comprehensive Pharmacology Reference*; Elsevier: Amsterdam, The Netherlands, 2007.

17. Rebec, G.V. Vitamin C and Glutamate Uptake: Implications for Huntington's Disease. In *Diet and Nutrition in Dementia and Cognitive Decline*; Academic Press: San Diego, CA, USA; Elsevier, Inc.: San Diego, CA, USA, 2015.

18. Cetecioglu, Z.; Atasoy, M. Biodegradation and Inhibitory Effects of Antibiotics on Biological Wastewater Treatment Systems. In *Toxicity and Biodegradation Testing. Methods in Pharmacology and Toxicology*; Humana Press: New York, NY, USA, 2018; pp. 29–55. [CrossRef]

19. Muriuki, C.W.; Home, P.G.; Raude, J.M.; Ngumba, E.K.; Munala, G.K.; Kairigo, P.K.; Gachanja, A.N.; Tuhkanen, T.A. Occurrence, distribution, and risk assessment of pharmerciuticals in wastewater and open surface drains of peri-urban areas: Case study of Juja town, Kenya. *Environ. Pollut.* **2020**, *267*, 115503. [CrossRef] [PubMed]

20. Rodriguez-Mozaz, S.; Vaz-Moreira, I.; Della Giustina, S.V.; Llorca, M.; Barceló, D.; Schubert, S.; Berendonk, T.U.; Michael-Kordatou, I.; Fatta-Kassinos, D.; Martinez, J.L.; et al. Antibiotic residues in final effluents of European wastewater treatment plants and their impact on the aquatic environment. *Environ. Int.* **2020**, *140*, 105733. [CrossRef] [PubMed]

21. Polianciuc, S.I.; Gurzău, A.E.; Kiss, B.; Ștefan, M.G.; Loghin, F. Antibiotics in the environment: Causes and consequences. *Med. Pharm. Rep.* **2020**, *93*, 231–240. [CrossRef] [PubMed]

22. Wajahat, R.; Yasar, A.; Khan, A.M.; Tabinda, A.B.; Bhatti, S.G. Ozonation and Photo-Driven Oxidation of Ciprofloxacin in Pharmaceutical Wastewater: Degradation Kinetics and Energy Requirements. *Pol. J. Environ. Stud.* **2019**, *28*, 1933–1938. [CrossRef]

23. Huang, A.; Yan, M.; Lin, J.; Xu, L.; Gong, H.; Gong, H. A Review of Processes for Removing Antibiotics from Breeding Wastewater. *Int. J. Environ. Res. Public Health* **2021**, *18*, 4909. [CrossRef]

24. Ncube, S.; Nuapia, Y.B.; Chimuka, L.; Madikizela, L.M.; Etale, A. Trace Detection and Quantitation of Antibiotics in a South African Stream Receiving Wastewater Effluents and Municipal Dumpsite Leachates. *Front. Environ. Sci.* **2021**, *9*, 365. [CrossRef]

25. Nantaba, F.; Wasswa, J.; Kylin, H.; Palm, W.-U.; Bouwman, H.; Kümmerer, K. Occurrence, distribution, and ecotoxicological risk assessment of selected pharmaceutical compounds in water from Lake Victoria, Uganda. *Chemosphere* **2019**, *239*, 124642. [CrossRef] [PubMed]

26. Berglund, B.; Khan, G.A.; Weisner, S.E.; Ehde, P.M.; Fick, J.; Lindgren, P.-E. Efficient removal of antibiotics in surface-flow constructed wetlands, with no observed impact on antibiotic resistance genes. *Sci. Total Environ.* **2014**, *476–477*, 29–37. [CrossRef] [PubMed]

27. Moreira, F.C.; Soler, J.; Alpendurada, M.; Boaventura, R.A.; Brillas, E.; Vilar, V.J. Tertiary treatment of a municipal wastewater toward pharmaceuticals removal by chemical and electrochemical advanced oxidation processes. *Water Res.* **2016**, *105*, 251–263. [CrossRef] [PubMed]

28. Lien, L.T.Q.; Hoa, N.Q.; Chuc, N.T.K.; Thoa, N.T.M.; Phuc, H.D.; Diwan, V.; Dat, N.T.; Tamhankar, A.J.; Lundborg, C.S. Antibiotics in Wastewater of a Rural and an Urban Hospital before and after Wastewater Treatment, and the Relationship with Antibiotic Use—A One Year Study from Vietnam. *Int. J. Environ. Res. Public Health* **2016**, *13*, 588. [CrossRef] [PubMed]

29. Yang, Q.; Gao, Y.; Ke, J.; Show, P.L.; Ge, Y.; Liu, Y.; Guo, R.; Chen, J. Antibiotics: An overview on the environmental occurrence, toxicity, degradation, and removal methods. *Bioengineered* **2021**, *12*, 7376–7416. [CrossRef]

30. Mahdavi, H.; Bagherifar, R. Cellulose acetate/SiO_2-poly(2-Acrylamido-2-methylpropane sulfonic acid) hybrid nanofiltration membrane: Application in removal of ceftriaxone sodium. *J. Iran. Chem. Soc.* **2018**, *15*, 2839–2849. [CrossRef]

31. Abramović, B.F.; Uzelac, M.M.; Finčur, N.L. Photocatalytic degradation of thiotriazinone, stable hydrolysis product of antibiotic ceftriaxone. *Acta Period. Technol.* **2019**, *50*, 1–11. [CrossRef]

32. Shipingana, L.N.N.; Shivaraju, H.P.; Yashas, S.R. Quantitative assessment of pharmaceutical drugs in a municipal wastewater and overview of associated risks. *Appl. Water Sci.* **2022**, *12*, 16. [CrossRef]

33. Faleye, A.; Adegoke, A.A.; Ramluckan, K.; Bux, F.; Stenström, T.A. Antibiotic Residue in the Aquatic Environment: Status in Africa. *Open Chem.* **2018**, *16*, 890–903. [CrossRef]

34. Bekele, L.K.; Gebeyehu, G.G. Application of Different Analytical Techniques and Microbiological Assays for the Analysis of Macrolide Antibiotics from Pharmaceutical Dosage Forms and Biological Matrices. *ISRN Anal. Chem.* **2012**, *2012*, 859473. [CrossRef]

35. Davani, B. *Pharmaceutical Analysis for Small Molecules*; John Wiley & Sons, Inc.: Hoboken, NJ, USA, 2017.

36. Dasgupta, A.; Krasowski, M.D. Application of chromatographic techniques for therapeutic drug monitoring. In *Therapeutic Drug Monitoring Data*; Academic Press: London, UK, 2020; pp. 53–63. [CrossRef]

37. Pauter, K.; Szultka-Młyńska, M.; Buszewski, B. Determination and Identification of Antibiotic Drugs and Bacterial Strains in Biological Samples. *Molecules* **2020**, *25*, 2556. [CrossRef] [PubMed]

38. Raeisi, A.; Ramezani, M.; Ravazadeh, H.; Taher, M.A. Chromatographic Behaviour of Antibiotics on Thin Layers of Zeolite. *J. Pharm. Res. Int.* **2019**, *30*, 1–8. [CrossRef]

39. Hancu, G.; Simon, B.; Kelemen, H.; Rusu, A.; Mircia, E.; Gyéresi, Á. Thin Layer Chromatographic Analysis of Beta-Lactam Antibiotics. *Adv. Pharm. Bull.* **2013**, *3*, 367–371. [CrossRef] [PubMed]

40. Asan, A.; Seddiq, N. A Simple Spectrophotometric Determination of Amoxicillin in Drug samples. *J. Turk. Chem. Soc.* **2022**, *9*, 423–432. [CrossRef]

41. De Paula, C.E.R.; Almeida, V.G.K.; Cassella, R.J. Determinação espectrofotométrica de cefalexina em formulações farmacêuticas explorando a sua reação de transferência de carga com a quinalizarina. *Quim. Nova* **2010**, *33*, 914–919. [CrossRef]

42. Moldoveanu, S.C.; David, V. Short Overviews of the Main Analytical Techniques Containing a Separation Step. In *Selection of the HPLC Method in Chemical Analysis*; Elsevier Inc.: Amsterdam, The Netherlands, 2017; pp. 55–85. [CrossRef]

43. Chew, Y.-L.; Khor, M.-A.; Lim, Y.-Y. Choices of chromatographic methods as stability indicating assays for pharmaceutical products: A review. *Heliyon* **2021**, *7*, e06553. [CrossRef] [PubMed]

44. Parys, W.; Dołowy, M.; Pyka-Pająk, A. Significance of Chromatographic Techniques in Pharmaceutical Analysis. *Processes* **2022**, *10*, 172. [CrossRef]

45. Ahuja, S. Derivatization in gas chromatography. *J. Pharm. Sci.* **1976**, *65*, 163–182. [CrossRef] [PubMed]

46. Lozano-Sánchez, J.; Borrás-Linares, I.; Sass-Kiss, A.; Segura-Carretero, A. Chromatographic Technique: High-Performance Liquid Chromatography (HPLC). In *Modern Techniques for Food Authentication*; Academic Press: London, UK, 2018; pp. 459–526.

47. Jena, A.K.A. HPLC: Highly Accessible Instrument in Pharmaceutical Industry for Effective Method Development. *Pharm. Anal. Acta* **2012**, *3*, 1–9. [CrossRef]

48. Locatelli, M.; Melucci, D.; Carlucci, G.; Locatelli, C. Recent hplc strategies to improve sensitivity and selectivity for the analysis of complex matrices. *Instrum. Sci. Technol.* **2012**, *40*, 112–137. [CrossRef]

49. Gallagher, T.; Riedel, S.; Kapcia, J.; Caverly, L.J.; Carmody, L.; Kalikin, L.M.; Lu, J.; Phan, J.; Gargus, M.; Kagawa, M.; et al. Liquid Chromatography Mass Spectrometry Detection of Antibiotic Agents in Sputum from Persons with Cystic Fibrosis. *Antimicrob. Agents Chemother.* **2021**, *65*, 2. [CrossRef] [PubMed]

50. Usman, M.R.; Prasasti, A.; Islamiah, S.; Firdaus, A.N.; Marita, A.W.; Fajriyah, S.; Yanti, E.F. Ceftriaxone Degradation by Titanium Dioxide (TiO2) Nanoparticles: Toxicity and Degradation Mechanism. *J. Kim. Val.* **2020**, *6*, 82–89. [CrossRef]

51. Nováková, L.; Svoboda, P.; Pavlík, J. Ultra-high performance liquid chromatography. In *Liquid Chromatography: Fundamentals and Instrumentation*, 2nd ed.; Elsevier: Amsterdam, The Netherlands, 2017; Volume 1, pp. 719–769.

52. Zhang, A.-H.; Wang, P.; Sun, H.; Yan, G.-L.; Han, Y.; Wang, X.-J. High-throughput ultra-performance liquid chromatography-mass spectrometry characterization of metabolites guided by a bioinformatics program. *Mol. BioSyst.* **2013**, *9*, 2259–2265. [CrossRef] [PubMed]

53. Rathod, R.H.; Chaudhari, S.R.; Patil, A.S.; Shirkhedkar, A.A. Ultra-high performance liquid chromatography-MS/MS (UHPLC-MS/MS) in practice: Analysis of drugs and pharmaceutical formulations. *Futur. J. Pharm. Sci.* **2019**, *5*, 6. [CrossRef]

54. Zhang, Y.; Li, X.Q.; Li, H.M.; Zhang, Q.H.; Gao, Y.; Li, X.J. Antibiotic residues in honey: A review on analytical methods by liquid chromatography tandem mass spectrometry. *TrAC Trends Anal. Chem.* **2018**, *110*, 344–356. [CrossRef]

55. Chen, F.; Cheng, Z.; Peng, Y.; Wang, Z.; Huang, C.; Liu, D.; Wang, B.; Pan, B.; Guo, W. A liquid chromatography-tandem mass spectrometry (LC-MS/MS)-based assay for simultaneous quantification of aldosterone, renin activity, and angiotensin II in human plasma. *J. Chromatogr. B* **2021**, *1179*, 122740. [CrossRef]

56. Keskar, M.R.; Jugade, R.M. Spectrophotometric Investigations of Macrolide Antibiotics: A Brief Review. *Anal. Chem. Insights* **2015**, *10*, ACI.S31857. [CrossRef]

57. Rachidi, M.; Elharti, J.; Digua, K.; Cherrah, Y.; Bouklouze, A. New Spectrophotometric Method for Azithromycin Determination. *Anal. Lett.* **2006**, *39*, 1917–1926. [CrossRef]

58. Rufino, J.L.; Fernandes, F.C.B.; Ruy, M.S.; Pezza, H.R.; Pezza, L. A simple spectrophotometric method for the determination of tetracycline and doxycycline in pharmaceutical formulations using chloramine-t. *Eclet. Quim.* **2018**, *35*, 139–145. [CrossRef]

59. Keskar, M.R.; Jugade, R.M. Spectrophotometric Determination of Cefixime Trihydrate in Pharmaceutical Formulations Based on Ion-Pair Reaction with Bromophenol Blue. *Anal. Chem. Insights* **2015**, *10*, 11–16. [CrossRef]

60. Omar, M.A.; Nagy, D.M.; Hammad, M.A.; Aly, A.A. Validated spectrophotometric methods for determination of certain aminoglycosides in pharmaceutical formulations. *J. Appl. Pharm. Sci.* **2013**, *3*, 151–161. [CrossRef]

61. Salem, H.; Askal, H. Colourimetric and AAS determination of cephalosporins using Reineck's salt. *J. Pharm. Biomed. Anal.* **2002**, *29*, 347–354. [CrossRef]

62. Choudhary, Y.S.; Jothi, L.; Nageswaran, G. Electrochemical Characterization. In *Spectroscopic Methods for Nanomaterials Characterization*; Elsevier: Amsterdam, The Netherlands, 2017; Volume 2, pp. 19–54.

63. Westbroek, P.; Priniotakis, G.; Kiekens, P. Electrochemical methods. In *Analytical Electrochemistry in Textiles*; Woodhead Publishing Limited: Cambridge, UK, 2005; pp. 37–69.

64. Svorc, L. Determination of Caffeine: A Comprehensive Review on Electrochemical Methods. *Int. J. Electrochem. Sci.* **2013**, *8*, 5755–5773.

65. Avramov, M.L.; Petrovi, S.D.; Mijin, Ž. Contribution to the Recent Advances in Electrochemical Analysis of Pharmaceuticals. In *Biomedical and Pharmaceutical Applications of Electrochemistry. Modern Aspects of Electrochemistry*; Springer International Publishing: Cham, Switzerland, 2016.

66. Aboul-Enein, H.Y.; Sibel, A. Ozkan: Electroanalytical Methods in Pharmaceutical Analysis and Their Validation. *Chromatographia* **2012**, *75*, 811. [CrossRef]
67. Ayankojo, A.G.; Reut, J.; Ciocan, V.; Öpik, A.; Syritski, V. Molecularly imprinted polymer-based sensor for electrochemical detection of erythromycin. *Talanta* **2019**, *209*, 120502. [CrossRef]
68. Da Cunha, C.E.P.; Rodrigues, E.S.B.; Fernandes Alecrim, M.; Thomaz, D.V.; Macêdo, I.Y.L.; Garcia, L.F.; de Oliveira Neto, J.R.; Moreno, E.K.G.; Ballaminut, N.; de Souza Gil, E. Voltammetric Evaluation of Diclofenac Tablets Samples through Carbon Black-Based Electrodes. *Pharmaceuticals* **2019**, *12*, 83. [CrossRef]
69. Shah, J.; Jan, M.R.; Shah, S.; Naeem, M. Spectrofluorimetric Protocol for Ceftriaxone in Commercial Formulation and Human Plasma After Condensation with Formaldehyde and Ethyl Acetoacetate. *J. Fluoresc.* **2011**, *21*, 2155–2163. [CrossRef]
70. Tariq, A.; Siddiqui, M.R.; Kumar, J.; Reddy, D.; Negi, P.S.; Chaudhary, M.; Srivastava, S.M.; Singh, R.K. Development and validation of high performance liquid chromatographic method for the simultaneous determination of ceftriaxone and vancomycin in pharmaceutical formulations and biological samples. *Sci. Asia* **2010**, *36*, 297–304. [CrossRef]
71. Tange, M.; Yoshida, M.; Nakai, Y.; Uchida, T. The Role of an Impurity in Ceftriaxone Sodium Preparation for Injection in Determining Compatibility with Calcium-Containing Solutions. *Chem. Pharm. Bull.* **2016**, *64*, 207–214. [CrossRef]
72. Diwan, V.; Tamhankar, A.J.; Khandal, R.K.; Sen, S.; Aggarwal, M.; Marothi, Y.; Iyer, R.V.; Sundblad-Tonderski, K.; Lundborg, C.S. Antibiotics and antibiotic-resistant bacteria in waters associated with a hospital in Ujjain, India. *BMC Public Health* **2010**, *10*, 414. [CrossRef]
73. Kratzer, A.; Liebchen, U.; Schleibinger, M.; Kees, M.G.; Kees, F. Determination of free vancomycin, ceftriaxone, cefazolin and ertapenem in plasma by ultrafiltration: Impact of experimental conditions. *J. Chromatogr. B* **2014**, *961*, 97–102. [CrossRef] [PubMed]
74. Kale, R.S.; Jain, H.K.; Ghode, P.D.; Mhaske, G.S.; Puri, M.V.; Raut, M.D.; Patil, H.S. An rphplc method for simultaneous estimation of Ceftriaxone sodium and sulbactam sodium in parenteral dosage form. *Int. J. Pharm. Pharm. Sci.* **2011**, *3*, 406–409.
75. Akl, M.A.; Ahmed, M.A.; Ramadan, A. Validation of an HPLC-UV method for the determination of ceftriaxone sodium residues on stainless steel surface of pharmaceutical manufacturing equipments. *J. Pharm. Biomed. Anal.* **2011**, *55*, 247–252. [CrossRef] [PubMed]
76. Shrivastava, S.M.; Singh, R.; Tariq, A.; Siddiqui, M.R.; Yadav, J.; Negi, P.S.; Chaudhary, M. A Novel High Performance Liquid Chromatographic Method for Simultaneous Determination of Ceftriaxone and Sulbactam in Sulbactomax. *Int. J. Biomed. Sci. IJBS* **2009**, *5*, 37–43. [PubMed]
77. Rehm, S.; Rentsch, K.M. LC-MS/MS method for nine different antibiotics. *Clin. Chim. Acta* **2020**, *511*, 360–367. [CrossRef]
78. Herrera-Hidalgo, L.; Gil-Navarro, M.; Penchala, S.D.; López-Cortes, L.; de Alarcón, A.; Luque-Márquez, R.; Gutiérrez-Valencia, A. Ceftriaxone pharmacokinetics by a sensitive and simple LC–MS/MS method: Development and application. *J. Pharm. Biomed. Anal.* **2020**, *189*, 113484. [CrossRef]
79. Wongchang, T.; Winterberg, M.; Tarning, J.; Sriboonvorakul, N.; Muangnoicharoen, S.; Blessborn, D. Determination of ceftriaxone in human plasma using liquid chromatography–tandem mass spectrometry. *Wellcome Open Res.* **2019**, *4*, 47. [CrossRef]
80. Mohamed, D.; Kamal, M. Enhanced HPLC-MS/MS method for the quantitative determination of the co-administered drugs ceftriaxone sodium and lidocaine hydrochloride in human plasma following an intramuscular injection and application to a pharmacokinetic study. *Biomed. Chromatogr.* **2018**, *32*, e4322. [CrossRef]
81. Ongas, M.; Standing, J.; Ogutu, B.; Waichungo, J.; Berkley, J.A.; Kipper, K. Liquid chromatography–tandem mass spectrometry for the simultaneous quantitation of ceftriaxone, metronidazole and hydroxymetronidazole in plasma from seriously ill, severely malnourished children. *Wellcome Open Res.* **2017**, *2*, 43. [CrossRef]
82. Shrestha, B.; Bhuyan, N.R.; Sinha, B.N. Simultaneous determination of Ceftriaxone and Tazobactam in injectables by UHPLC method. *Pharm. Methods* **2013**, *4*, 46–51. [CrossRef]
83. Sun, H.; Xing, H.; Tian, X.; Zhang, X.; Yang, J.; Wang, P. UPLC-MS/MS Method for Simultaneous Determination of 14 Antimicrobials in Human Plasma and Cerebrospinal Fluid: Application to Therapeutic Drug Monitoring. *J. Anal. Methods Chem.* **2022**, *2022*, 7048605. [CrossRef] [PubMed]
84. Yu, X.; Tang, X.; Zuo, J.; Zhang, M.; Chen, L.; Li, Z. Distribution and persistence of cephalosporins in cephalosporin producing wastewater using SPE and UPLC–MS/MS method. *Sci. Total Environ.* **2016**, *569-570*, 23–30. [CrossRef]
85. Salman, A.T. RP-HPLC Estimation of Ceftriaxone Sodium in Pharmaceuticals. *Egypt. J. Chem.* **2021**, *64*, 4901–4906. [CrossRef]
86. De Aléssio, P.V.; Kogawa, A.C.; Salgado, H.R.N. Quality of Ceftriaxone Sodium in Lyophilized Powder for Injection Evaluated by Clean, Fast, and Efficient Spectrophotometric Method. *J. Anal. Methods Chem.* **2017**, *2017*, 7530242. [CrossRef] [PubMed]
87. Ethiraj, R.; Thiruvengadam, E.; Sampath, V.S.; Vahid, A.; Raj, J. Development and Validation of Stability Indicating Spectroscopic Method for Content Analysis of Ceftriaxone Sodium in Pharmaceuticals. *Int. Sch. Res. Not.* **2014**, *2014*, 278173. [CrossRef]
88. Abu, T.M.M.; Ghithan, J.; Abu-Taha, M.I.; Darwish, S.M.; Abu-Hadid, M.M. Spectroscopic approach of the interaction study of ceftriaxone and human serum albumin. *J. Biophys. Struct. Biol.* **2014**, *6*, 1–12. [CrossRef]
89. Pasha, C.; Narayana, B. A simple method for the spectrophotometric determination of cephalosporins in pharmaceuticals using variamine blue. *Eclét. Quím.* **2008**, *33*, 41–46. [CrossRef]
90. Gunasekaran, S.; Charles, J. Spectral measurements and qualitative analysis of ceftriaxone and cefotaxime. *Asian J. Chem.* **2008**, *20*, 1343–1356.

91. Manimekalai, P.; Dhanalakshmi, R.; Manavalan, R. Preparation and characterization of ceftriaxone sodium encapsulated chitosan nanoparticles. *Int. J. Appl. Pharm.* **2017**, *9*, 10. [CrossRef]

92. Feng, Y.-C.; Ni, Z.; Hu, C.-Q. Variable selection in near infrared spectroscopy for quantitative models of homologous analogs of cephalosporins. *J. Innov. Opt. Health Sci.* **2014**, *7*, 1450005. [CrossRef]

93. Shah, J.; Jan, M.R.; Shah, S. Inayatullah Development and validation of a spectrofluorimetric method for the quantification of ceftriaxone in pharmaceutical formulations and plasma. *Luminescence* **2013**, *28*, 516–522. [CrossRef] [PubMed]

94. Dafale, N.A.; Semwal, U.P.; Agarwal, P.K.; Sharma, P.; Singh, G.N. Quantification of ceftriaxone sodium in pharmaceutical preparations by a new validated microbiological bioassay. *Anal. Methods* **2012**, *4*, 2490–2498. [CrossRef]

95. Aléssio, P.V.; Salgado, H.R.N. Development and Validation of a Successful Microbiological Agar Assay for Determination of Ceftriaxone Sodium in Powder for Injectable Solution. *Pharmaceutics* **2012**, *4*, 334–342. [CrossRef] [PubMed]

96. Solangi, A.; Memon, S.; Mallah, A.; Memon, N.; Khuhawar, M.Y.; Bhanger, M.I. Determination of ceftriaxone, ceftizoxime, paracetamol, and diclofenac sodium by capillary zone electrophoresis in pharmaceutical formulations and in human blood serum. *Turk. J. Chem.* **2010**, *34*, 921–934. [CrossRef]

97. Crini, G.; Lichtfouse, E. Advantages and disadvantages of techniques used for wastewater treatment. *Environ. Chem. Lett.* **2019**, *17*, 145–155. [CrossRef]

98. Vymazal, J. Constructed Wetlands for Wastewater Treatment. *Water* **2010**, *2*, 530–549. [CrossRef]

99. Li, Y.; Zhu, G.; Ng, W.J.; Tan, S.K. A review on removing pharmaceutical contaminants from wastewater by constructed wetlands: Design, performance and mechanism. *Sci. Total Environ.* **2014**, *468–469*, 908–932. [CrossRef] [PubMed]

100. Ávila, C.; Nivala, J.; Olsson, L.; Kassa, K.; Headley, T.; Mueller, R.A.; Bayona, J.M.; García, J. Emerging organic contaminants in vertical subsurface flow constructed wetlands: Influence of media size, loading frequency and use of active aeration. *Sci. Total Environ.* **2014**, *494–495*, 211–217. [CrossRef]

101. Ilyas, H.; Masih, I.; van Hullebusch, E.D. Pharmaceuticals' removal by constructed wetlands: A critical evaluation and meta-analysis on performance, risk reduction, and role of physicochemical properties on removal mechanisms. *J. Water Health* **2020**, *18*, 253–291. [CrossRef]

102. Choi, Y.-J.; Kim, L.-H.; Zoh, K.-D. Removal characteristics and mechanism of antibiotics using constructed wetlands. *Ecol. Eng.* **2016**, *91*, 85–92. [CrossRef]

103. Guan, Y.; Wang, B.; Gao, Y.; Liu, W.; Zhao, X.; Huang, X.; Yu, J. Occurrence and Fate of Antibiotics in the Aqueous Environment and Their Removal by Constructed Wetlands in China: A review. *Pedosphere* **2017**, *27*, 42–51. [CrossRef]

104. Cetecioglu, Z.; Ince, B.; Azman, S.; Gokcek, N.; Coskun, N.; Ince, N.C.A.O. Determination of Anaerobic and Anoxic Biodegradation Capacity of Sulfamethoxasole and the Effects on Mixed Microbial Culture. In *Biodegradation-Engineering and Technology*; Intech: Rijeka, Croatia, 2013. [CrossRef]

105. Chen, J.; Liu, Y.-S.; Zhang, J.-N.; Yang, Y.-Q.; Hu, L.-X.; Yang, Y.-Y.; Zhao, J.-L.; Chen, F.-R.; Ying, G.-G. Removal of antibiotics from piggery wastewater by biological aerated filter system: Treatment efficiency and biodegradation kinetics. *Bioresour. Technol.* **2017**, *238*, 70–77. [CrossRef] [PubMed]

106. Wang, R.; Feng, F.; Chai, Y.; Meng, X.; Sui, Q.; Chen, M.; Wei, Y.; Qi, K. Screening and quantitation of residual antibiotics in two different swine wastewater treatment systems during warm and cold seasons. *Sci. Total Environ.* **2019**, *660*, 1542–1554. [CrossRef]

107. Zheng, W.; Zhang, Z.; Liu, R.; Lei, Z. Removal of veterinary antibiotics from anaerobically digested swine wastewater using an intermittently aerated sequencing batch reactor. *J. Environ. Sci.* **2018**, *65*, 8–17. [CrossRef] [PubMed]

108. Xu, Z.; Song, X.; Li, Y.; Li, G.; Luo, W. Removal of antibiotics by sequencing-batch membrane bioreactor for swine wastewater treatment. *Sci. Total Environ.* **2019**, *684*, 23–30. [CrossRef]

109. Akbari, M.Z.; Xu, Y.; Lu, Z.; Peng, L. Review of antibiotics treatment by advance oxidation processes. *Environ. Adv.* **2021**, *5*, 100111. [CrossRef]

110. Cuerda-correa, E.M.; Alexandre-franco, M.F.; Fern, C. Advanced Oxidation Processes for the Removal of Antibiotics from Water. An Overview. *Water* **2020**, *12*, 102. [CrossRef]

111. Saharan, V.K.; Pinjari, D.V.; Gogate, P.R.; Pandit, A.B. Advanced Oxidation Technologies for Wastewater Treatment: An Overview. In *Industrial Wastewater Treatment, Recycling and Reuse*; Butterworth-Heinemann: Oxford, UK, 2014; pp. 141–191.

112. Pandis, P.K.; Kalogirou, C.; Kanellou, E.; Vaitsis, C.; Savvidou, M.G.; Sourkouni, G.; Zorpas, A.A.; Argirusis, C. Key Points of Advanced Oxidation Processes (AOPs) for Wastewater, Organic Pollutants and Pharmaceutical Waste Treatment: A Mini Review. *ChemEngineering* **2022**, *6*, 8. [CrossRef]

113. Miyata, M.; Ihara, I.; Yoshid, G.; Toyod, K.; Umetsu, K. Electrochemical oxidation of tetracycline antibiotics using a Ti/IrO2 anode for wastewater treatment of animal husbandry. *Water Sci. Technol.* **2011**, *63*, 456–461. [CrossRef]

114. Zhou, Y.; Hu, B.; Zhuang, X.; Qiu, J.; Xu, T.; Zeng, M.; He, X.; Yu, G. Investigation on Mechanism of Tetracycline Removal from Wastewater by Sinusoidal Alternating Electro-Fenton Technique. *Sustainability* **2022**, *14*, 2328. [CrossRef]

115. Mahdi, M.H.; Mohammed, T.J.; A Al-Najar, J. Advanced Oxidation Processes (AOPs) for treatment of antibiotics in wastewater: A review. *IOP Conf. Ser. Earth Environ. Sci.* **2021**, *779*, 012109. [CrossRef]

116. Saleh, T.A.; Gupta, V.K. An Overview of Membrane Science and Technology. In *Nanomaterial and Polymer Membranes*; Elsevier: Amsterdam, The Netherlands, 2016; pp. 1–23. [CrossRef]

117. Nqombolo, A.; Mpupa, A.; Moutloali, R.; Nomngongo, P. Wastewater Treatment Using Membrane Technology. In *Wastewater and Water Quality*; Yonar, T., Ed.; IntechOpen: London, UK, 2018; pp. 29–40.

118. Sethy, N.K.; Arif, Z.; Sista, K.S.; Mishra, P.K.; Kumar, P.; Kushwaha, A.K. Advances in Remediation of Water Pollution Advances in Membrane Technology Used in the Wastewater Treatment Process. In *Pollutants and Water Management: Resources, Strategies and Scarcity*; Singh, P., Singh, R., Singh, V.K., Bhadouria, R., Eds.; John Wiley & Sons Ltd.: Hoboken, NJ, USA, 2021.
119. Deegan, A.M.; Shaik, B.; Nolan, K.; Urell, K.; Oelgemöller, M.; Tobin, J.; Morrissey, A. Treatment options for wastewater effluents from pharmaceutical companies. *Int. J. Environ. Sci. Technol.* **2011**, *8*, 649–666. [CrossRef]
120. Ezugbe, E.O.; Rathilal, S. Membrane Technologies in Wastewater Treatment: A Review. *Membranes* **2020**, *10*, 89. [CrossRef] [PubMed]
121. Liu, P.; Zhang, H.; Feng, Y.; Yang, F.; Zhang, J. Removal of trace antibiotics from wastewater: A systematic study of nanofiltration combined with ozone-based advanced oxidation processes. *Chem. Eng. J.* **2014**, *240*, 211–220. [CrossRef]
122. Lan, L.; Kong, X.; Sun, H.; Li, C.; Liu, D. High removal efficiency of antibiotic resistance genes in swine wastewater via nanofiltration and reverse osmosis processes. *J. Environ. Manag.* **2018**, *231*, 439–445. [CrossRef]
123. Badi, M.Y.; Azari, A.; Pasalari, H.; Esrafili, A.; Farzadkia, M. Modification of activated carbon with magnetic Fe_3O_4 nanoparticle composite for removal of ceftriaxone from aquatic solutions. *J. Mol. Liq.* **2018**, *261*, 146–154. [CrossRef]
124. Reynoso, E.; Spesia, M.B.; García, N.A.; Biasutti, M.A.; Criado, S. Riboflavin-sensitized photooxidation of Ceftriaxone and Cefotaxime. Kinetic study and effect on Staphylococcus aureus. *J. Photochem. Photobiol. B: Biol.* **2015**, *142*, 35–42. [CrossRef]
125. Hashemi, S.Y.; Badi, M.Y.; Pasalari, H.; Azari, A.; Arfaeinia, H.; Kiani, A. Degradation of Ceftriaxone from aquatic solution using a heterogeneous and reusable $O3/UV/Fe_3O_4@TiO_2$ systems: Operational factors, kinetics and mineralisation. *Int. J. Environ. Anal. Chem.* **2020**, 1–17. [CrossRef]
126. Shokri, M.; Isapour, G.; Shamsvand, S.; Kavousi, B. Photocatalytic degradation of ceftriaxone in aqueous solutions by immobilized TiO2 and ZnO nanoparticles: Investigating operational parameters. *J. Mater. Environ. Sci.* **2016**, *7*, 2843–2851.
127. Tutunaru, B.; Samide, A.; Iordache, S.; Tigae, C.; Simionescu, A.; Popescu, A. Ceftriaxone Degradation in the Presence of Sodium Halides Investigated by Electrochemical Methods Assisted by UV-Vis Spectrophotometry. *Appl. Sci.* **2021**, *11*, 1376. [CrossRef]
128. Takdastan, A.; Sadeghi, H.; Dobaradaran, S.; Ma, L.; Sorooshian, A.; Ravanbakhsh, M.; Niari, M.H. Synthesis and characterization of γ-Fe_2O_3 encapsulated NaY zeolites as solid adsorbent for degradation of ceftriaxone through heterogeneous catalytic advanced oxidation processes. *J. Iran. Chem. Soc.* **2019**, *17*, 725–734. [CrossRef]

Article

Assessment of Levonorgestrel Leaching in a Landfill and Its Effects on Placental Cell Lines and Sperm Cells

Ramiro Ríos-Sossa [1], Juan José García-Londoño [1], Daniel Gil-Ramírez [1], Arley Camilo Patiño [2], Walter D. Cardona-Maya [3], Juan Carlos Quintana-Castillo [4] and Jhon Fredy Narváez-Valderrama [1,*]

[1] Grupo de Investigación Ingeniar, Facultad de Ingeniería, Corporación Universitaria Remington, Calle 51 No. 51-27, Medellín 050010, Colombia; ramiro.rios@udea.edu.co (R.R.-S.); juan.garcia.3750@miremington.edu.co (J.J.G.-L.); danielh.gil@udea.edu.co (D.G.-R.)

[2] Grupo Toxicología, Alternativas Terapéuticas y Alimentarias BIOPOLIMER, Facultad de Ciencias Farmacéuticas y Alimentarias, Universidad de Antioquia, Calle 70 No. 52-21, Medellín 050010, Colombia; arley.patino@udea.edu.co

[3] Grupo Reproducción, Facultad de Medicina, Universidad de Antioquia, Calle 70 No. 52-21, Medellín 050010, Colombia; wdario.cardona@udea.edu.co

[4] Grupo de Investigación Infettare, Universidad Cooperativa de Colombia, Calle 50A 41-20, Medellín 050012, Colombia; juan.quintanac@ucc.edu.co

* Correspondence: jhon.narvaez@uniremington.edu.co

Abstract: The Buenavista landfill is located east of the city of Medellín, but it has a slope steeper than 30% and is less than 600 m away from the Piedras River, possibly influencing the quality of the drinking water in the city. Many complex residues are disposed of in this landfill, including pharmaceuticals and personal care products (PPCPs) such as levonorgestrel (LNG), which may reach water bodies via runoff and leaching. We assessed the levels of LNG in the effluent of an upflow anaerobic sludge blanket (UASB) reactor from the Buenavista landfill by uHPLC–DAD, as well as the endocrine disruptor effect of LNG on placental cell lines (BeWo) and human sperm cells. Additionally, the potential leaching of LNG was assayed under laboratory conditions using soil layers that were sampled from the Buenavista landfill. LNG was detected at levels of 315 $\mu g \cdot L^{-1}$ in the effluents of the UASB reactor. Thus, the UASB reactor is not an efficient treatment method for the removal of recalcitrant pollutants. Additionally, we found that a layer of soil used as a cover material may adsorb more than 90% of LNG pollutants, but small amounts may still be leached, which means that a cover material is not a strong enough barrier to fully prevent the leaching of LNG. Finally, our results show that the leachate fraction decreased the levels of β-human chorionic gonadotropin, but not sperm motility or viability. Thus, leached LNG could trigger reproduction disorders, but further studies should be carried out to investigate its potential effects in more detail.

Keywords: cover material; soil sorption; leaching; endocrine disruptor; pharmaceutical; personal care products

Citation: Ríos-Sossa, R.; García-Londoño, J.J.; Gil-Ramírez, D.; Patiño, A.C.; Cardona-Maya, W.D.; Quintana-Castillo, J.C.; Narváez-Valderrama, J.F. Assessment of Levonorgestrel Leaching in a Landfill and Its Effects on Placental Cell Lines and Sperm Cells. *Water* **2022**, *14*, 871. https://doi.org/10.3390/w14060871

Academic Editors: François Gagné, Stefano Magni and Valerio Matozzo

Received: 14 December 2021
Accepted: 8 March 2022
Published: 10 March 2022

Publisher's Note: MDPI stays neutral with regard to jurisdictional claims in published maps and institutional affiliations.

1. Introduction

Pharmaceutical and personal care products (PPCPs) have been classified as emerging pollutants because of their environmental impact. Long-term exposure to these substances may affect a wide range of species, even at low concentrations [1]. Furthermore, the high persistence of PPCPs and their low removal from wastewater may increase their impact on aquatic ecosystems [2]. Traces of PPCPs have been found in drinking water supplies, which is a public health concern [3]. The implications of the toxicity of PPCPs in water include endocrine disruption (ED), which affects the equilibrium of hormone levels and leads to metabolic, neurological, and reproductive consequences in exposed species [4].

Levonorgestrel (LNG) is a pharmaceutical product widely used in Colombia as a contraceptive method; thus, complex effects similar to those of other PPCPs were found [5].

This substance is known to interact not only with progesterone receptors (PR), but also with another estrogen receptor, which may trigger complex pathways in non-targeted organisms [6]. LNG was detected in very low levels in water, leading to disruptions in the gonadotropin expression of aquatic species, such as teleost fish [7]. However, the information available on the occurrence and fate of this substance in aquatic ecosystems is limited. LNG is a steroidal molecule with strong molecular bonds, and thus, LNG is recalcitrant to degradation processes [8]. Therefore, LNG shows environmental persistence because natural conditions do not completely remove this substance. For instance, LNG is not removed by solar radiation at wavelengths close to 320 nm, and even energy radiation (UVC-265) reduces only 80% of this substance [9]. Therefore, more complex photodegradation should be applied for its removal in treatment plants.

Many PPCPs, including LNG, are disposed of in landfills, which could be a source of water pollution due to leaching [10]. The leaching that occurs in this landfill is a risk for long-term exposure in many people, but the effects are still unknown. For instance, some landfill leachates containing heavy metals induced genotoxic and cytotoxic effects in the peripheral blood erythrocytes extracted from Wistar strain rats, which showed the impact of landfills on biota [11]. Furthermore, the leachate may contain more complex substances, including PPCPs. However, leaching impacts may be reduced when the interaction of pollutants with a soil layer that is applied for covering waste residues in landfills (cover material) increases. Thus, the selection of an appropriate cover material decreases soilborne pollutants due to the soil layer working as a natural barrier. In previous research, Yang et al. found that hydrophobic partitioning is the main pathway for the sorption of progestins in soil [12]. For instance, the organic fractions in the soil layer may affect LNG leaching from landfills. The leached fraction of LNG may trigger ED in biota, among other complex implications in exposed organisms. For example, it has been found that levels between 3.3 and 40 ng·L^{-1} of LNG can reduce fish reproduction by the masculinization of females, changes in gonad histology, and disruption to hormone levels [13]. However, LNG toxicity in human reproduction has not been deeply studied. In a previous report, we showed that LNG and its photodegraded fractions decrease human chorionic gonadotropin (β-hCG) hormone levels in placental cell lines. This may cause problems to occur during gestational development [9].

The Buenavista landfill, which is located in Antioquia, Colombia, has been found to contain expired PPCPs, including painkillers, contraceptives, and antibiotics, among others. Thus, this landfill could be a source of PPCPs leaching into the Piedras River, which plays an essential role in the supply of drinking water to the city of Medellín. However, an appropriate cover material to reduce the transport of leachate pollutants, as well as treatment, may reduce their impact in natural water. To this end, an upflow anaerobic sludge blanket (UASB) reactor is used for the treatment of leachates produced by the Buenavista landfill. However, the UASB reactor, among other conventional treatments, is not sufficient for persistent organic pollutant removal because they are extremely resistant to biological degradation processes, and more specific advanced oxidation processes should be included for the treatment of pollutants [14].

The aim of this study was to assay the levels of LNG from the effluent of a UASB reactor used in the Buenavista landfill to evaluate possible transport by leaching from cover materials. Additionally, we studied the effects of the cover material used in this landfill on LNG adsorption to approximate a better material selection for reducing leaching. Finally, we studied the endocrine disruptor effect of LNG on the BeWo cell line and human sperm cells by in vitro assay. The results may be relevant for future studies on the assessment of risk due to the long-term exposure of PPCPs disposed of in landfills and their impact on bodies of water.

2. Materials and Methods

2.1. Chemicals and Materials

A stock solution of LNG was prepared at a concentration of 10.5 mg·L^{-1} from the commercial pharmaceutical Cerciorat$^®$ containing 1.5 mg per pill (98–102% USP29). Then, seven pills were macerated and diluted in Milli-Q water. The stock solution was stored at $-20\,°C$. Solvents, including methanol and acetonitrile (HPLC grade), fetal bovine serum (FBS), and Ham-F10 medium, were purchased from Sigma-Aldrich (San Luis, MO, USA). The water was purified in a Thermo Scientific$^®$ Barnstead 50131217 GenPure TM UV/UF for Type I Milli-Q water (deionized). The buffer solution for the mobile phase was prepared with 0.05% formic acid in deionized water. All solvents were filtered through cellulose nitrate (with pore sizes of 0.2 μm) and degassed in an ultrasonic bath for 10 min at 25 °C. Finally, the antibodies for β-hCG and progesterone analysis were provided by Medife S.A.S (the provider for Roche Colombia).

2.2. Soil Layer Sampling and Monitoring of LNG in the Effluent of the UASB Reactor from the Buenavista Landfill

Different soil layers used as cover materials were sampled from the Buenavista landfill, located in La Union city, Antioquia, at the following coordinates: 5°57′31.37″ N 75°19′41.76″ W (latitude and longitude, respectively). These soil layers were identified at different depths from a complete soil profile that was mixed to obtain the cover material to bury the solid waste. The soil properties of the layers differed in terms of color, texture, structure, and thickness. Three layers were identified: Layer 1 (L1), Layer 2 (L2), and Layer 3 (L3). Samples of the soil layers were individually taken and stored at 4 °C until physicochemical analysis. They were sterilized by autoclaving at 120 °C for 20 min in order to reduce the microbial activity and thus avoid biodegradation. A sample of leachate was taken from the effluent of the UASB reactor in the Buenavista landfill to assay the LNG levels. This sample was filtered through a 0.2 μm membrane and analyzed using an ultra-high-performance liquid chromatography–diode array detector (uHPLC–DAD). The LNG was confirmed by the UV spectrum at its retention time, which was in agreement with the standard UV spectrum presented in the literature. The sample was spiked via the standard addition method.

2.3. Leaching Analysis of LNG from the Soil Layers under Laboratory Conditions

The experiment was carried out inside packed columns of polyvinyl chloride (PVC) using each of the soil layers in duplicate (with a diameter of 2.54 cm and a length of 30 cm). Then, 50 mL of the LNG stock solution was passed through the columns of PVC. The volume of the standard solution was calculated according to the average precipitation around the sampling site per area. Next, 50 mL of deionized water was used in the soil layers as a negative control. All leachate fractions were collected after 24 h in conical flasks to measure the volume, and they were centrifuged at 3000 rpm for 10 min at 25 °C for cleaning. Finally, 1 mL of the sample was injected into the uHPLC–DAD system using the recommended method.

2.4. Validation Parameters for LNG Analysis by uHPLC–DAD

We analyzed 10.5 mg.L^{-1} of the stock solution of LNG by uHPLC–DAD (Thermo Scientific Dionex UltiMate 3000) at a wavelength of 245 nm. The LNG was identified using the UV spectrum at its retention time. The LNG was separated on a C18 Hypersil Gold column (5 μm × 150 × 4.6 mm). A volume of 20 μL of the LNG solution was injected and separated using a mobile phase with a ratio of 85:15 (*v:v*) of buffer–acetonitrile, by applying a flow of 0.450 mL/min^{-1}. After 10 min, the mobile phase was modified at a ratio of methanol–acetonitrile of 50:50 (*v:v*), under a flow of 0.500 mL/min^{-1}. The mobile phase was set back to initial conditions (85:15 (*v:v*) of buffer–acetonitrile). Finally, a calibration curve was plotted between 1 and 10 ppm, and the validation parameters, such as the limit of detection (LOD), the limit of quantification (LOQ), and recovery were estimated. See the Supplementary Materials (SM) for more details.

2.5. Assessment of Leached Fractions in Human Sperm Cells

The experimental in vitro assays included five fresh semen samples from healthy individuals, collected by masturbation in a sterile container after sexual abstinence for 3 to 5 days. The normozoospermia for semen parameters is presented in Supplementary Materials SM1. The sperm motility and viability were quantified for each seminal sample following the guidelines established in the manual for seminal analysis published by the World Health Organization [15] before and after incubating an aliquot of $4–5 \times 10^6$ sperm cells with 100 μL of the leachate samples from the effluents of the UASB reactor for three hours. Both sperm parameters were quantified at the initial time (0 h) and 3 h after incubation (once every hour). The negative control was incubated with only 100 μL of PBS, while the leachate sample was incubated with 100 μL of samples from the UASB reactor. Additionally, the direct effect of the LNG was tested with 100 μL of the stock solution.

2.6. Effects of Leachate Sample from the UASB Reactor on BeWo Cells and Their Production of β-hCG

The BeWo cell line was supplied by ATCC® CCL-98™, and it was cultured using Roswell Park Memorial Institute (RPMI)-1640 medium with 2% fetal bovine serum under a 5% CO_2 atmosphere at 37 °C and 95% humidity (confluence > 70% and viability > 90%). A negative control was prepared with 100 μL of PBS in 5 mL of medium containing 5×10^5 cells and similar conditions. It was tested with 100 μL of the leachate sample from the UASB reactor. Additionally, 100 μL of the LNG stock solution was assayed. All experiments were carried out in duplicate for 24 h. A colorimetric assay (MTT assay) was carried out for cell viability according to Lomonte et al. [16]. For more details, see the Supplementary Materials (Supplementary Materials SM2). The analysis of the levels of β-hCG hormone in the medium was assayed after 24 h by ElectroChemiLuminescence using the COBAS® 411 Integrated System (HITACHI) analyzer for the immunoassay test. A total of 1 mL of the medium was taken out from each experimental well and added to a conical Eppendorf tube for the analysis of progesterone and β-hCG using the COBAS® 411 device. The UV–vis absorption and photoluminescence were recorded on the COBAS® 411 by the HITACHI provider. This device was supplied by the Hospital Marco Fidel Suarez (Bello, Colombia).

2.7. Statistical Analysis and Graphs

All graphs were plotted using GraphPad Prism 7.0 software (GraphPad Software Inc., San Diego, CA, USA). For linearity, homoscedasticity analysis was carried out, and the R square was found. The limit of detection (LOD) and the limit of quantification (LOQ) were determined by the signal–noise method, and five blanks were analyzed by uHPLC–DAD via the method presented above. Finally, the ED effects, due to changes in β-hCG and sperm, were defined using case–control methods via the application of an unpaired *t*-test analysis.

2.8. Ethical Statements

The BeWo cell line experiments do not require ethical review and approval because this line is obtained as a commercial immortalized from placental choriocarcinoma (ATCC® CCL-98) which did not include human experimentation. However, ethical approval for the semen analysis was obtained from the Bioethics Committee of the School of Medicine at Universidad de Antioquia (F-CBI-012, 14 February 2019), and the patients signed an informed consent form.

3. Results

3.1. Validation Parameters for LNG Analysis by uHPLC–DAD

The chromatography method allowed for the separation of the LNG at a retention time of 17.63 min. The quality parameters, for quantification purposes, are presented in Table 1.

Table 1. Data on LNG for qualification and quantification purposes.

LNG	R_t (min)	r^2	LOD (ppm)	LOQ (ppm)	Linearity range (ppm)	Linear equation	$Log\ K_{ow}$ [17]
	17.63	0.998	0.09	0.30	0.3–10.5	$y = 0.7271x + 0.6011$	3.48

The UV wavelength for LNG detection and quantification was 254 nm. The *n*-octanol-water partition coefficient ($Log\ K_{ow}$) is included.

The linearity was analyzed by homoscedasticity and r^2, which produced a good fit in a linear model, and the slope was significantly non-zero (p-value < 0.05) For more details, see Supplementary Materials SM3.

According to the uHPLC–DAD method, the LOQ appeared to be sensitive. However, some integrated techniques for sample extraction may increase its sensitivity. For example, Huang et al. applied dispersive liquid–liquid microextraction for LNG analysis in natural water, obtaining a sensitivity 100 times higher [18].

3.2. Analysis of LNG in the Effluent of the UASB Reactor from the Buenavista Landfill

The UV spectrum obtained at a retention time of 17.63 was in agreement with the spectrum found in the literature; therefore, the sample showed the presence of LNG (see Supplementary Materials SM3). Additionally, the standard addition appeared to increase at the same time in the spiked samples. Therefore, the leachate sample from the Buenavista landfill showed small amount of LNG at 315 $\mu g \cdot L^{-1}$, which was calculated by linear regression (see Figure 1). A similar method was developed to determine the levels of LNG in tap water, lake water, and river water samples using uHPLC–DAD [18]. However, the authors applied dispersive liquid–liquid microextraction for analysis to increase the sensitivity of the methods. In our work, we analyzed samples by direct injection because we consider that more polluted water (leached water), and thus higher levels of pollutants, do not require a more difficult extraction process.

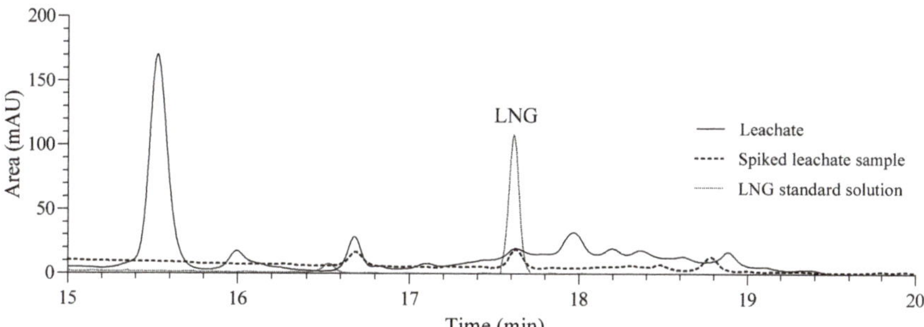

Figure 1. LNG analysis by uHPLC–DAD. The dashed line shows the analysis of the LNG standard. The black line shows the analysis of the leachate sample, and the horizontal dashed line shows the spiked sample.

According to the detected levels, this pharmaceutical product may be transported through the cover material used in landfills. This substance has been detected in wastewater and drinking water treatment plants worldwide at levels between 3.6 and 78.95 $ng \cdot L^{-1}$ [13,19]. Although the $Log\ K_{ow}$ presented for LNG in Table 1 indicates a possible lipophilic interaction, the soil is the final barrier that can trap molecules inside. The soil sorption and interaction are presented in Section 3.3 in order to explain the leaching process.

Consequently, the risk assessment for LNG exposure may be considered for public health studies because people can ingest low levels of these substances that are not entirely removed, as was previously found [20]. Progestins may accumulate in exposed organisms because of their lipophilic properties. For instance, LNG was reported to accumulate in the blood plasma of rainbow trout, reaching levels of 8.5–12 ng·mL^{-1} when the species was exposed to treated wastewater effluent (at a concentration of 1 ng·L^{-1} of LNG) for 14 days [21]. This concentration exceeds the plasma concentrations of the therapeutic doses used in humans (2.4 ng·mL^{-1}).

On the other hand, LNG may bioaccumulate and biomagnify in all tropic chains. For instance, this substance can cause severe consequences in the gestation process and embryonic development of zebrafish by bioaccumulation [22].

3.3. Assessment of Leaching Processes for LNG in Soil Layers Used as Cover Materials in the Buenavista Landfill

To approximate a possible explanation as to why LNG may be found in the effluent of the UASB reactor located at the Buenavista landfill, a leaching experiment was carried out. First, physicochemical analysis of the soil layers was performed. See Table 2 for details.

Table 2. Soil properties. CEC: cation exchange capacity.

Property	Layer 1 (L1)	Layer 2 (L2)	Layer 3 (L3)
Soil texture	Clay sandy loam	Sandy loam	Loam
% Clay	50	36	44
% Silt	18	26	44
% Sand	32	38	12
pH	5.6	5.1	5.1
Org. matter (%)	4.1	0.15	0.12
Al (cmol kg^{-1})	–	1.4	2
Ca (cmol kg^{-1})	0.2	0.1	0.1
p	1.0	2.0	8
CEC (cmol kg^{-1})	0.2	1.5	2.1
Soil water infiltration (%)	75.6	42.6	31

Overall, the soil layers showed high clay and silt content, which is related to the high soil sorption of PPCPs [23]. Although low amounts of organic material were found in the soil layers, lipophilic interactions may occur (Table 2). For instance, the humic acid in organic matter may increase the soil–pollutant interaction [24]. Additionally, the cation exchange capacity (CEC) in the soil allows for electrostatic interactions with pollutants due to the charges and ionizable states of molecules.

The water infiltration in the soil was found to be between 31% and 75.6%. The highest level of infiltration was found in L1, which may play an essential role in the transport of pollutants by soil–water partitioning. For instance, intensive rainfall events in the landfill may increase the leaching of pollutants [10].

Second, the leaching of LNG was tested in all soil layers. According to Figure 2a, more than 90% of LNG pollutants were adsorbed in all layers used as a cover material, but a higher level of soil sorption of LNG was found in L2 and L3. Some physicochemical characteristics in soil layers may increase the LNG–soil interaction. For instance, the lipophilic properties of LNG may induce more sorption in organic fractions and decrease the water interaction. Thus, higher amounts of LNG may remain for a long period of time in cover material. For instance, according to Ahmed et al., nonpolar substances are adsorbed into soil more than polar substances due to lipophilic affinity [25]. To understand the influences of the soil physicochemical properties of pollutant sorption, a correlation plot is provided in Figure 2. Some properties, such as the CEC, water infiltration, and silty fraction showed a good correlation (Pearson $r > 0.99$, p-value = 0.1910).

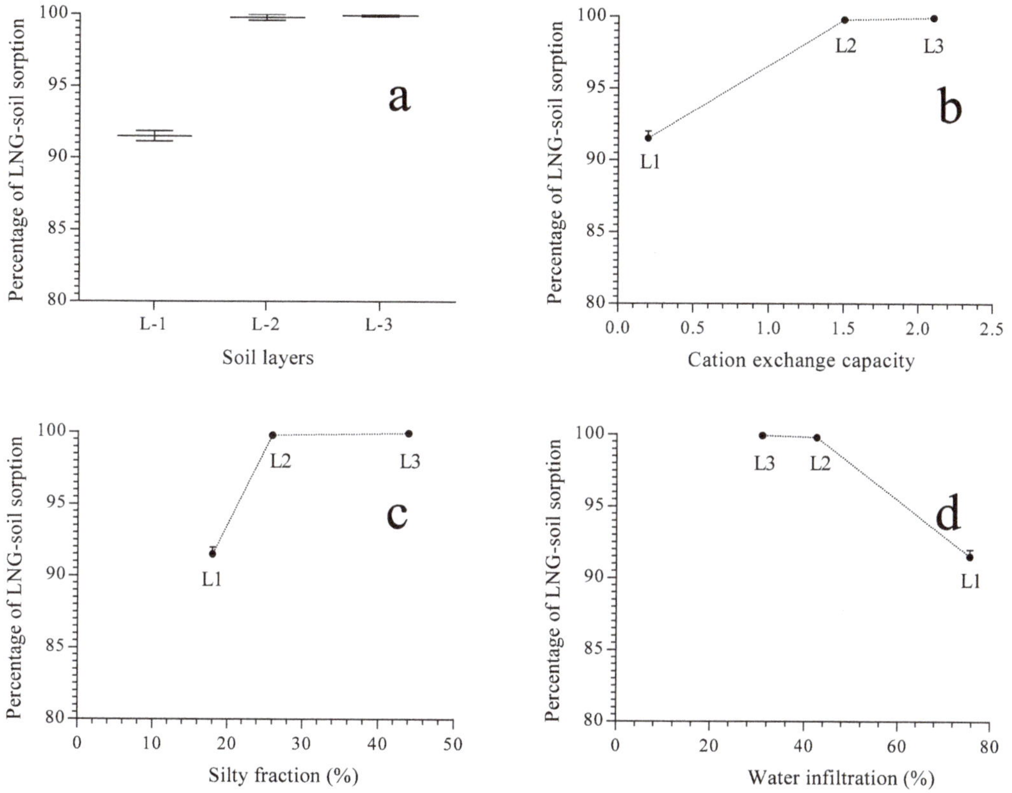

Figure 2. Analysis of soil sorption of LNG by leaching analysis: (**a**) leaching of LNG from soil layers used as cover materials; (**b**) effects of water infiltration on soil layer sorption; (**c**) effects of the silty fraction on soil layer sorption; (**d**) effects of CEC on soil layer sorption.

The water infiltration measured in the soil layers was indirectly correlated with the soil sorption of LNG (Figure 2b). Therefore, the leaching of pollutants depends on water infiltration, which occurs due to hydrogen bonds, polar interactions, and even liquid–solid partitioning. LNG shows a hydroxyl group that may interact with water by hydrogen bonds, but due to the high *Log K_{ow}*, only a lower amount can remain in water at equilibrium, which explains why less than 8% of the LNG was transported through the column. For instance, some LNG ketone and hydroxy groups may interact in soil, increasing LNG uptake in soils, and inducing hydrogen bonding and pore-filling [12]. Unfortunately, desorption of synthetic progestin may be induced by rainfall, and thus, a low amount may be transported by leaching. Therefore, selecting an appropriate cover material and ensuring adequate operating conditions is critical to reducing LNG leaching from landfills.

The silty fraction showed a direct relationship with the soil sorption of LNG (see Figure 2c). Therefore, a higher silty fraction may increase the soil sorption of steroid molecules because the size particles may uptake substances with a higher molecular weight, operating as size exclusion chromatography. According to Yang et al., the relatively high surface area of soils and sediment favored the uptake of some estrogens with similar molecular structures of LNG, such as altrenogest, drospirenone, and medroxyprogesterone acetate [26]. However, the soil layers studied in this paper contain small amounts of clay, which may play an important role in potentializing the soil sorption of pollutants due to the small grain size [27].

Similarly, the CEC showed a direct relationship with the soil sorption of LNG (see Figure 2d). Thus, higher values of CEC increased its soil sorption. According to Tang et al., the equilibrium time for the soil sorption of LNG took a shorter amount of time at lower CEC values because LNG may fill the vacant sites in soils more quickly in order to reach saturation [2].

According to our results, the grain size, the CEC, and water infiltration may play essential roles in the leaching fractions of LNG. Some soil-borne LNG levels may induce chronic toxicity with long-term exposure in aquatic organisms and possibly overexposed humans. However, more data for the correlation between the soil sorption of LNG in cover materials and the soil properties should be included in future studies.

3.4. Assessment of Leachate Sample in Placental Cell Lines (BeWo) and Sperm Cells

Both the LNG and leached LNG from soil layers under laboratory conditions did not show a change in the conventional sperm parameters, including progressive motility (Figure 3a), non-progressive motility (Figure 3b), immotile sperm cells (Figure 3c), and viability (Figure 3d) compared to the negative controls.

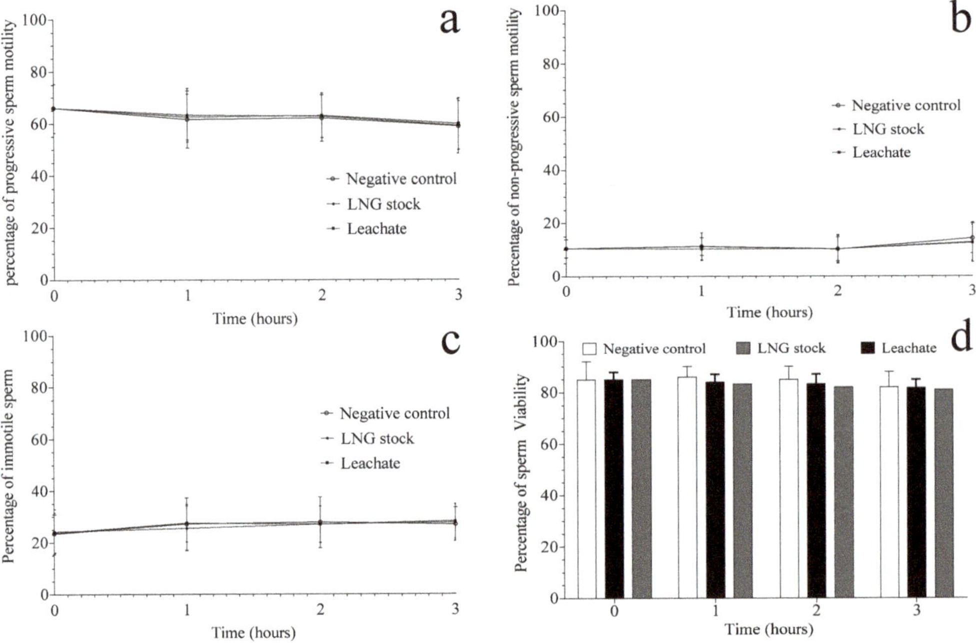

Figure 3. Sperm motility test and viability. Experimental data ($n = 5$): (**a**) Percentage of sperm progressive motility; (**b**) Percentage of sperm non-progressive motility; (**c**) percentage of immotile sperm cells; and (**d**) Percentage of sperm viability.

The findings of this study show that the sperm motility and viability of seminal samples were not affected when they were in contact with LNG. However, chronic exposure to this substance, even at low concentrations, may alter some species' reproductive capacities. For instance, low amounts of LNG may result in reproductive disorders in male zebrafish in progestin-contaminated aquatic environments due to LNG causing a significant decrease in the plasma concentrations of 11-ketotestosterone (11-KT) or estradiol (E2) in males exposed to between 10 and 100 ng·L^{-1} of LNG [28]. Additionally, previous studies have shown a reduction in sperm kinematics and capacitation in ovine spermatozoa when exposed to estrogens such as 17β-estradiol and progesterone [29]. On the other hand, more long-term exposure to LNG may affect sperm kinematics in male fathead minnow (Pimephales promelas) [30].

The BeWo cell line was applied to assess possible changes in β-hCG production because this hormone is related to corpus luteum rescue in trophoblastic cell implantation. Significant differences (SD) between the negative control and the stock solution of LNG or the leachate sample were found by applying an unpaired *t*-test. Our results show that LNG and leached fractions decreased β-hCG in the cell medium after BeWo cell line exposure (see Figure 4). This could be a result of the ED effect because the MTT assay did not show differences in viability during the test. This effect is related to changes in β-hCG production and not to acute toxicity due to the reduction in cell numbers.

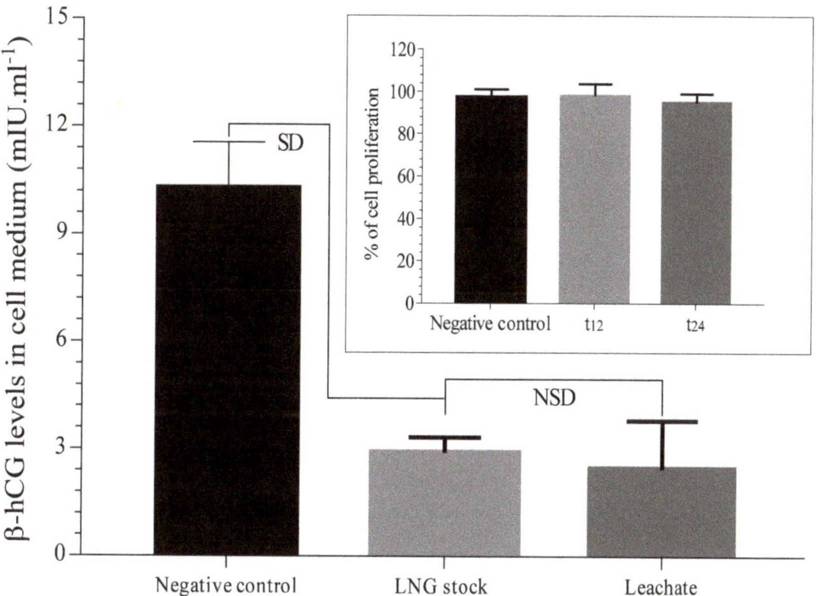

Figure 4. Study of LNG reproduction effects and leached LNG from different soil layers by analysis of β-hCG levels in the medium of the placental cell line. The MTT viability test is presented in the small box. Significant differences (SD; *p*-value < 0.05); non-significant differences (NSD; *p*-value > 0.05).

In a previous study, we found that low levels of LNG and its degradation products may induce a similar reduction in β-hCG levels after the exposure of placental lines to those fractions [9].

On the other hand, previous reports have shown that LNG may work as an ED in the reproductive functions of fish and amphibians, which causes the gene expression of pituitary gonadotropins and gonadal steroidogenic enzymes [7,31]. Furthermore, these alterations induce estrogenic effects in fathead minnows (males and females) exposed to different concentrations of LNG (up to 3124 ng·L⁻¹) for 28 days [32].

According to the biological activity of progestins, it is known that they act intracellularly through the binding and activation of the progesterone receptor (PR), which serves as a transcription factor to induce genomic effects [33]. These nuclear receptors are evolutionarily conserved in vertebrate animals and, thus, indirect exposure in water at low levels could induce similar pathways [4].

3.5. A Possible Implication of Leached LNG in Medellín, Colombia

The Buenavista landfill is located 600 m away from the Piedras River. This landfill has a slope steeper than 30% and, thus, runoff and leaching may occur. Therefore, the quality of the river may be impacted by pollutants transported from the landfill (for more details, see Supplementary Materials SM4). For instance, we detected LNG in effluents of the UASB reactor in the Buenavista landfill, and this substance may reach the water body. Therefore, the levels of LNG should be monitored in this river. A percentage of water from the Piedras River is pumped to the La Fe reservoir, one of the most important drinking water reservoirs for the Valle del Aburrá in Antioquia (Colombia), supplying potable water to more than two million people. The natural water is purified in La Ayurá, a drinking water treatment plant (DWTP) located in Medellín before its distribution. Thus, the exposure risk for people may depend on five factors, including: (1) the final disposal of PPCPs in the landfill; (2) the cover material sorption; (3) the leaching treatment at the UASB reactor; (4) natural degradation; and (5) purification in the DWTP (La Ayurá). However, regarding the first factor, we found many empty pharmaceutical blisters and plastic containers in the Buenavista landfill, which shows that this place is the final destination for many PPCPs. Secondly, we found that cover materials may allow for the leaching of LNG at values lower than 8% (under laboratory conditions). Furthermore, the levels of LNG in the effluents of the UASB reactor indicate that the cover material interaction and even the leaching treatment do not sufficiently reduce the levels of LNG. Recalcitrant pollutants, such as PPCPs, are not efficiently removed from wastewater treatment plants (WWTPs) by conventional processes. For instance, Botero-Coy et al. detected levels of antibiotics above $1\ \mu g \cdot L^{-1}$ in the effluents of WWTPs in cities such as Bogotá, Medellín, and Florencia, which means that these pollutants are not completely removed by conventional methods [34]. Thus, the design of treatment plants for leaching water and wastewater should include advanced oxidation, among other tertiary treatment processes for the removal of pollutants.

Natural degradation is not efficient enough to remove LNG in natural water. In a previous report, we found that solar energy for the photolysis of LNG does not have a strong enough effect to enable its degradation under natural conditions [9]. Thus, LNG fractions may remain in the water for a long time. However, biodegradation should be considered in future studies. Although information about the biodegradation and biotransformation of LNG as well other contraceptives is still unknown, its half-life in non-sterile soil is ten times higher than in sterile soil and, thus, microorganisms may play an important role in its natural degradation [35]. According to our results, LNG was detected in the effluent of the UASB reactor; through runoff, this fraction could reach the natural water in the Piedras River. Thus, the exposure risk for people in Medellín may depend on the DWTP, which applied coagulation and chlorination during treatment. However, these conventional methods are not efficient for the removal of pollutants, as discussed above. Thus, LNG exposure could have some endocrine effects on the biota and exposed population. Although the effects of short-term exposure were not found in sperm cells by in vitro assay in this study, long-term exposure to LNG should be considered because of the risks of continuous exposure at low levels. According to Elisabeth Carlsen et al., the quantity of semen has decreased by fifty percent in the last 50 years from $113 \times 10^6\ mL^{-1}$ [36]. Although the exact causes of this decrease have not been confirmed, complex xenobiotic exposure may be related to the reduction in sperm count. For instance, recently, more PPCPs have been found in bodies of water, and reproduction disorders in exposed aquatic species have increased [1]. According to the US Environmental Protection Agency (EPA), steroidal hormones, including LNG, may induce feminization, among other reproductive dysfunctions, in non-target organisms [37].

Finally, in this study, we found that leached fractions of LNG decreased the levels of β-hCG in a placental cell line, which indicates possible effects on implantation pathways. Therefore, more studies should be carried out on human ED due to low levels of unintentional exposure to contraceptives that may leach from landfills. Expired products may be disposed of in the Buenavista landfill, and soil sorption may be the final barrier preventing pollutants from ending up in surface water. However, we found that the cover material allowed the passage, through the soil layer, of small amounts of freely dissolved concentration of LNG, which may affect the endocrine systems of exposed species, even in humans. Future studies should focus on reducing the disposal of PPCPs in landfills and developing an efficient cover material for the sorption of pollutants to avoid exposure to soil-borne LNG.

4. Conclusions

Many expired or partially consumed PPCPs are disposed of in landfills. In these places, the soil layers used as cover materials influence the sorption capacity of these pollutants. This interaction may affect the transport of these substances, allowing them to reach water bodies, inducing imbalances in exposed ecosystems and, consequently, increasing the risk to public health. The soil layers used in the Buenavista landfill contain high amounts of clay and silt, which increase LNG–soil interactions. However, LNG is not completely adsorbed and, as a result, it was detected in leachate samples. The samples that contained LNG and the LNG stock solution affected the production of β-hCG in placental cell lines. Therefore, indirect exposure to LNG may cause problems during gestational development.

On the other hand, the levels of LNG detected in the leachate samples did not affect the seminal parameters of sperm cells. However, studies on the effects of chronic toxicity on the andrological pathways should be carried out to assess the outcomes of long-term exposure to LNG in drinking water. This was a preliminary study carried out to understand the influence of the Buenavista landfill as a source of PPCPs in the Piedras River due to leaching.

Supplementary Materials: The following supporting information can be downloaded at: https://www.mdpi.com/article/10.3390/w14060871/s1, SM1. Normal sperm values, SM2. MTT assay, SM3. UV spectrum of LNG, SM4. Analysis of PPCPs in the landfill.

Author Contributions: Conceptualization, A.C.P. and J.C.Q.-C.; methodology, R.R.-S., W.D.C.-M. and J.J.G.-L.; validation, D.G.-R.; formal analysis, writing—original draft preparation and supervision, J.F.N.-V. All authors have read and agreed to the published version of the manuscript.

Funding: The materials and methods were supported by grants from Ministerio de Ciencia y Tecnología MINCIENCIAS (Project code: 1365777757707). Additionally, this paper was supported by Corporación Universitaria Remington and Universidad Cooperativa de Colombia (Project code: 4000000098).

Institutional Review Board Statement: Informed consent was obtained from all subjects involved in the study.

Informed Consent Statement: The study was conducted in accordance with the Declaration of Helsinki, and approved by the Ethic Committee of the Faculty of Medicine-University of Antioquia (Acta de aprobación No. 003, date: 14-02-2019)." for studies involving humans. Written informed consent has been obtained from the patients to publish this paper.

Data Availability Statement: Not applicable.

Acknowledgments: The authors would like to thank the Ministerio de Ciencia y Tecnología MINCIENCIAS for funding the project entitled "Potencial de bioacumulación de agroquímicos y contaminantes persistentes en una cuenca del oriente antioqueño: Evaluación de un problema de salud pública". The authors also thank the authors of the research project entitled "Ensayo de disrupción endocrina de micro-contaminantes acumulados en prótesis mamarias de silicona por medio de passive dosing y expresión de la hormona β-hCG en la línea celular BeWo" which was funding by Universidad Cooperativa de Colombia and Corporación Universitaria Remington. Additionally, the authors thank the Hospital Marco Fidel Suarez (Bello-Antioquia) for analyzing the levels of β-hCG in the samples and for their support.

Conflicts of Interest: The authors declare that there is no conflict of interest with regard to the published results.

References

1. Narvaez, J.F.; Jimenez, C. Pharmaceutical Products in the Environment: Sources, Effects and Risks. *Vitae* **2012**, *19*, 93–108.
2. Tang, T.; Shi, T.; Li, D.; Xia, J.; Hu, Q.; Cao, Y. Adsorption properties and degradation dynamics of endocrine-disrupting chemical levonorgestrel in soils. *J. Agric. Food Chem.* **2012**, *60*, 3999–4004. [CrossRef] [PubMed]
3. Padhye, L.P.; Yao, H.; Kung'U, F.T.; Huang, C.-H. Year-long evaluation on the occurrence and fate of pharmaceuticals, personal care products, and endocrine disrupting chemicals in an urban drinking water treatment plant. *Water Res.* **2014**, *51*, 266–276. [CrossRef] [PubMed]
4. Christen, V.; Hickmann, S.; Rechenberg, B.; Fent, K. Highly active human pharmaceuticals in aquatic systems: A concept for their identification based on their mode of action. *Aquat. Toxicol.* **2010**, *96*, 167–181. [CrossRef]
5. Bonneuil, N.; Medina, M. Between Tradition and Modernity: The Transition of Contraception Use in Colombia. *Rev. Desarro. Soc.* **2009**, *2*, 119–151. [CrossRef]
6. Africander, D.; Verhoog, N.; Hapgood, J.P. Molecular mechanisms of steroid receptor-mediated actions by synthetic progestins used in HRT and contraception. *Steroids* **2011**, *76*, 636–652. [CrossRef]
7. Kroupova, H.K.; Trubiroha, A.; Lorenz, C.; Contardo-Jara, V.; Lutz, I.; Grabic, R.; Kocour, M.; Kloas, W. The progestin levonorgestrel disrupts gonadotropin expression and sex steroid levels in pubertal roach (Rutilus rutilus). *Aquat. Toxicol.* **2014**, *154*, 154–162. [CrossRef]
8. Olivera, E.R.; Luengo, J.M. Steroids as Environmental Compounds Recalcitrant to Degradation: Genetic Mechanisms of Bacterial Biodegradation Pathways. *Genes* **2019**, *10*, 512. [CrossRef]
9. Narvaez, J.F.; Grant, H.; Gil, V.C.; Porras, J.; Sanchez, J.C.B.; Duque, L.F.O.; Sossa, R.R.; Quintana-Castillo, J.C. Assessment of endocrine disruptor effects of levonorgestrel and its photoproducts: Environmental implications of released fractions after their photocatalytic removal. *J. Hazard. Mater.* **2019**, *371*, 273–279. [CrossRef]
10. Yu, X.; Sui, Q.; Lyu, S.; Zhao, W.; Wu, D.; Yu, G.; Barcelo, D. Rainfall Influences Occurrence of Pharmaceutical and Personal Care Products in Landfill Leachates: Evidence from Seasonal Variations and Extreme Rainfall Episodes. *Environ. Sci. Technol.* **2021**, *55*, 4822–4830. [CrossRef]
11. Torres-González, O.R.; Sánchez-Hernández, I.M.; Flores-Soto, M.E.; Chaparro-Huerta, V.; Soria-Fregozo, C.; Hernández-García, L.; Padilla-Camberos, E.; Flores-Fernández, J.M. Landfill Leachate from an Urban Solid Waste Storage System Produces Genotoxicity and Cytotoxicity in Pre-Adolescent and Young Adults Rats. *Int. J. Environ. Res. Public Health* **2021**, *18*, 11029. [CrossRef]
12. Yang, X.; Lin, H.; Zhang, Y.; He, Z.; Dai, X.; Zhang, Z.; Li, Y. Sorption and desorption of seven steroidal synthetic progestins in five agricultural soil-water systems. *Ecotoxicol. Environ. Saf.* **2020**, *196*, 110586. [CrossRef] [PubMed]
13. Fent, K. Progestins as endocrine disrupters in aquatic ecosystems: Concentrations, effects and risk assessment. *Environ. Int.* **2015**, *84*, 115–130. [CrossRef] [PubMed]
14. Klavarioti, M.; Mantzavinos, D.; Kassinos, D. Removal of residual pharmaceuticals from aqueous systems by advanced oxidation processes. *Environ. Int.* **2009**, *35*, 402–417. [CrossRef] [PubMed]
15. WHO. *WHO Laboratory Manual for the Examination and Processing of Human Semen*, 6th ed.; World Health Organization: Geneva, Switzerland, 2021.
16. Lomonte, B.; Gutiérrez, J.; Romero, M.; Núñez, J.; Tarkowski, A.; Hanson, L. An MTT-based method for the in vivo quantification of myotoxic activity of snake venoms and its neutralization by antibodies. *J. Immunol. Methods* **1993**, *161*, 231–237. [CrossRef]
17. Oropesa, A.L.; Guimaraes, L. Occurrence of Levonorgestrel in Water Systems and Its Effects on Aquatic Organisms: A Review. *Rev. Environ. Contam. Toxicol.* **2021**, *254*, 57–84.
18. Huang, Y.; Lin, H.; Song, R.; Tian, Y.; Zhang, Z. Optimization of dispersive liquid-liquid microextraction for analysis of levonorgestrel in water samples using uniform design. *Anal. Methods* **2011**, *3*, 857–864. [CrossRef]
19. Kot-Wasik, A.; Jakimska, A.; Sliwka-Kaszynska, M. Occurrence and seasonal variations of 25 pharmaceutical residues in wastewater and drinking water treatment plants. *Environ. Monit. Assess.* **2016**, *188*, 661. [CrossRef]
20. Yu, Q.; Geng, J.; Zong, X.; Zhang, Y.; Xu, K.; Hu, H.; Deng, Y.; Zhao, F.; Ren, H. Occurrence and removal of progestagens in municipal wastewater treatment plants from different regions in China. *Sci. Total Environ.* **2019**, *668*, 1191–1199. [CrossRef]
21. Fick, J.; Lindberg, R.H.; Parkkonen, J.; Arvidsson, B.; Tysklind, M.; Larsson, D.G.J. Therapeutic levels of levonorgestrel detected in blood plasma of fish: Results from screening rainbow trout exposed to treated sewage effluents. *Environ. Sci. Technol.* **2010**, *44*, 2661–2666. [CrossRef]
22. Hua, J.; Han, J.; Guo, Y.; Zhou, B. The progestin levonorgestrel affects sex differentiation in zebrafish at environmentally relevant concentrations. *Aquat. Toxicol.* **2015**, *166*, 1–9. [CrossRef] [PubMed]
23. Ma, X.; Liu, X.; Ding, S.; Su, S.; Gan, Z. Sorption and leaching behavior of bithionol and levamisole in soils. *Chemosphere* **2019**, *224*, 519–526. [CrossRef] [PubMed]
24. Grathwohl, P. Influence of Organic Matter from Soils and Sediments from Various Origins on the Sorption of Some Chlorinated Aliphatic Hydrocarbons: Implications on Koc Correlations. *Environ. Sci. Technol.* **1990**, *24*, 1687–1693. [CrossRef]
25. Ahmed, A.A.; Thiele-Bruhn, S.; Aziz, S.G.; Hilal, R.H.; Elroby, S.A.; Al-Youbi, A.O.; Leinweber, P.; Kühn, O. Interaction of polar and nonpolar organic pollutants with soil organic matter: Sorption experiments and molecular dynamics simulation. *Sci. Total Environ.* **2015**, *508*, 276–287. [CrossRef] [PubMed]
26. Yang, X.; Dai, X.; Zhang, Y.; Lin, H.; Wang, J.; He, Z.; Li, Y. Sorption, desorption, and transformation of synthetic progestins in soil and sediment systems. *Geoderma* **2020**, *362*, 114141. [CrossRef]

27. Lambert, J.-F. 7-Organic pollutant adsorption on clay minerals. In *Developments in Clay Science*; Elsevier: Amsterdam, The Netherlands, 2018; Volume 9.

28. Jianghuan, H.J.H.; Yongyong, G.; Bingsheng, Z. Reproductive Toxicity in Male Zebrafish after Long-term Exposure to Low Concentrations of Progestin Levonorgestrel. *Asian J. Ecotoxicol.* **2015**, *10*, 100–107.

29. Alfradique, V.A.P.; Batista, R.I.T.P.; Souza-Fabjan, J.M.G.; Côrtes, L.R.; Bragança, G.M.; de Souza, C.V.; da Costa, L.C.; Brandão, F.Z. Supplementation of 17beta-estradiol and progesterone in the co-culture medium of bovine oviductal epithelial cells and ovine spermatozoa reduces the sperm kinematics and capacitation. *Reprod. Biol.* **2018**, *18*, 368–379. [CrossRef]

30. Frankel, T.; Yonkos, L.; Ampy, F.; Frankel, J. Exposure to levonorgestrel increases nest acquisition success and decreases sperm motility in the male fathead minnow (Pimephales promelas). *Environ. Toxicol. Chem.* **2018**, *37*, 1131–1137. [CrossRef]

31. Lorenz, C.; Contardo-Jara, V.; Trubiroha, A.; Krüger, A.; Viehmann, V.; Wiegand, C.; Pflugmacher, S.; Nützmann, G.; Lutz, I.; Kloas, W. The synthetic gestagen Levonorgestrel disrupts sexual development in Xenopus laevis by affecting gene expression of pituitary gonadotropins and gonadal steroidogenic enzymes. *Toxicol. Sci.* **2011**, *124*, 311–319. [CrossRef]

32. Overturf, M.D.; Huggett, D.B. Responses to various exposure durations of levonorgestrel during early-life stages of fathead minnows (*Pimephales promelas*). *Aquat. Toxicol.* **2015**, *161*, 33–40. [CrossRef]

33. Stanczyk, F.Z.; Hapgood, J.; Winer, S.; Mishell, D.R. Progestogens used in postmenopausal hormone therapy: Differences in their pharmacological properties, intracellular actions, and clinical effects. *Endocr. Rev.* **2013**, *34*, 171–208. [CrossRef] [PubMed]

34. Botero-Coy, A.M.; Martínez-Pachón, D.; Boix, C.; Rincón, R.J.; Castillo, N.; Arias-Marín, L.; Manrique-Losada, R.; Torres-Palma, R.; Moncayo-Lasso, A.; Hernández, F. 'An investigation into the occurrence and removal of pharmaceuticals in Colombian wastewater'. *Sci. Total Environ.* **2018**, *642*, 842–853. [CrossRef] [PubMed]

35. Zhang, Q.; Wang, C.; Liu, W.; Qu, J.; Liu, M.; Zhang, Y.; Zhao, M. Degradation of the potential rodent contraceptive quinestrol and elimination of its estrogenic activity in soil and water. *Environ. Sci. Pollut. Res. Int.* **2014**, *21*, 652–659. [CrossRef] [PubMed]

36. Carlsen, E.; Giwercman, A.; Keiding, N.; Skakkebaek, N.E. Evidence for decreasing quality of semen during past 50 years. *BMJ* **1992**, *305*, 609–613. [CrossRef]

37. Crisp, T.M.; Clegg, E.D.; Cooper, R.L. *Special Report on Environmental Endocrine Disruption An Effects Assessment and Analysis*; U.S. Environmental Protection Agency: Washington, DC, USA, 1997; p. 11.

Article

Distribution and Transfer of Antibiotic Resistance Genes in Coastal Aquatic Ecosystems of Bohai Bay

Lei Jia [1], Hao Liu [1], Na Zhao [2], Qiuxia Deng [2], Chunhua Zhu [2] and Bo Zhang [1,2,*]

[1] Tianjin Fisheries Research Institute, Tianjin 300200, China; tianjinbohaisuo@163.com (L.J.); 2793773@163.com (H.L.)
[2] Southern Marine Science and Engineering Guangdong Laboratory-Zhanjiang, Fisheries College, Guangdong Ocean University, Zhanjiang 524000, China; zhaona1985@163.com (N.Z.); qiuxiadeng1992@163.com (Q.D.); hu860025@163.com (C.Z.)
* Correspondence: zb611273@163.com

Abstract: Antibiotic resistance genes (ARGs) are abundant in diverse ecosystems and the resistome may constitute a health threat for humans and animals. It is necessary to uncover ARGs and the accumulation mechanisms from different environmental sources. Various habitats, such as soil, seawater and fish intestines, could overflow a considerable amount of ARGs and the horizontal transfer of ARGs may occur in these environments. Thus, we assessed the composition and abundance of ARGs in seawater, soil and intestinal tracts of *Cynoglossus semilaevis* collected from different sites in Bohai Bay (China), including a natural area and three fish farms, through a high-throughput qPCR array. In total, 243 ARGs were uncovered, governing the resistance to aminoglycoside, multidrug, beta-lactamase, macrolide lincosamide streptomycin B (MLSB), chloramphenicol, sulfonamide, tetracycline, vancomycin and other antibiotics. The action mechanisms of these ARGs were mainly antibiotic deactivation, efflux pump and cellular protection. Importantly, similar ARGs were detected in different samples but show dissimilar enrichment levels. ARGs were highly enriched in the fish farms compared to the natural sea area, with more genes detected, while some ARGs were detected only in the natural sea area samples, such as bacA-02, tetL-01 and ampC-06. Regarding sample types, water samples from all locations shared more ARGs in common and held the highest average level of ARGs detected than in the soil and fish samples. Mobile genetic elements (MGEs) were also detected in three sample types, in the same trend as ARGs. This is the first study comparing the resistome of different samples of seawater, soil and intestines of *C. semilaevis*. This study contributes to a better understanding of ARG dissemination in water sources and could facilitate the effective control of ARG contamination in the aquatic environment.

Keywords: antibiotic resistance genes; seawater; soil; fish intestines; horizontal transfer

Citation: Jia, L.; Liu, H.; Zhao, N.; Deng, Q.; Zhu, C.; Zhang, B. Distribution and Transfer of Antibiotic Resistance Genes in Coastal Aquatic Ecosystems of Bohai Bay. *Water* **2022**, *14*, 938. https://doi.org/10.3390/w14060938

Academic Editors: François Gagné, Stefano Magni and Valerio Matozzo

Received: 29 January 2022
Accepted: 14 March 2022
Published: 17 March 2022

1. Introduction

Antibiotic resistance genes (ARGs) are genes governing the expression of enzymes involved in the degradation or in the chemical modification and subsequent inactivation of antibiotics [1]. The emergence of bacterial strains resistant to many antibiotics could be called into questions when using these antibiotics to fight against infectious diseases [2,3]. In the face of this public health problem, it is necessary to define the elements responsible for the emergence of these multiresistant bacteria.

It has now been established that the environment represents a reservoir of resistant bacteria or ARGs that could be transmissible to humans and animals [4]. The occurrence of ARGs in hospital settings, fish farms, organic waste products, wastewater treatment plants and soils/sediments has also been investigated [5–11]. Many organic waste products are used as fertilizer on agricultural soils [12] and treated wastewater or river water into which they are discharged can be used to irrigate crops [13]. Certain agricultural practices can, therefore, potentially lead to the environmental dissemination of ARGs and the contamination of animals, crops, soils and water resources [14,15]. For example, previous studies have reported that pond water in fish farms and shrimp farms contained ciprofloxacin, penicillin G, rifampicin and vancomycin-related ARGs [16]. The persistence of ARGs in the soils of fish farms has been strongly compared to the surrounding sediments [6].

Aquatic ecosystems constitute a big reservoir of ARGs. It was previously reported that aquaculture systems have a significant impact on the accumulation and spread of antibiotic-resistant bacteria (ARB) and ARGs in freshwater and sediment [16,17]. Fish may ingest ARB from the aquatic environment for their unique living environments, so the fish intestine would be an optimal niche for conjugal ARG transfer. In previous studies, it was reported that the guts of aquatic animals such as zebrafish contained a high abundance and diversity of bacteria, which indicated them as significant potential recipients of ARGs through horizontal gene transfer (HGT) and was beneficial to the spread of ARGs in water environments [18,19]. Moreover, the unique living environments are able to cause aquatic animals to ingest ARB in water easily [19]. There were some data regarding the abundance and the transfer of ARGs among water, sediment (soil) and organisms in aquatic systems [20,21], but the research topic needs an in-depth study. Here, we discuss the abundance of ARGs in aquaculture environments as well as fish ecosystems and the occurrence of the transfer of ARGs among these micro-systems.

The Chinese tongue sole (*Cynoglossus semilaevis*) is a marine fish species widely spread in the Bohai Sea and cultured in the area; however, not enough attention has been paid to the distributions of ARGs related to this fish till now. Based on the above, this study intends to investigate the different classes of ARGs present in the intestinal tract of *C. semilaevis*, culture water and soils in the same area. We also aimed to investigate the exchange and transfer mechanisms of ARGs between these ecosystems. This work hopes to provide a better understanding of the abundance and dissemination of ARGs in aquatic ecosystems.

2. Materials and Methods

2.1. Study Sites and Sample Collection

The sampling sites selected for this study are located in Bohai Bay, which is situated in the west of the Bohai Sea, in northeast China (Figure 1). The four sampling sites were as follows: P0 (117.935° E, 39.0415° N), P1 (117.626544° E, 38.852927° N), P2 (117.912634° E, 39.304451° N) and P3 (117.964958° E, 39.264878° N). There were three types of samples, including intestinal tracts of *C. semilaevis* (A), sea water (B) and soil (C), collected from the natural sea area or fish farms. Fish samples A0–A3 (*C. semilaevis*) were intestinal tracts collected using sterile plastic bags, followed by separation with sterile forceps and homogenization in a sterile mortar with liquid nitrogen. A volume of 5 L of water samples so-called B0–B3 were collected from 30 cm below the surface with a stainless steel sampler, and then transferred to sterile polyethylene bottles. For the extraction of DNA, 200 mL of each sample B was filtered with vacuum filtration apparatus with a sterile membrane filter (0.22 μm pore diameter), which was aseptically stored at −80 °C. About 100 g of dry soil samples C0–C3 was collected from a depth of 0–10 cm with the help of a grab sampler and gently mixed using sterile polyethylene bottles. For the fish farm area, soil at the drain outlet was sampled. Soil samples were then freeze-dried, ground with a mortar and sieved through a 100-mesh screen. All samples were collected in three replicates, rapidly stored in a portable ice box, which were then transported to the laboratory and the pretreatment was performed within 1 day (Table 1).

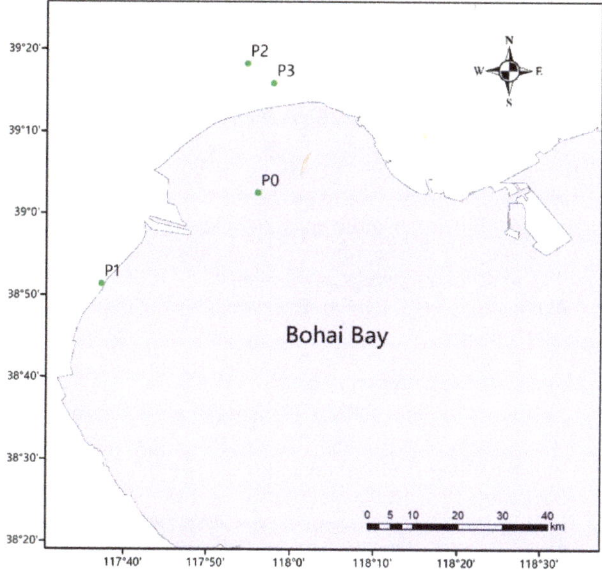

Figure 1. Map of the sampling location and sampling sites: The four sampling sites were P0 (117.935° E, 39.0415° N), P1 (117.626544° E, 38.852927° N), P2 (117.912634° E, 39.304451° N) and P3 (117.964958° E, 39.264878° N) in the west of the Bohai Sea, in northeast China.

Table 1. Samples from different sites.

Site	Type	Sample		
		Intestinal Tracts of Fish	**Sea Water**	**Soil**
P0	Natural sea area	A0	B0	C0
P1	Fish farm	A1	B1	C1
P2	Fish farm	A2	B2	C2
P3	Fish farm	A3	B3	C3

2.2. DNA Extraction and Purification

According to the manufacturer's instructions and previous studies [4,22], total DNA was extracted from samples A, B and C using the DNeasy PowerWater Kit (QIAGEN Sciences, Germantown, MD, USA), the DNeasy PowerSoil Kit (QIAGEN Sciences, Germantown, MD, USA) and the QIAamp DNA Stool Mini Kit (QIAGEN Sciences, Germantown, MD, USA), respectively. A Nanodrop 1000™ spectrophotometer (Thermo Scientific, Wilmington, DE, USA) was used to analyze the DNA quality and concentration.

2.3. High-Throughput Quantitative PCR

Detection of ARGs was performed using the Wafergen Smartchip ultra-high-throughput fluorescent quantitative PCR system according to the method of Zhou [4], by controlling the total DNA concentration of bacteria in different sampling media identically. In this experiment, 296 pairs of primers were set up, including 285 pairs of primers for ARGs, 9 pairs of primers for MGEs, 1 pair of primers for the clinical class 1 integron-integrase gene and 1 pair of 16s rRNA internal reference primers to normalize the abundance of ARGs in different sample types [22]. The PCR mixture was introduced into a microwell chip using a nanoscale multi-sample spotter (MSND) 296 (assays) × 16 (samples) mode, followed by qPCR on the cycler. One no-template control (NTC) with three repetitions was set for each assay in each chip for each sample. The reaction system included 1 × LightCycler® 480 SYBR Green I Master, 500 nM of each primer and 2 ng/μL DNA template. The total reaction volume was 100 nL. PCR reaction conditions were as follows: predenaturation at 95 °C, 10 min, 40 cycles of 95 °C for 30 s; 60 °C for 30 s; dissolution curves. The qPCR results were analyzed through the qPCR software of the instrument. CT = 31 was set as the detection domain, and two or more of the three replicates were detected, with the deviation < 20%. All 16s rRNA internal reference genes of all samples with concentration $\geq$10 ng/μL in this project were detected, and NTC was not amplified, which proved that the experimental results were credible. The absolute copy number of each gene was calculated indirectly based on the relative copy number. The formula used for calculation was as follows: absolute copy number of target gene = 16 s rRNA gene absolute copy numbe $\times$ $2^{-\Delta CT}$, which means the absolute abundance of each gene was recorded as the product of the relative copy number and the 16S rRNA's absolute copy number, where in $\Delta CT = CT$ (gene)-CT (16 s). It is worth noting that CT (gene) and CT (16 s) were from the same sample.

2.4. Statistical Analysis

Distribution of ARG type in each sample was visualized via Circos graphs, which were carried out by Circos software (http://circos.ca/; accessed on 17 October 2021) online, while Heatmap graphs were produced using RStudio with pheatmap package. The $2^{-\Delta CT}$ abundance table was used to calculate the Pearson Correlation Coefficient and the Bray–Curtis distance. Adonis and Anosim analyses were then conducted using the Bray–Curtis distance matrix. Ordinary least square (OLS) graphs were drawn based on the $2^{-\Delta CT}$ abundance table. The correlation analysis among ARGs and MEGs was performed using the Hmisc package in R. One-way ANOVA test in SPSS version 26.0 was used to analyze the statistical differences of quantity of ARGs between fish farms and the adjacent natural sea area.

3. Results

3.1. Distribution, Classification and Abundance of ARGs in Different Environments Associated with C. semilaevis Farming Industry

The results of the high-throughput PCR indicated that a variety of ARGs could be found in different kinds of samples (Figure S1 in Supplementary Materials). The 243 detected ARGs included genes of eight distinct and predominant classes, conferring resistance to antibiotics such as aminoglycosides, multidrug, beta-lactamases, macrolide lincosamide streptomycin B (MLSB), chloramphenicol, sulfonamide, tetracycline and vancomycin, with a few of them related to other antibiotics (Figure S1 in Supplementary Materials). The proportion of the eight classes in each sample depended on both the sample type and location.

In *C. semilaevis* samples, ARGs were mostly abundant in A2 (P2), followed by A1 (P1) and A3 (P3). The lowest number of ARGs was found in A0 (P0). The copy number of almost all of the ARGs was higher in the A2 sample too (Figure S1A in Supplementary Materials). In A2, ARGs related to aminoglycosides (26 genes), beta-lactamases (26 genes), MLSB (23 genes), multidrug (20 genes) and tetracycline (14 genes) were the most abundant, among which the aph3ia (aminoglycoside), tetk (Tetracycline), ermK (MLSB) and bl3_cpha (beta-lactamase) genes were the ARGs with the highest copy numbers (Figure S1A in Supplementary Materials). ARGs responsible for resistance to chloramphenicol could not be detected in A1–A3, while ARGs related to aminoglycosides and tetracycline resistance were absent in A0. In general, in type A samples, the abundant ARGs were multidrug, MLSB, beta-lactamases, aminoglycoside and tetracycline-related ones.

In sea water samples, tetracycline, aminoglycosides and multidrug-related ARGs were the most abundant in B1 (P1), while tetracycline, aminoglycosides and MLSB in B2 (P2) (Figure S1B in Supplementary Materials). In B3, ARGs involved in the resistance to beta-lactamase, multidrug and tetracycline were the most abundant (Figure S1B in Supplementary Materials). The number of ARGs was lower in B0 (natural seawater collected from P0) compared to the culture water samples (Figure S1B in Supplementary Materials), among which multidrug, tetracycline and aminoglycosides-related ARGs were dominant. In type

B samples, tetracycline, aminoglycoside, multidrug, beta-lactamases and MLSB-related ones were the main ARGs.

ARGs were detected in the soil samples C0-C3, among which multidrug, beta-lactamases, aminoglycoside, MLSB and tetracycline-related ARGs were in the majority (Figure S1C in Supplementary Materials). These ARGs were abundant in C1 (P1) and C2 (P2) compared to other soil samples (Figure S1 in Supplementary Materials).

3.2. Differential Distribution of ARGs between Fish Farms and the Adjacent Natural Sea Area

Based on the statistical analysis with the one-way ANOVA test in SPSS version 26.0, the comparison between samples from the fish farms (P1, P2 and P3) and the natural sea area (P0) indicated that ARGs were highly enriched in the fish farms, with more genes detected (Figure S1, Table S1 in Supplementary Materials). In addition, ARGs with a high abundance were recorded in P2. Some ARGs were detected only in the samples from the natural sea area, such as bacA-02, tetL-01 and ampC-06. Water samples from four locations shared more ARGs in common.

3.3. Mechanisms of ARGs in Different Type of Samples

The mechanisms of ARGs detected in all type of samples were antibiotic deactivation, cellular protection and efflux pump, with little other or unknown mechanisms (Figure S1 in Supplementary Materials). These ARG mechanisms were predominant in the A2 sample, but they were found in lesser extent in the other fish samples. In soil samples, antibiotic deactivation and efflux pump were the most representative mechanisms, especially in C1 and C2.

3.4. Transfer Potential of ARG Types between Fish, Soil and Water Samples

A comparative analysis was performed between samples from the gut of *C. semilaevis* (A0-A3), water (B0-B3) and soil (C0-C3) to evaluate the possibility of HGT between the different biomes. The correlation of ARGs detected from different samples showed a good cluster in the sample type (Figure 2), and samples from fish (A) were distinguished from water (B) and soil (C) (Adonis: *p*-value = 0.017; Anosim: *p*-value = 0.011). The results (Table S1 in Supplementary Materials) indicated that the average level of ARGs detected in fish samples was lower compared to the soil samples and the numbers of ARGs in soil samples were lower compared to the water samples. Relatively pristine environments (A0 and B0) revealed fewer ARGs than other fish farm (Table S1 in Supplementary Materials). Clustered A1 and A2 separated from A0; B1, 2 and 3 separated from B0 (Figure 2). The Venn diagram showing the distribution of ARGs among fish (A), water (B) and soil (C) samples was depicted (Figure 3). A total of 366 ARGs, all classes combined, was screened from the water samples, while 340 ARGs were screened from the soil samples and only 221 ARGs were obtained from the fish samples. Some individual ARGs followed this trend. For example, 62 ARGs encoding for resistances to aminoglycosides were found in water samples, while only 46 and 30 ARGs were recorded for the soil and fish samples, respectively. In addition, 10 ARGs encoding for resistance to chloramphenicol were obtained from water samples, while 6 and 1 from the soil and fish samples, respectively. The same trend was observed for ARGs encoding for tetracycline, as well as mobile genetic elements (MGEs). CIntI-1 (class1), IS613, Tp614, intI-1 (clinic), tnpA01, tnpA02, tnpA03, tnpA04, tnpA05 and tnpA07 were detected, and cIntI-1 (class1), intI-1 (clinic) tnpA04 and tnpA07 were the most common MGEs among the three sample types. Nevertheless, some disparities were observed. ARGs encoding for β-lactam, multidrug and sulfonamide were more abundant in soil samples than the water and fish samples, in this order. Additionally, the MLSB-associated ARGs were uniformly distributed in the water (32), soil (42) and fish (42) samples.

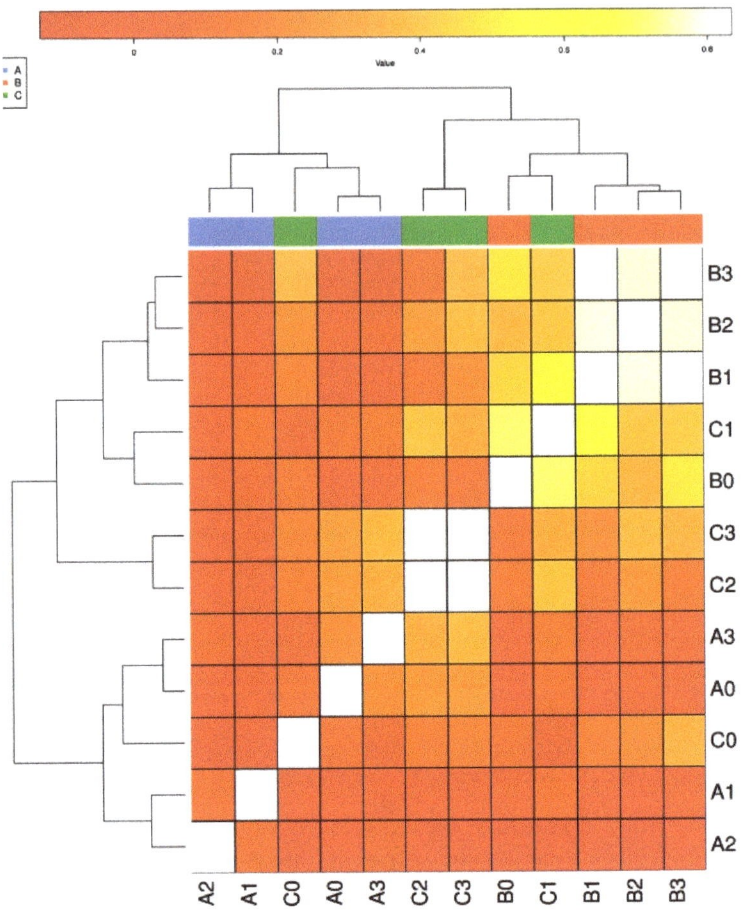

Figure 2. Correlation of ARGs between different samples: detected correlation of ARGs between different samples from the gut of *C. semilaevis* (A0–A3), water samples (B0–B3) and soil samples (C0–C3) to evaluate the possibility of HGT between the different biomes.

MGEs were in direct proportion with ARGs, and the integrase genes and transposases were in the same trend (Figure S2 in Supplementary Materials). The Spearman correlation analysis among ARGs indicated a significant correlation among all the ARGs across the studied samples (Figure S3 in Supplementary Materials). The correlation coefficient R was > 0.7, and the p-value was < 0.05. The integrase gene intI-1 (clinic) was significantly and positively correlated with ARGs related to multidrug, MLSB and tetracycline resistance., while the other integrase gene cIntI-1 (class1) was significantly and positively correlated with ARGs related to aminoglycosides and tnpA07. Transposases exhibited more correlations with ARGs. IS613 was significantly and positively correlated with vancomycin ARGs, while Tp614 with MLSB, beta-lactamase, aminoglycoside, multidrug and other ARGs were not. TnpA01 was significantly and positively correlated with MLSB, while tnpA02 mainly with tetracycline and beta-lactamase, tnpA03 with MLSB, multidrug, vancomycin and aminoglycosides and tnpA07 were significantly and positively correlated with aminoglycoside ARGs. TnpA04 and tnpA05 interacted more tightly with ARGs related to tetracycline and beta-lactamase.

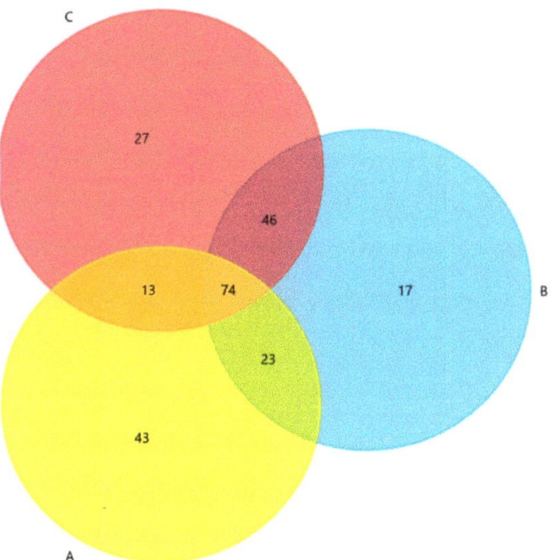

Figure 3. Venn diagram indicating the abundance of shared and dissimilar ARGs in the three types of samples: Three types of samples were fish samples (**A**), water samples (**B**) and soil samples (**C**). There were 366 ARGs in total, 74 of them shared in the three types of samples, 23 of them shared between fish samples and water samples, 46 of them shared between water samples and soil samples and 13 of them shared between fish samples and soil samples.

4. Discussion

Studies have shown that approximately 30% to 90% of used antibiotics promote the production and spread of ARGs [23], but the abundance and dissemination of ARGs in the environment and their mechanisms are far from being completely elucidated. The present study aimed to explore the distribution of ARGs in aquatic microecosystems. Eight classes of ARGs, namely, aminoglycoside, multidrug, beta-lactamase, MLSB, chloramphenicol, sulfonamide, tetracycline and vancomycin ARGs, which are all common antibiotics, were those predominantly abundant in the aquatic environment. The abundance of these ARGs differed according to the microsystem and sampling site, mostly abundant in the soils, indicating an accumulation in the sediments. Regardless of the type of samples, the mechanisms adopted by the ARGs were the same and were represented by antibiotic deactivation, cellular protection and efflux pump. These results were the first look into the distribution and spread of ARGs related to the *C. semilaevis* breeding industry.

The results of the high-throughput analysis of the resistome revealed that a high level of ARGs could be found in the soil, water and intestinal tract of *C. semilaevis*. These results signposted that fish farms constitute a great reservoir of ARGs, conforming with previous research demonstrating that fish-rearing ponds, lakes, bays [24–26] and shrimp ponds [16,17] were all enriched in ARGs. In addition, the bacterial community structures also differed, which might affect the composition of ARGs [27–29]. The aminoglycoside-related ARGs were more abundant in the water samples, and this was corroborated by previous findings indicating that aminoglycoside ARGs are abundant and persistent in wastewater plants [9] and drinking waters [30]. The presence of aminoglycoside ARGs in the fish intestinal tract and soil samples was also supported by studies conveying the persistence of aminoglycoside ARGs in the gut of fish and zoo animals [31] and plenty of studies reporting the presence of aminoglycoside ARGs in the soils of diverse types [6,12,13,26,32]. Our findings also indicated that β-lactam ARGs were enriched in soil and water samples, but were found in a lesser extent in the fish intestinal tract. This observation is supported by previous works in the Tibetan environment, sewage sludge field application soils, agricultural soils and parks [9,30,33–37]. The abundance of chloramphenicol ARGs was relatively low in our samples, but its occurrence was in line with previous works which indicated the presence of chloramphenicol ARGs in feedlots, Baltic sea fish farm soils, water effluents and pristine remote Antarctic soils [6,8,32,38]. MLSB-related ARGs were uniformly distributed in the fish intestinal tract, water and soil samples. Studies reporting MLSB ARGs in fish intestinal tract are limited, but these ARGs have been reported in water effluents and soils treated with sewage sludge [8,26]. A particular enrichment of multidrug ARGs was recorded in the soil samples, followed by water and fish intestinal tract samples. These ARGs have been also reported in soil and water samples by other researchers [6,16,26,30,36,39]. Similarly, sulfonamide, tetracycline and vancomycin-resistant ARGs were all found in the three types of samples, especially tetracycline, which corroborated previous findings [6,8,30,40,41]. Our study revealed a diversity of ARGs with different degrees of abundance in water, the intestinal tract of *C. semilaevis* and soil in the aquatic settings and adjacent natural sea areas.

Disparities were found in ARGs detected from the same type of samples collected from diverse geographical locations. The level of ARGs was lower in the samples from natural sources. In daily production, antibiotics are usually applied through the splashing of the factory culture ponds in the fish farm, mainly florfenicol, quinolones (oxoquine), streptomycin, neomycin (neomycin sulfate and tetracycline), doxycycline hydrochloride, sulfonamide antibiotics (sulfadiazine and sulfamethoxazole) and enrofloxacin, related to the ARGs detected here. A frequent outside release of the culture water may decrease the concentration of antibiotics and ARB in it, which could explain the decreased level of ARGs in the water samples. The discharge of effluent water in the soil may also explain the increased accumulation of ARGs in soil samples. Indeed, previous studies indicated that the effluent water could be released into the natural world, infiltrating into the ground or being dumped into the sea, which leads to the accumulation of antibiotic-resistant bacteria in the soil; thus, causing the enrichment of soil in ARGs, which is equivalent to ARG transfer and enrichment from water to soil [29]. From the geographical map of our sampling, the three fish farms were all near the sea and the water for fish farming came from the sea, a large part of which would flow back in. Therefore, the transfer of ARGs may occur from the coastal fish farm to the sea (including natural seawater, sea mud and marine organisms such as fish species). Accordingly, inflow water was also proved to be a major source of trout farming contamination with salmonella and multidrug-resistant bacteria [27]. Thus, a cycle accumulation pattern seems to exist.

MGEs involved in the mobility of ARGs are an important factor [33,36,42,43] and were detected in all three types of samples, such as intI-1 (clinic) tnpA04 and tnpA07. Wild fish in a peri-urban river might be recipients and communicators of ARGs in the water environment and ARGs might transfer between fish and water using bacteria as a spreading medium, with nine ARGs and one MGE continuously shared in both sample types [44]. Notably, a differentially ecological niche of the host resulted in the various resistomes of the lower vertebrate, among which the resistomes of seawater fish shared more similarities and were characterized by a high abundance, distinct from other observed habitats, and ARGs in seawater fish were colocalized with mobile elements with a high frequency, suggesting that they were likely spread through HGT [45]. On the whole, these data revealed the significant diversity and heterogeneity of ARGs in lower vertebrates, indicating that these wild species potentially play an essential role in the global spread of ARGs [44,45].

The HGT of ARGs among environmental bacteria in different taxa is proved to be an important pathway that disseminates resistance and, subsequently, acquires resistance through human pathogens and commensals. Since transposition-related genes and integrases were found in the studied samples, HGT might occur between bacteria by transposition or integration in the aquatic environment. This threes in corroboration with previous reports, confirming that ARGs are introduced by DNA transposition into plasmid and subsequent drug selection in the aquaculture sites [46]. Further studies should be carried out to verify these HGT mechanisms.

In this study, mechanisms driven by ARGs were mainly antibiotic deactivation, cellular protection and efflux pump. Efflux pump is the major mechanism of multidrug resistance in bacteria and acts by interfering with the synthesis of nucleic acid or the inhibition of protein synthesis [6,13,17,47]. Antibiotic deactivation has also been reported in previous studies on urban park soils with reclaimed water irrigation [2,13]. Cellular protection has been proved as a mechanism driven by ARGs screened from farmed fish feces [6], which corroborated our results. Antibiotic deactivation was the most abundant ARG mechanism in the fish intestinal extract, water and soil samples, which indicated that the deactivation of antibiotics may be the main mechanism adopted by resistant bacteria in the aquatic environment. The three mechanisms were also reported by Muziasari [6]. Thus, finding ways to counteract this mechanism could be the key to overcome antibiotic resistance, especially for animals left in these environments.

This study investigated the distributions, diversity, abundance and potential transfer mechanism of ARGs in *C. semilaevis*-related samples; however, these results still need to be verified by a larger-scale sample test. This work also outlined the crisis of ARG transmission from fish farms to adjacent sea areas.

5. Conclusions

Our study indicated that there were abundant ARGs in the aquatic ecosystem and that there was a transfer of these ARGs into the water, which, subsequently, led to the serious enrichment of aquatic sediments in ARGs. This study allowed not only the knowledge of ARGs in the *C. semilaevis*-related niche, but also the understanding of ARG dissemination in aquatic ecosystems. Therefore, the investigation and in-depth study of the distribution and spread of ARGs could allow us to control the spread of ARGs in the environment and minimize the risk to animal and human health.

Supplementary Materials: The following supporting information can be downloaded at: https://www.mdpi.com/article/10.3390/w14060938/s1. Figure S1: distribution and abundance of detected ARGs, Figure S2: correlation between trends of ARGs and MEGs in three types of samples, Figure S3: correlations among the detected ARGs and MEGs, Table S1: Abundance and distribution of ARG types in different samples. Table S2: The information of primers used in the study.

Author Contributions: Conceptualization, B.Z.; methodology, H.L.; validation, Q.D. and H.L.; formal analysis, L.J.; investigation, N.Z.; data curation, N.Z.; writing—review and editing, L.J.; project administration, C.Z.; funding acquisition, L.J. All authors have read and agreed to the published version of the manuscript.

Funding: This research was funded by grants from the Tianjin Natural Science Foundation (17JC-QNJC15000), China Agriculture Research System of MOF and MARA (CARS-47-Z01) and the modern industrial technology system in Tianjin ITTMRS2022001.

Institutional Review Board Statement: The animal study protocol was approved by the Institutional Review Board of Tianjin Fisheries Research Institute (protocol code 2019003 and date of approval is 2 April 2019).

Informed Consent Statement: Not applicable.

Data Availability Statement: The data presented in this study are available on request from the corresponding author.

Acknowledgments: The authors would like to acknowledge the support of the client project team and their collaborative effort with the wider project team, which included: the Tianjin Fisheries Research Institute; Guangdong Ocean University; Southern Marine Science and Engineering Guangdong Laboratory-Zhanjiang.

Conflicts of Interest: The authors declare no conflict of interest.

References

1. Lin, H.; Zhang, J.; Chen, H.; Wang, J.; Sun, W.; Zhang, X.; Yang, Y.; Wang, Q.; Ma, J. Effect of temperature on sulfonamide antibiotics degradation, and on antibiotic resistance determinants and hosts in animal manures. *Sci. Total Environ.* **2017**, *607–608*, 725–732. [CrossRef] [PubMed]
2. Pontes, D.S.; de Araujo, R.S.A.; Dantas, N.; Scotti, L.; Scotti, M.T.; de Moura, R.O.; Mendonca-Junior, F.J.B. Genetic mechanisms of antibiotic resistance and the role of antibiotic adjuvants. *Curr. Top. Med. Chem.* **2018**, *18*, 42–74. [CrossRef] [PubMed]
3. Loftie-Eaton, W.; Crabtree, A.; Perry, D.; Millstein, J.; Baytosh, J.; Stalder, T.; Robison, B.D.; Forney, L.J.; Top, E.M. Contagious Antibiotic Resistance: Plasmid Transfer among Bacterial Residents of the Zebrafish Gut. *Appl. Environ. Microbiol.* **2021**, *87*, e02735-20. [CrossRef] [PubMed]
4. Zhou, Z.C.; Feng, W.Q.; Han, Y.; Zheng, J.; Chen, T.; Wei, Y.Y.; Gillings, M.; Zhu, Y.G.; Chen, H. Prevalence and transmission of antibiotic resistance and microbiota between humans and water environments. *Environ. Int.* **2018**, *121*, 1155–1161. [CrossRef] [PubMed]
5. Hassan, M.; Brede, D.A.; Diep, D.B.; Nes, I.F.; Lotfipour, F.; Hojabri, Z. Efficient inactivation of multi-antibiotics resistant nosocomial enterococci by purified Hiracin bacteriocin. *Adv. Pharm. Bull.* **2015**, *5*, 393–401. [CrossRef]
6. Muziasari, W.I.; Pitkänen, L.K.; Sørum, H.; Stedtfeld, R.D.; Tiedje, J.M.; Virta, M. The resistome of farmed fish feces contributes to the enrichment of antibiotic resistance genes in sediments below Baltic Sea fish farms. *Front. Microbiol.* **2016**, *7*, 2137.
7. Calero-Cáceres, W.; Méndez, J.; Martín-Díaz, J.; Muniesa, M. The occurrence of antibiotic resistance genes in a Mediterranean river and their persistence in the riverbed sediment. *Environ. Pollut.* **2017**, *223*, 384–394. [CrossRef]
8. Szekeres, E.; Baricz, A.; Chiriac, C.M.; Farkas, A.; Opris, O.; Soran, M.L.; Andrei, A.S.; Rudi, K.; Balcázar, J.L.; Dragos, N.; et al. Abundance of antibiotics, antibiotic resistance genes and bacterial community composition in wastewater effluents from different Romanian hospitals. *Environ. Pollut.* **2017**, *225*, 304–315. [CrossRef]
9. An, X.L.; Su, J.Q.; Li, B.; Ouyang, W.Y.; Zhao, Y.; Chen, Q.L.; Cui, L.; Chen, H.; Gillings, M.R.; Zhang, T.; et al. Tracking antibiotic resistome during wastewater treatment using high throughput quantitative PCR. *Environ. Int.* **2018**, *117*, 146–153. [CrossRef]
10. Lerminiaux, N.A.; Cameron, A.D.S. Horizontal transfer of antibiotic resistance genes in clinical environments. *Can. J. Microbiol.* **2019**, *65*, 34–44. [CrossRef]
11. Bouki, C.; Venieri, D.; Diamadopoulos, E. Detection and fate of antibiotic resistant bacteria in wastewater treatment plants: A review. *Ecotoxicol. Environ. Saf.* **2013**, *91*, 1–9. [CrossRef] [PubMed]
12. Liu, P.; Jia, S.; He, X.; Zhang, X.; Ye, L. Different impacts of manure and chemical fertilizers on bacterial community structure and antibiotic resistance genes in arable soils. *Chemosphere* **2017**, *188*, 455–464. [CrossRef] [PubMed]
13. Wang, F.H.; Qiao, M.; Su, J.Q.; Chen, Z.; Zhou, X.; Zhu, Y.G. High throughput profiling of antibiotic resistance genes in urban park soils with reclaimed water irrigation. *Environ. Sci. Technol.* **2014**, *48*, 9079–9085. [CrossRef] [PubMed]
14. Larrañaga, O.; Brown-Jaque, M.; Quirós, P.; Gómez-Gómez, C.; Blanch, A.R.; Rodríguez-Rubio, L.; Muniesa, M. Phage particles harboring antibiotic resistance genes in fresh-cut vegetables and agricultural soil. *Environ. Int.* **2018**, *115*, 133–141. [CrossRef] [PubMed]
15. Muurinen, J.; Stedtfeld, R.; Karkman, A.; Pärnänen, K.; Tiedje, J.; Virta, M. Influence of manure application on the environmental resistome under finnish agricultural practice with restricted antibiotic use. *Environ. Sci. Technol.* **2017**, *51*, 5989–5999. [CrossRef]
16. Pham, T.T.H.; Rossi, P.; Dinh, H.D.K.; Pham, N.T.A.; Tran, P.A.; Ho, T.; Dinh, Q.T.; De Alencastro, L.F. Analysis of antibiotic multi-resistant bacteria and resistance genes in the effluent of an intensive shrimp farm (Long An, Vietnam). *J. Environ. Manag.* **2018**, *214*, 149–156. [CrossRef]
17. Zhao, Y.; Zhang, X.X.; Zhao, Z.; Duan, C.; Chen, H.; Wang, M.; Ren, H.; Yin, Y.; Ye, L. Metagenomic analysis revealed the prevalence of antibiotic resistance genes in the gut and living environment of freshwater shrimp. *J. Hazard Mater.* **2018**, *350*, 10–18. [CrossRef]
18. Almeida, A.R.; Tacão, M.; Soares, J.; Domingues, I.; Henriques, I. Tetracycline-resistant bacteria selected from water and zebrafish after antibiotic exposure. *Int. J. Environ. Res. Public Health* **2021**, *18*, 3218. [CrossRef]
19. Fu, J.; Yang, D.; Jin, M.; Liu, W.; Zhao, X.; Li, C.; Zhao, T.; Wang, J.; Gao, Z.; Shen, Z.; et al. Aquatic animals promote antibiotic resistance gene dissemination in water via conjugation: Role of different regions within the zebra fish intestinal tract, and impact on fish intestinal microbiota. *Mol. Ecol.* **2017**, *26*, 5318–5333. [CrossRef]
20. Wang, L.; Su, H.; Hu, X.; Xu, Y.; Xu, W.; Huang, X.; Li, Z.; Cao, Y.; Wen, G. Abundance and removal of antibiotic resistance genes (ARGs) in the rearing environments of intensive shrimp aquaculture in South China. *J. Environ. Sci. Health B* **2019**, *54*, 211–218. [CrossRef]
21. Cycoń, M.; Mrozik, A.; Piotrowska-Seget, Z. Antibiotics in the soil environment-degradation and their impact on microbial activity and diversity. *Front. Microbiol.* **2019**, *10*, 338. [CrossRef] [PubMed]
22. Zhu, Y.G.; Zhao, Y.; Li, B.; Huang, C.L.; Zhang, S.Y.; Yu, S.; Chen, Y.S.; Zhang, T.; Gillings, M.R.; Su, J.Q. Continental-scale pollution of estuaries with antibiotic resistance genes. *Nat. Microbiol.* **2017**, *2*, 16270. [CrossRef] [PubMed]
23. Ma, J.; Cui, Y.; Li, A.; Zou, X.; Ma, C.; Chen, Z. Antibiotics and antibiotic resistance genes from wastewater treated in constructed wetlands. *Ecol. Eng.* **2022**, *177*, 106548. [CrossRef]
24. Calero-Cáceres, W.; Melgarejo, A.; Colomer-Lluch, M.; Stoll, C.; Lucena, F.; Jofre, J.; Muniesa, M. Sludge as a potential important source of antibiotic resistance genes in both the bacterial and bacteriophage fractions. *Environ. Sci. Technol.* **2014**, *48*, 7602–7611. [CrossRef] [PubMed]

25. Korzeniewska, E.; Harnisz, M. Relationship between modification of activated sludge wastewater treatment and changes in antibiotic resistance of bacteria. *Sci. Total Environ.* **2018**, *639*, 304–315. [CrossRef] [PubMed]

26. Xie, W.Y.; McGrath, S.P.; Su, J.Q.; Hirsch, P.R.; Clark, I.M.; Shen, Q.; Zhu, Y.G.; Zhao, F.J. Long-term impact of field applications of sewage sludge on soil antibiotic resistome. *Environ. Sci. Technol.* **2016**, *50*, 12602–12611. [CrossRef]

27. Antunes, P.; Campos, J.; Mourão, J.; Pereira, J.; Novais, C.; Peixe, L. Inflow water is a major source of trout farming contamination with Salmonella and multidrug resistant bacteria. *Sci. Total Environ.* **2018**, *642*, 1163–1171. [CrossRef] [PubMed]

28. Gu, J.; Zhang, L.; Wang, X.; Lu, C.; Liu, J.; Liu, Y.; Li, L.; Peng, J.; Xue, M. High-throughput analysis of the effects of different fish culture methods on antibiotic resistance gene abundances in a lake. *Environ. Sci. Pollut. Res. Int.* **2019**, *26*, 5445–5453. [CrossRef]

29. Yuan, J.; Ni, M.; Liu, M.; Zheng, Y.; Gu, Z. Occurrence of antibiotics and antibiotic resistance genes in a typical estuary aquaculture region of Hangzhou Bay, China. *Mar. Pollut. Bull.* **2019**, *138*, 376–384. [CrossRef]

30. Ma, L.; Li, B.; Jiang, X.T.; Wang, Y.L.; Xia, Y.; Li, A.D.; Zhang, T. Catalogue of antibiotic resistome and host-tracking in drinking water deciphered by a large scale survey. *Microbiome* **2017**, *5*, 154. [CrossRef]

31. Bello González Tde, J.; Zuidema, T.; Bor, G.; Smidt, H.; van Passel, M.W. Study of the aminoglycoside subsistence phenotype of bacteria residing in the gut of humans and zoo animals. *Front. Microbiol.* **2015**, *6*, 1550. [CrossRef] [PubMed]

32. Van Goethem, M.W.; Pierneef, R.; Bezuidt, O.K.I.; Van De Peer, Y.; Cowan, D.A.; Makhalanyane, T.P. A reservoir of 'historical' antibiotic resistance genes in remote pristine Antarctic soils. *Microbiome* **2018**, *6*, 40. [CrossRef] [PubMed]

33. Chen, B.; Yuan, K.; Chen, X.; Yang, Y.; Zhang, T.; Wang, Y.; Luan, T.; Zou, S.; Li, X. Metagenomic analysis revealing antibiotic resistance genes (ARGs) and their genetic compartments in the Tibetan environment. *Environ. Sci. Technol.* **2016**, *50*, 6670–6679. [CrossRef] [PubMed]

34. Chen, Q.; An, X.; Li, H.; Su, J.; Ma, Y.; Zhu, Y.G. Long-term field application of sewage sludge increases the abundance of antibiotic resistance genes in soil. *Environ. Int.* **2016**, *92–93*, 1–10. [CrossRef] [PubMed]

35. Cheng, J.H.; Tang, X.Y.; Cui, J.F. Effect of long-term manure slurry application on the occurrence of antibiotic resistance genes in arable purple soil (entisol). *Sci. Total Environ.* **2019**, *647*, 853–861. [CrossRef]

36. Gou, M.; Hu, H.W.; Zhang, Y.J.; Wang, J.T.; Hayden, H.; Tang, Y.Q.; He, J.Z. Aerobic composting reduces antibiotic resistance genes in cattle manure and the resistome dissemination in agricultural soils. *Sci. Total Environ.* **2018**, *612*, 1300–1310. [CrossRef]

37. Han, X.M.; Hu, H.W.; Shi, X.Z.; Wang, J.T.; Han, L.L.; Chen, D.; He, J.Z. Impacts of reclaimed water irrigation on soil antibiotic resistome in urban parks of Victoria, Australia. *Environ. Pollut.* **2016**, *211*, 48–57. [CrossRef]

38. He, L.Y.; Liu, Y.S.; Su, H.C.; Zhao, J.L.; Liu, S.S.; Chen, J.; Liu, W.R.; Ying, G.G. Dissemination of antibiotic resistance genes in representative broiler feedlots environments: Identification of indicator ARGs and correlations with environmental variables. *Environ. Sci. Technol.* **2014**, *48*, 13120–13129. [CrossRef]

39. Kang, W.; Zhang, Y.J.; Shi, X.; He, J.Z.; Hu, H.W. Short-term copper exposure as a selection pressure for antibiotic resistance and metal resistance in an agricultural soil. *Environ. Sci. Pollut. Res. Int.* **2018**, *25*, 29314–29324. [CrossRef]

40. Zhao, X.; Wang, J.; Zhu, L.; Ge, W.; Wang, J. Environmental analysis of typical antibiotic-resistant bacteria and ARGs in farmland soil chronically fertilized with chicken manure. *Sci. Total Environ.* **2017**, *593–594*, 10–17. [CrossRef]

41. Lin, H.; Chapman, S.J.; Freitag, T.E.; Kyle, C.; Ma, J.; Yang, Y.; Zhang, Z. Fate of tetracycline and sulfonamide resistance genes in a grassland soil amended with different organic fertilizers. *Ecotoxicol. Environ. Saf.* **2019**, *170*, 39–46. [CrossRef]

42. Chen, Q.L.; An, X.L.; Zhu, Y.G.; Su, J.Q.; Gillings, M.R.; Ye, Z.L.; Cui, L. Application of struvite alters the antibiotic resistome in soil, rhizosphere, and phyllosphere. *Environ. Sci. Technol.* **2017**, *51*, 8149–8157. [CrossRef] [PubMed]

43. Xiang, Q.; Chen, Q.L.; Zhu, D.; An, X.L.; Yang, X.R.; Su, J.Q.; Qiao, M.; Zhu, Y.G. Spatial and temporal distribution of antibiotic resistomes in a peri-urban area is associated significantly with anthropogenic activities. *Environ. Pollut.* **2018**, *235*, 525–533. [CrossRef] [PubMed]

44. Zhou, Z.C.; Lin, Z.J.; Shuai, X.Y.; Zheng, J.; Meng, L.X.; Zhu, L.; Sun, Y.J.; Shang, W.C.; Chen, H. Temporal variation and sharing of antibiotic resistance genes between water and wild fish gut in a peri-urban river. *J. Environ. Sci.* **2021**, *103*, 12–19. [CrossRef] [PubMed]

45. Chen, Y.M.; Holmes, E.C.; Chen, X.; Tian, J.H.; Lin, X.D.; Qin, X.C.; Gao, W.H.; Liu, J.; Wu, Z.D.; Zhang, Y.Z. Diverse and abundant resistome in terrestrial and aquatic vertebrates revealed by transcriptional analysis. *Sci. Rep.* **2020**, *10*, 18870. [CrossRef]

46. Nonaka, L.; Maruyama, F.; Onishi, Y.; Kobayashi, T.; Ogura, Y.; Hayashi, T.; Suzuki, S.; Masuda, M. Various pAQU plasmids possibly contribute to disseminate tetracycline resistance gene tet(M) among marine bacterial community. *Front. Microbiol.* **2014**, *5*, 152. [CrossRef]

47. Chen, B.; He, R.; Yuan, K.; Chen, E.; Lin, L.; Chen, X.; Sha, S.; Zhong, J.; Lin, L.; Yang, L.; et al. Polycyclic aromatic hydrocarbons (PAHs) enriching antibiotic resistance genes (ARGs) in the soils. *Environ. Pollut.* **2017**, *220*, 1005–1013. [CrossRef]

Journal of
Xenobiotics

Article

Isolation and Quantification of Polystyrene Nanoplastics in Tissues by Low Pressure Size Exclusion Chromatography

François Gagné

Aquatic Contaminants Research Division, Environment and Climate Change Canada, Montréal, QC H2Y 2E7, Canada; francois.gagne@ec.gc.ca

Abstract: Ecotoxicity investigations of plastic nanoparticles (NPs) should pay more attention to their ability to pass barriers, accumulate, and initiate toxicity in cells. The purpose of this study was to develop a simple size exclusion chromatography (SEC) methodology to measure plastic NPs in biological tissues. A SEC column was prepared using a high-resolution gel for large macromolecules to separate plastic NPs from the protein/lipid pools in tissues. It was necessary to prepare the samples in high salt and non-ionic detergent (0.5 M NaCl and 0.2% Tween-20) and apply 0.2% Tween-20 containing 14 mM NaCl for the elution buffer to limit proteins adsorption to NPs. This methodology was able to resolve 50 and 100 nm polystyrene NPs from the protein/lipid pools in tissue homogenates. The fluorescent dye neutral red (NR) was also used for transparent NPs. Moreover, a sample fractionation step was also proposed for plastic NPs concentration using a salting-out methodology with saturated NaCl (5 M) and acetonitrile. Polystyrene NPs partition in acetonitrile, which were further analyzed by SEC. This methodology was tested in two case studies with clams collected in a high boat traffic (harbor) area and with caged freshwater mussels downstream of a large urban area. Although the present methodology was developed with polystyrene NPs it should be amenable to other plastic polymers that react with the NR fluorescent probe.

Keywords: polystyrene; nanoparticle; size exclusion chromatography; mussels

Citation: Gagné, F. Isolation and Quantification of Polystyrene Nanoplastics in Tissues by Low Pressure Size Exclusion Chromatography. *J. Xenobiot.* **2022**, *12*, 109–121. https://doi.org/10.3390/jox12020010

Academic Editors: Tiziana Cappello and Eider Bilbao

Received: 1 March 2022
Accepted: 5 May 2022
Published: 9 May 2022

Publisher's Note: MDPI stays neutral with regard to jurisdictional claims in published maps and institutional affiliations.

1. Introduction

The observation that plastic materials are found at a global scale represents a major contamination problem, thus raising concerns about their long-term impacts on ecosystems. Indeed, plastic wastes reach millions of tons per year, and not all are recycled back into our economy, contaminating both terrestrial and aquatic environments. Microplastics (5 mm to 1 µm) are found in consumer products such as cosmetics, find their way into wastewater treatment plants—ill equipped to remove them—and are released into the aquatic environment [1,2]. To make this situation worse, the environment breakdown of plastics materials by abiotic and biotic processes will release an exponential number of particles as the size decreases, reaching the nanoscales (<100 nm). Plastic nanoparticles (NPs) are defined as particles with sizes generally <100 nm but some include the 100–1000 nm range as well. Small NPs (<100 nm) can become readily bioavailable towards organisms and are absorbed in cells [3,4].

Plastic NPs are considered toxic for organisms due to their ability to bioaccumulate in tissues, even passing the blood–brain barrier, leading to oxidative stress and tissue damage [5,6]. A recent meta-analysis on plastic NPs revealed that these NPs significantly decrease survival, and disrupt the behavior and reproduction of fish and invertebrates [7]. NPs are also known to produce oxidative stress and series of biophysical changes (viscosity and altered fractal organization of proteins/enzymes) in aquatic organisms, leading to toxicity [8,9]. Plastics NPs could also act as vectors by absorbing existing toxic compounds such as bisphenol A, polyaromatic hydrocarbons, and heavy metals [10]. This could modify the bioavailability and exacerbate the toxic impacts of these existing environmental pollutants. In addition to oxidative stress, polystyrene NPs were also genotoxic to *Mytilus galloprovincialis* mussels, most notably in gills [11], suggesting long-term negative effects. Methods to quantify plastic NPs in tissues are scarce, which limits our understanding about their occurrence in the environment, distribution in tissues, and health impacts. Plastics are readily stained by the solvatochromic dye, neutral red (NR), often used to detect the presence of both micro and plastic NPs [6,11]. However, NR also reacts to nonpolar lipids in cells, which could interfere with the assay. Hence, there is a need to find other means to determine plastic NPs in biological tissues involving hyphenated procedures. More recently, a more specific fluorescent assay for polystyrene NPs in tissues was proposed using molecular rotor probes [12,13]. The presence of plastic NPs in the crowded environment of cells could disrupt the protein networks and lipid dynamics producing organized (nematic-like) liquid crystals, protein condensation/denaturation, and changes in metabolic pathways [8]. A methodology to study the interactions of plastic NPs in cells with respect to proteins and lipids would be valuable to help us better understand the uptake and toxicity of NP at the molecular level.

The purpose of this study was therefore to develop a low-pressure size exclusion chromatography (SEC) methodology for polystyrene NPs in tissue homogenate fraction. The method should be available to resolve NPs from the protein and lipids pool in the cytoplasm and permit some degree of quantification. The proposed method was first developed with fluorescently labeled polystyrene NPs, albumin, and in spiked homogenate fractions. The resulting method was optimized to analyze plastic NPs directly in marine and freshwater bivalve tissues in two real-life case studies.

2. Materials and Methods

2.1. Sample Preparation

Tissue samples were dissected on ice, weighted and placed in three volumes of 140 mM NaCl containing 10 mM Hepes-NaOH pH 7.4, 1 mM EDTA and 1 mM dithiothreitol. The tissues were homogenized using a Teflon pestle tissue grinder (5 passes) and centrifuged at $3000 \times g$ for 20 min. The supernatant was collected, filtered on the 0.45 μm pore membrane and stored at -85 °C until analysis. A fractionation procedure was also studied based on salting-out methodology. The homogenate sample (300 μL) was mixed with 400 μL of saturated NaCl (5 M) and 300 μL of 100% acetonitrile (ACN) for 10 min and centrifuged at $3000 \times g$ for 5 min. Polystyrene NPs readily partitions in the upper ACN layer and was collected for chromatographic analysis.

2.2. Size Exclusion Chromatography

Plastic NPs were analyzed by low-pressure size exclusion chromatography using the Bio-logic system (Biorad, Ontario, Canada). The column (40 cm × 1 cm) was filled with Sephacryl S-500 gel and equilibrated with 0.2% Tween-20 and 14 mM NaCl at pH 7.4 at a flow rate of 1 mL/min. Sephacryl S-500 is a high-resolution gel designed for the separation of large macromolecules between 2×10^7 to 4×10^4 g/mole size range and suitable to resolve plastic NPs from the protein/lipid pool. Conductivity (NaCl and other salts) and absorbance at 280 nm (proteins) were continuously monitored. One mL fractions (0.75 mL/min) were collected for NR staining (10 μM final concentration). Clear and fluorescently labeled (yellow–green Fluoresbrite) polystyrene nanoparticles of 50 and 100 nm were purchased at Polyscience Inc. (Warrington, PA, USA) and were used for column calibration along with NaCl and albumin (66 kDa).

Calibration was achieved using fluorescently labeled polystyrene NPs in the absence/presence of biological sample (proteins) to determine the optimal separation conditions. As plastic NPs tend to adsorb proteins and lipids, forming a corona at the surface, various loading and elution buffers were tested to find optimal conditions at the levels of detergent type (ionic vs. nonionic detergents), concentration, solvent (acetonitrile, acetone, and ethanol) and salt concentration (NaCl). It was found that the addition of albumin and tissue homogenate extracts (proteins/lipids) interacted with nanoparticles in low salt

(<10 mM NaCl) and detergent concentrations (<0.05%). Incubating the homogenates fractions in high salts (0.7 M NaCl) in the presence of nonionic detergent (0.05–0.2% Tween-20) in the sample buffer and nonionic detergent (0.2% Tween-20 in 14 mM NaCl) for the elution buffer eliminated nanoparticle–protein interactions and facilitated separation of plastic NPs from the proteins/lipids pool.

Homogenate fractions were either directly injected in the column or previously fractionated by a salting-out methodology using saturated NaCl/ACN. For the direct injection, the homogenate supernatants ($3000 \times g$ 10 min) were thawed on ice and mixed with one volume of 1.4 M NaCl and 0.2% Tween-20 buffered at pH 7.4 (with 1 M K_2HPO_4) for 5 min. As explained above, the addition of salts and nonionic detergents were necessary to keep proteins from adhering to the nanoparticle's surface and to exclude suspended plastic particles (density). For the salting-out step, 300 µL of the homogenate was mixed with 400 µL of 5 M NaCl and 400 µL of ACN. The sample was mixed and centrifuged at $3000 \times g$ for 10 min as described above. The ACN upper phase was mixed with 0.1 volume of 1.4 M NaCl/Tween-20 0.1% and injected to the chromatography column (0.25 mL) using the same elution buffer (0.2% Tween-20, 14 mM NaCl). A volume of 0.25 mL of transparent or fluorescent plastic nanoparticles (50 and 100 nm), albumin (66 kDa), and 0.5 M NaCl (total volume) was also injected in the column for calibration. The flow rate was 0.75 mL/min and 1 mL fractions were collected to a total volume of 30 mL. The absorbance at 280 nm (proteins and polystyrene) and conductivity (mS) were continuously measured.

The collected fractions were stained with 10 µM NR dye (from a 100 µM stock solution prepared in 0.2% Tween-20 and 10% ethanol in 14 mM NaCl) and fluorescence was measured at 485 nm excitation and 530 nm (40 nm bandpass) (Turner Biosystems, Sunnyvale, CA, USA). The blank consisted of the elution buffer, and increasing concentrations (0.05–1 µg/mL) of 50 nm diameter transparent polystyrene NPs (Polyscience Inc., Warrington, PA, USA) were used for calibration. The elution volume of the analyte (Ve) was determined based on NR fluorescence, and total volume (Vt) of the column was determined by NaCl conductivity spike from the sample buffer. The void volume (Vo) was based on the supplier information (around 24% of the total column volume = 7.5 mL). The elution profiles data were expressed as Ve/Vt, which permits comparisons with other column volumes.

2.3. Case Studies

The proposed methodology was used to screen for plastic NPs in two real-life case studies. The first case study consisted of sampling wild *Mya arenaria* clam populations from a clean or reference site in the Saint-Lawrence estuary (Baie du Moulin à Baude, Québec, Canada) and polluted site (harbor) at Baie Sainte Catherine supporting regular boat tourism activities (Québec, Canada). The polluted site was located some 8 km south from the reference site in the Saint-Lawrence River estuary. A total of 30 clams between 4–8 cm longitudinal shell length were collected by hand and frozen at $-20\ ^\circ$C for transportation to the laboratory. At the laboratory, the clams ($N = 4$) were thawed on ice and the soft tissues homogenized in a tissue grinder apparatus in ice-cold homogenization buffer as described above. The homogenates were analyzed by chromatography in each individual (homogenates were not pooled). The second case study consisted of experimentally caged *Elliptio complanata* mussels collected at a pristine lake (Lac des Écorces, Québec, Canada) and 5 km downstream of a municipal wastewater dispersion plume of a largely populated city (population of 1.8 million, City of Montréal, Québec, Canada). The mussels (mean size of 6 ± 0.5 cm, $N = 10$ per cage) were exposed for 3 months during the summer of 2017 (July to September) in cages consisting of 1×0.25 m cylindered nets fixed on cement blocks (5–10 kg) and immersed between 2–3 m depth. At the end of the exposure period, the mussels were processed in the same way as the clams, with the exception that the digestive glands of mussels were used instead of whole tissues and kept at $-80\ ^\circ$C.

2.4. Data Analysis

Sample preparation and column optimisation experiments were repeated three times and the mean value with standard deviation was used as the descriptive statistics. The chromatographic elution profiles in absorbance, conductivity, and NR fluorescence data were normalized at 1 to facilitate data visualization and reporting: (sample reading—blank (elution buffer))/(maximal reading—blank). A number of four individuals were collected for *Mya arenaria* clams and *Elliptio complanata* caging experiments for the chromatography analyses with either direct injection or after NaCl/ACN fractionation. Graphical and statistical analysis were performed using the SYSTAT software package (version 13).

3. Results and Discussion

The proposed methodology makes use of a high-resolution gel (Sephacryl S500) designed to resolve large macromolecules between 20 million and 400,000 daltons (g/mol). The upper protein size range is in the order of circa 750 kDa containing many thousands of amino acids, thus the gel matrix should be able to resolve plastic NPs larger from the protein/lipid pool in cells. Since SEC separates particles based on the Stokes radius rather than weight, the following discussion will be provided in size rather than mass. It is estimated that the protein size range is between 3 and 700 kDa where large 500 kDa globular proteins theoretically occupy a radius of 5.2 nm, which is well within the separation range of the gel matrix [14] and sets the theoretical lower size range of the column. Hence, the gel matrix offers the possibility to separate large plastic nanoparticles (10–1000 nm) from the protein pool in the sample without the need to extract/purify nanoparticles before running the column (Table 1). Large filamentous proteins (e.g., actinomyosin) should precipitate during centrifugation with appropriate speeds depending of the tissues. In the present study with molluscs, homogenisation with blender or Teflon pestle followed by $3000 \times g$ centrifugation appeared satisfactory. To validate these assumptions, the reported diameter of albumin is about 4 nm (mass of 66 kDa) [15] and should separate well from 50 or 100 nm NPs. A 40 cm $\times$ 1 cm column was filled with Sephacryl S-500 gel for separation of pure fluorescently labeled polystyrene nanoplastics (50 and 100 nm mean diameter), albumin, and NaCl (Figure 1). The total volume (Vt = 25 mL) was calculated with a conductivity (NaCl) peak, which permeates completely the gel beads, and the theoretical void volume (Vo) of the column was at Vo/Vt = 0.3 according to the supplier's information for Sephacryl S500 (Table 1). The 100 and 50 nm NPs were resolved at Ve/Vt = 0.4 and 0.45 respectively, albumin eluted at Ve/Vt of 0.75, and NaCl at Ve/Vt = 1 using the 0.2% Tween-20–14 mM NaCl elution buffer. The addition of the non-denaturing detergent (Tween-20) was necessary to maintain separation of albumin from nonspecific hydrophobic interactions of polystyrene NPs. Indeed, NPs adsorb proteins forming a corona [16] and could change the elution volume (Ve) for polystyrene NPs. This was observed by lowering the detergent concentration to 0.01%, which increased the Ve/Vo to 0.60 for the fluorescent NPs in the presence of 1 mg/mL albumin (results not shown). The addition of denaturing detergent such as SDS (0.1%) denatured albumin and eluted at the void volume Vo/Vt = 0.30 and overlapped upper size range of plastic nanoparticles (>100 nm). A significant and linear relationship between the size (log size in nm) and the Ve/Vt of the samples was obtained with r = 0.99 (Figure 1B). This relationship sets the upper and lower size range of NPs between 161–11 nm respectively.

We tested this chromatographic method with transparent polystyrene NPs (50 nm diameter) alone and in *Mya arenaria* soft tissues homogenates ($3000 \times g$ supernatant) using NR staining for plastic particles (Figure 2). The solvatochromic NR dye is well-recognized as a fluorescent dye for plastics but also stains lipids [12,17]. Transparent polystyrene NPs eluted at Ve/Vt = 0.45 based on NR fluorescence (Figure 2A). The addition of soft tissues fractions revealed two peaks at 280 nm (Figure 2B): one at Ve/Vo corresponding to polystyrene NPs and the other at Ve/Vo = 0.9–0.95 corresponding to the protein/lipids pool. NR staining revealed also two major peaks overlapping with 280 nm absorbance corresponding to polystyrene NPs at Ve/Vo = 0.44 and the protein/lipid pool at Ve/Vo = 0.9.

The elution profile of tissue fractions alone did not show any peaks at Ve/Vo < 0.7 in keeping with the resolution range of the gel matrix. The NR dye is a well-recognized fluorescent stain for many types of plastics such as polyethylene, polystyrene, nylon, polyethylene terephthalate, and polypropylene at the micro- and nanoscales, thus could be used with these plastic polymers [12,18]. The average length of phospholipids is about 2 nm [19] giving a Ve/Vo = 0.8–0.9 overlapping to the second NR peak, hence not suitable for measuring small plastic NPs (<10 nm). The presence of the detergent Tween-20 in the elution buffer also serves to solubilize membranes vesicles in the extracts removing larger lipid vesicles/structures in the sample.

Table 1. Characteristics of the size exclusion column.

	Ve/Vt [1]	Comments
Direct injection		
100 nm NP	0.40	NR staining
50 nm NP	0.45	NR staining
Albumin (A280)	0.75	
Protein pool (A280)	0.7–0.95	
NaCl/lipids	0.95–1	Conductivity and NR staining
NaCl/ACN fractionation		
100 nm NP	0.76	Polystyrene osmotic shrinkage
50 nm NP	0.80	Polystyrene osmotic shrinkage
A280 peak	0.9	Not protein origin
NR peak (lipids)	0.95	Phospholipids
NaCl	1	

[1] Chromatography column: 40 cm × 1 cm (31.4 mL total volume), sample buffer 0.70 M NaCl-0.05% Tween-20 and elution buffer 14 mM NaCl-0.2% Tween-20. Ve: elution volume, Vt: total volume (NaCl). The void volume Vo was Vo/Vt = 0.3 according to the suppliers' information for Sephacryl S500.

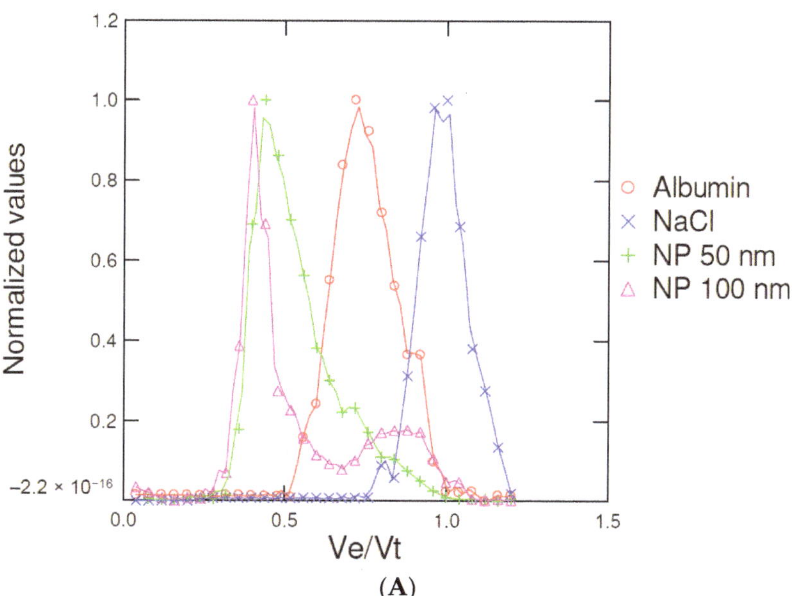

Figure 1. *Cont.*

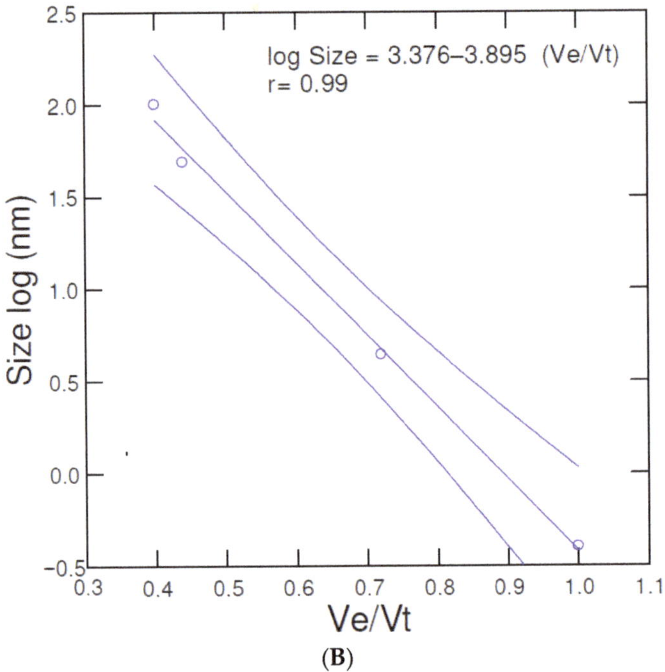

Figure 1. Gel chromatographic analysis of fluorescently labeled plastic NPs. Elution profile of polystyrene NPs (**A**) using 100 nm, 50 nm polystyrene NPs (fluorescence), albumin (absorbance 280 nm, MW = 60 kDa), and NaCl (conductivity: mS); calibration of the size exclusion chromatography column (**B**).

This chromatographic procedure was used to screen for plastic NPs in two real-life case studies, one involving wild *Mya arenaria* clams collected at a harbor/marina and the second with caged *Elliptio complanata* mussels at downstream sites of a major urban area, both generally associated to plastics pollution [20]. The first case consisted of feral *Mya arenaria* clams collected at a reference site under no direct source of pollution and a harbour site supporting intense commercial and touristic boating activities (Figure 3A,B). Clam tissues collected at the reference site showed one major UV absorbance (280 nm) peak at Ve/Vt = 0.8 corresponding to the protein pool, one NR peak at Ve/Vt = 0.8–0.9 corresponding to hydrophobic proteins and lipids, and a conductivity peak at Ve/Vt = 1 corresponding to salts/NaCl (Figure 3A). Clams collected at the contaminated harbor site showed two major UV and NR peaks at Ve/Vo = 0.4 and 0.8 (Figure 3B) corresponding to large-size compounds consistent with UV-absorbing plastics (polystyrene and polyethylene terephthalate) followed by the protein/lipid pool peak. Large hydrophilic carbohydrate-based polymers (sugars, glycogen), if present, do not absorb at 280 nm and should not interact with the NR dye. The second case concerned exposure of caged mussels exposed to pristine lake (reference) and downstream of a highly populated city of circa 3 million inhabitants in the Saint-Lawrence river (Figure 3C,D). Sediments in the Saint-Lawrence river were recently reported contaminated by microplastics [20] and prompted investigation on plastic NPs in this study. The major form of plastic was polyethylene (UV absorbance negative and NR positive) and reached densities of 1.4×10^5 microbeads·m^{-2}. *Elliptio complanata* digestive gland homogenates caged at the reference lake revealed one major band at Ve/Vt = 0.75 and one small band at the void volume (Figure 3C). One conductivity peak was observed at the Ve/Vt = 1 corresponding to dissolved salts with no evidence of conducting large-size molecules (i.e., absence of peaks at Ve/Vt < 0.8). The NR fluorescence peak generally followed the UV peak with a major band at Ve/Vt = 0.8 of the protein/lipid pool. In mussels caged downstream of a large city, the UV signal was distributed over the elution profile (Figure 3D) with a maxima at Ve/Vt = 0.75. The same conductivity peak at Ve/Vo = 1 was obtained, suggesting no presence of conductive large-molecular-size compounds (e.g., elemental nanoparticles). In the case of NR staining, we observed two major peaks at Ve/Vt = 0.6–0.65 and Ve/Vt = 0.35–0.5, corresponding to compounds of sizes ranging from 11 to 120 nm. This suggests that exposure of mussels to urban activities leads to a complex distribution pattern of NR-stained materials outside the protein and lipids pools in tissues. This is in keeping with the reported levels of microplastics in sediments in the Saint-Lawrence river [20].

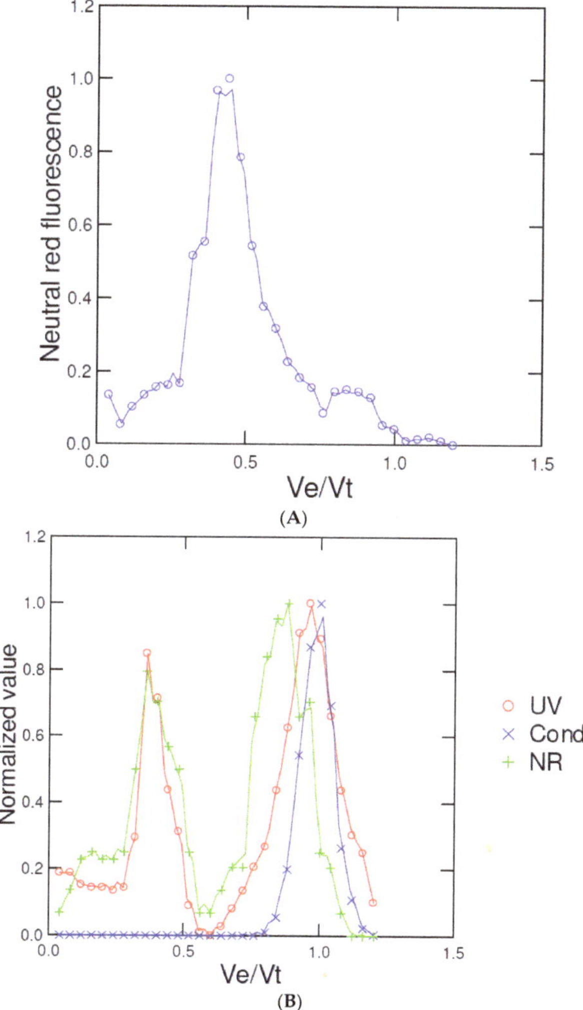

Figure 2. NP addition in mussel tissue homogenates. Pure transparent 50 nm polystyrene NPs at 10 µg/mL (**A**) and spiked mussel tissue homogenates (**B**), centrifuged at 3000× *g* for 10 min, the supernatant mixed with one volume of 1.5 M NaCl containing 0.1% Tween-20 and injected into the column. The eluted samples (1 mL) were collected and were analyzed for absorbance at 280 nm, conductivity and neutral red (NR) staining.

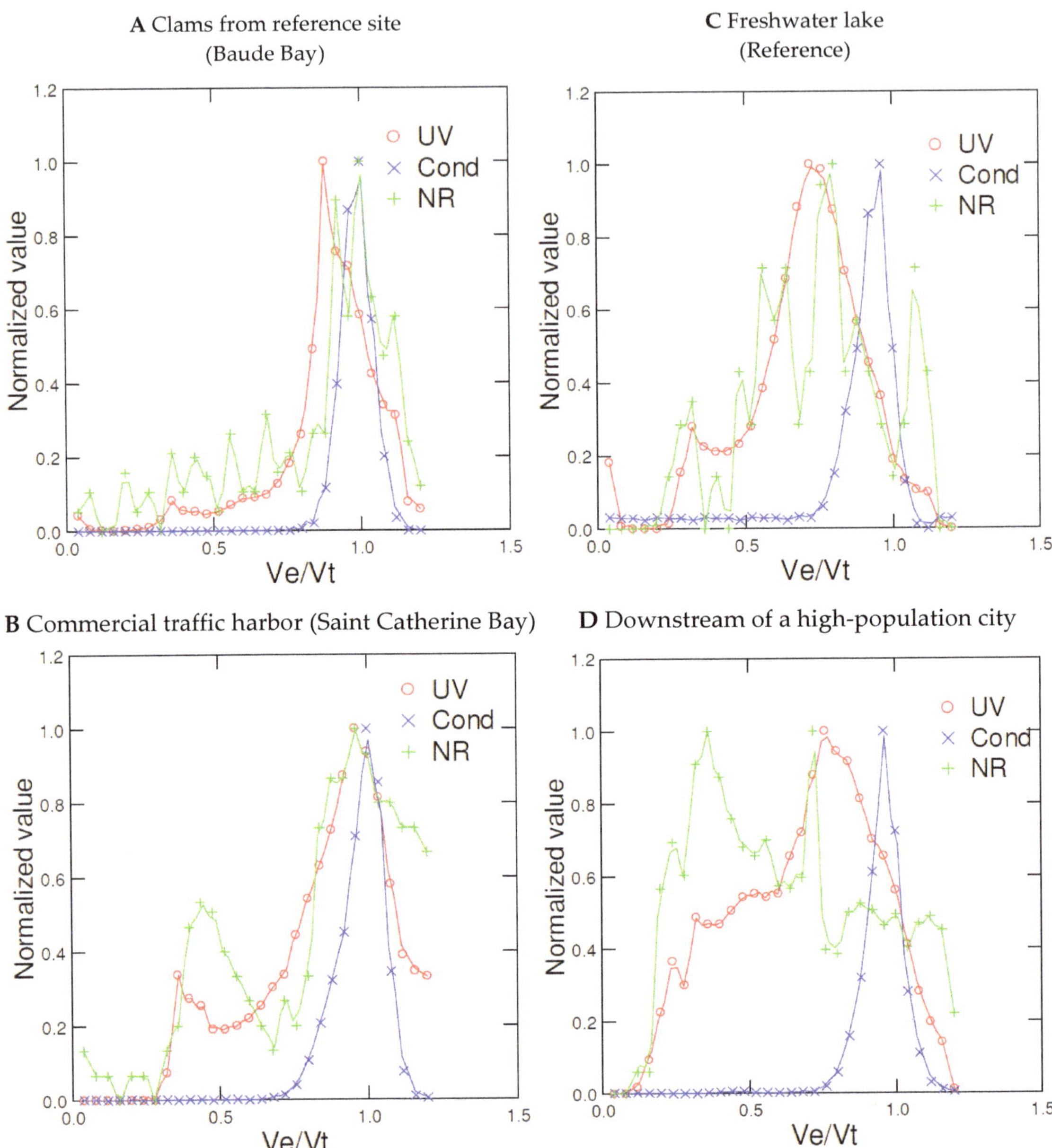

Figure 3. Chromatographic profile marine and freshwater bivalves exposed to anthropogenic pollution. Representative chromatographic profiles of *Mya arenaria* clam tissue homogenates from a reference site (**A**), a harbor supporting intense boat traffic (**B**), *Elliptio complanata* mussel digestive glands caged at reference lake (**C**), and downstream of a large-population city (**D**) for 3 months during the summer of 2017. The homogenates were centrifuged at $3000 \times g$ for 10 min, the supernatant mixed with one volume of 1.5 M NaCl-0.1% Tween-20 and directly injected to the size exclusion chromatography column. The data represented the measurements at each 1 mL fraction and normalized to 1 for viewing.

Direct sample analysis of homogenate fractions offers the advantage of measuring other endpoints (enzyme activities, proteins) or specific metabolites in addition to the presence of nanoplastic materials. In the attempt to increase the sensitivity of this methodology and as a means to remove potential interfering large protein filaments, a salting-out step with saturated NaCl and acetonitrile (NaCl/ACN) was used to extract and concentrate NPs. Mussel homogenates were spiked with increasing amounts of fluorescent polystyrene NPs, fractionated with NaCl/ACN methodology, and centrifuged at $3000\times g$ for 5 min. The ACN upper phase was mixed with 0.1 volume of 1.4 M NaCl (for total volume calibration of the column) and resolved by SEC with the same elution buffer. Because polystyrene absorbs at 280 nm, the absorbance readings of the elution profile were included (Figure 4A). The data revealed that UV absorbance increased linearly but at Ve/Vt = 0.75–0.8 with the added NPs. In an unspiked sample, low absorbance at 280 nm was measured at Ve/Vt = 0.95–1, suggesting the presence of nonprotein UV-absorbing compounds—perhaps nonpolar metabolites such as tyrosine or phenylalanine amino acids/peptides. Lipids would elute near this volume but would not absorb at 280 nm. The fluorescence peak of the fluorescent NPs eluted at Ve/Vt = 0.75 and peak height/area increased linearly with added NP in the homogenates (Figure 4B). The elution of polystyrene at these fractions (Ve/Vt = 0.75–0.8) suggests that polystyrene NPs shrunk in size with ACN. Polystyrene is completely soluble in ACN or acetone and was shown to shrink in the presence of organic solvents. Polystyrene shrinks in apolar environments from increased electrostatic interactions between the polystyrene polymer layers [21,22]. Hence, the salting-out step displaces polystyrene nanoparticles from water to the ACN phase and leads to osmotic shrinkage. This phenomenon is readily observed with soft polystyrene polymers but seemingly less in hard plastic polymers.

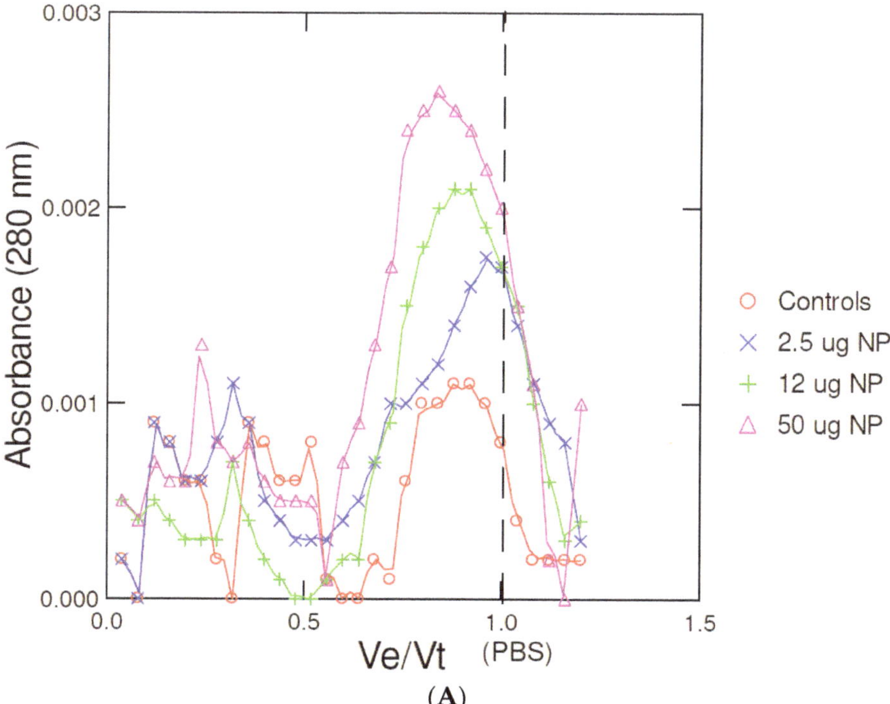

Figure 4. *Cont.*

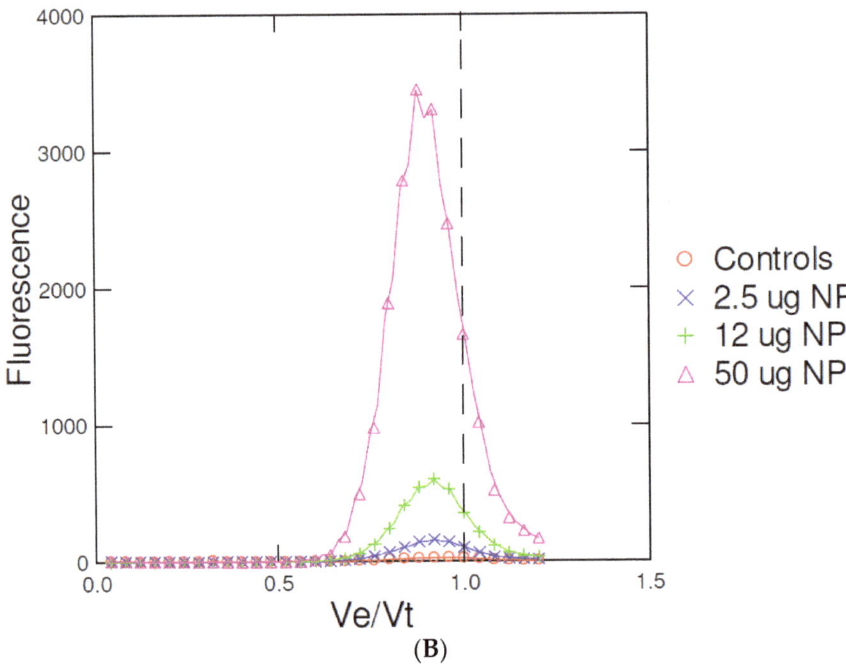

Figure 4. Calibration of the NaCl/acetonitrile fractionation method. Mussel homogenates were spiked with increasing quantities of fluorescently-labeled polystyrene NPs, fractionated using the salting-out step, and the acetonitrile upper phase injected in the SEC column. The elution buffer consisted of 0.2% tween-20 and 14 mM NaCl at pH 7.4. Absorbance at 280 nm (**A**) and fluorescence (**B**) were taken at each 1 mL volume of the eluate. The flow rate was 0.75 mL/min.

This approach was also tested on clam and mussel homogenates at sites under anthropogenic pollution/activity, which are likely sources of plastic pollution [23,24]. The eluted fractions of the Sephacryl S500 column were stained with 10 μM NR dye to detect plastic (apolar) material (Figure 5). In mussels caged for 3 months at downstream and rain runoff sites, NR staining was increased at Ve/Vt = 0.75–0.85 compared to reference Lake values (Figure 5A). In control mussels, the NR peak was Ve/Vt approximately 1 co-eluted with salts, suggesting the presence of other nonpolar low molecular weights compounds soluble in ACN. Moreover, the NR positive and staining peak was also detected at Ve/Vt = 0.3–0.4, suggesting the presence of larger plastic nanomaterials in the 100-nm size range. In *Mya arenaria* clam homogenates collected at a polluted site, the same pattern was observed, i.e., the appearance of NR staining at Ve/Vt = 0.75–0.80 at the polluted site and NR staining at Ve/Vt approximately 1 at the reference site (Figure 5B). Interestingly, the presence of large plastic compounds near the void volume of the column (at 160 nm) was not observed in this case.

The present methodology proposed a convenient and cost-effective means to isolate plastic NPs in tissues based on NR staining. Other, more specific, fluorescent stains could be used, such as molecular rotor probes (9-(2,2-dicyanovinyl)julolidine) that are more specific to plastic materials [12,13]. These approaches could be used as a screening tool to separate and detect plastic NPs but this methodology should be confirmed by more specific methods such as single particle plasma–mass spectrometry or pyrolysis gas chromatography–mass spectrometry [25,26]. The isolated fractions could be further analyzed for metals (e.g., arsenic, copper) or organic compounds to detect *Trojan horse* interactions i.e., xenobiotics sorbed to plastic NPs. They could also be measured using FT–infrared spectroscopy directly or after further ultrafiltration [27]. Nevertheless, the SEC methodology using NR or molecular rotor probes represents a convenient tool for the screening of biological tissues for NPs. However, the proposed methodology is convenient and easy to implement, and more accessible to low-budget laboratories interested in animals exposed to plastic NPs.

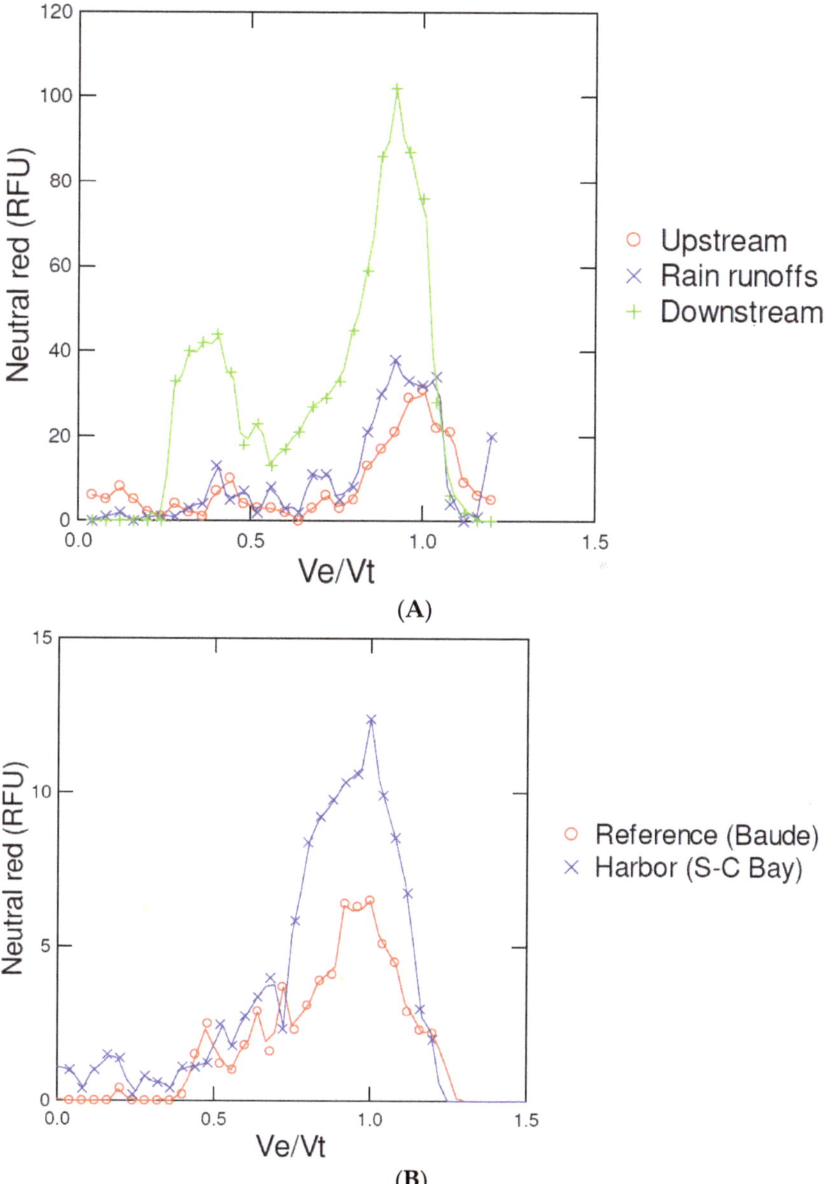

Figure 5. Chromatographic profile of freshwater mussels *Elliptio complanata* caged for 3 months at downstream a large population city (Montréal, Québec, Canada) (**A**) and in wild *Mya arenaria* clams at a polluted harbor, Saint Catherine (S-C) Bay (**B**). Digestive gland homogenates were prepared were extracted using the salting out step and injected to the SEC column using 0.2% Tween-20 and 14 mM NaCl elution buffer. The data represents NR fluorescence at each 1 mL fraction.

Funding: The project was funded under the Saint-Lawrence Action Plan of Environment and Climate Change Canada.

Institutional Review Board Statement: Not applicable.

Informed Consent Statement: Not applicable.

Data Availability Statement: Supplementary or raw data are available at ECCC upon demand.

Acknowledgments: This work was funded by the Aquatic Contaminants Research Division of Environment and Climate Change Canada.

Conflicts of Interest: The author declares no conflict of interests.

References

1. Fendall, L.S.; Sewell, M.A. Contributing to marine pollution by washing your face: Microplastics in facial cleansers. *Mar. Pollut. Bull.* **2009**, *58*, 1225–1228. [CrossRef]
2. Cole, M.; Lindeque, P.; Halsband, C.; Galloway, T.S. Microplastics as contaminants in the marine environment: A review. *Mar. Pollut. Bull.* **2011**, *62*, 2588–2597. [CrossRef] [PubMed]
3. Ng, E.L.; Huerta Lwanga, E.; Eldridge, S.M.; Johnston, P.; Hu, H.W.; Geissen, V.; Chen, D. An overview of microplastic and nanoplastic pollution in agroecosystems. *Sci. Total Environ.* **2018**, *627*, 1377–1388. [CrossRef] [PubMed]
4. Schell, T.; Rico, A.; Vighi, M. Occurrence, fate and fluxes of plastics and microplastics in Terrestrial and Freshwater Ecosystems. *Rev. Environ. Contam. Toxicol.* **2020**, *250*, 1–43. [PubMed]
5. Sökmen, T.O.; Sulukan, E.; Türkoğlu, M.; Baran, A.; Özkaraca, M.; Ceyhun, S.B. Polystyrene nanoplastics (20 nm) are able to bioaccumulate and cause oxidative DNA damages in the brain tissue of zebrafish embryo (Danio rerio). *Neurotoxicology* **2020**, *77*, 51–59. [CrossRef] [PubMed]
6. Auclair, J.; Quinn, B.; Peyrot, C.; Wilkinson, K.J.; Gagné, F. Detection, biophysical effects, and toxicity of polystyrene nanoparticles to the cnidarian Hydra attenuate. *Environ. Sci. Poll. Res. Int.* **2020**, *27*, 11772–11781. [CrossRef]
7. Han, Y.; Lian, F.; Xiao, Z.; Gu, S.; Cao, X.; Wang, Z.; Xing, B. Potential toxicity of nanoplastics to fish and aquatic invertebrates: Current understanding, mechanistic interpretation, and meta-analysis. *J. Hazard. Mater.* **2022**, *427*, 127870. [CrossRef]
8. Gagné, F. Ecotoxicology of altered fractal organization in cells. *Am. J. Biomed. Sci. Res.* **2020**, *8*, 498–502. [CrossRef]
9. Auclair, J.; Peyrot, C.; Wilkinson, K.J.; Gagné, F. Biophysical effects of polystyrene nanoparticles on *Elliptio complanata* mussels. *Environ. Sci. Pollut. Res. Int.* **2020**, *27*, 25093–25102. [CrossRef]
10. Campanale, C.; Massarelli, C.; Savino, I.; Locaputo, V.; Uricchio, V.F. A Detailed Review Study on Potential Effects of Microplastics and Additives of Concern on Human Health. *Int. J. Environ. Res. Public Health* **2020**, *17*, 1212. [CrossRef]
11. Erni-Cassola, G.; Gibson, M.I.; Thompson, R.C.; Christie-Oleza, J.A. Lost, but found with Nile Red: A novel method for detecting and quantifying small microplastics (1 mm to 20 µm) in Environmental Samples. *Environ. Sci. Technol.* **2017**, *51*, 13641–13648. [CrossRef] [PubMed]
12. Gagné, F. Detection of polystyrene nanoplastics in biological tissues with a fluorescent molecular rotor probe. *J. Xenobiotics* **2019**, *9*, 8147–8149. [CrossRef] [PubMed]
13. Moraz, A.; Breider, F. Detection and Quantification of Nonlabeled Polystyrene Nanoparticles Using a Fluorescent Molecular Rotor. *Anal. Chem.* **2022**, *45*, 14976–14984. [CrossRef] [PubMed]
14. Erickson, H.P. Size and Shape of Protein Molecules at the Nanometer Level Determined by Sedimentation, Gel Filtration, and Electron Microscopy. *Biol. Proced. Online* **2009**, *11*, 32. [CrossRef]
15. Diem, K.; Lentner, C. *Scientific Tables*, 7th ed.; Ciba-Geigy Limited: Basle, Germany, 1971.
16. Canesi, L.; Balbi, T.; Fabbri, R.; Salis, A.; Damonte, G.; Volland, M.; Blasco, J. Biomolecular coronas in invertebrate species: Implications in the environmental impact of nanoparticles. *NanoImpact* **2017**, *8*, 89–98. [CrossRef]
17. Greenspan, P.; Mayer, E.P.; Fowler, S.D. Nile Red: A Selective Fluorescent Stain for Intracellular Lipid Droplets. *J. Cell Biol.* **1985**, *100*, 965–973. [CrossRef]
18. Gagné, F.; Auclair, J.; André, C. Polystyrene nanoparticles induce anisotropic changes in subcellular fraction of the digestive system of freshwater mussels. *Curr. Top. Toxicol.* **2019**, *19*, 43–48.
19. Milo, R.; Phillips, R. Size and geometry. In *Cell Biology by Numbers*; GS Garland Science; Francis and Taylor Group: New York, NY, USA, 2015; Chapter 1; 395p, ISBN 978-0-8153-4537-4.
20. Castañeda, R.A.; Avlijas, S.; Simard, M.A.; Ricciardi, A. Microplastic pollution in St. Lawrence River sediments. *Can. J. Fish. Aquat. Sci.* **2014**, *71*, 1767–1771. [CrossRef]
21. Latreille, P.-L.; Adibnia, V.; Nour, A.; Rabanel, J.M.; Lalloz, A.; Arlt, J.; Poon, W.C.K.; Hildgen, P.; Martinez, V.A.; Banquy, X. Spontaneous shrinking of soft nanoparticles boosts their diffusion in confined media. *Nat. Commun.* **2019**, *10*, 4294. [CrossRef]
22. Jeon, S.; Granick, S. Polystyrene latex nanoparticles shrink when polyelectrolyte of the same charge is added. *Macromolecules* **2004**, *37*, 2919–2923. [CrossRef]

23. Singh, S.; Kumar Naik, T.S.S.; Anil, A.G.; Dhiman, J.; Kumar, V.; Dhanjal, D.S.; Aguilar-Marcelino, L.; Singh, J.; Ramamurthy, P.C. Micro (nano) plastics in wastewater: A critical review on toxicity risk assessment, behaviour, environmental impact and challenges. *Chemosphere* **2022**, *290*, 133169. [CrossRef] [PubMed]

24. Gagné, F.; Auclair, J.; Quinn, B. Detection of polystyrene nanoplastics in biological samples based on the solvatochromic properties of Nile red: Application in Hydra attenuata exposed to nanoplastics. *Environ. Sci. Poll. Res.* **2019**, *26*, 33524–33531. [CrossRef] [PubMed]

25. Jiménez-Lamana, J.; Marigliano, L.; Allouche, J.; Grassl, B.; Szpunar, J.; Reynaud, S. A Novel Strategy for the Detection and Quantification of Nanoplastics by Single Particle Inductively Coupled Plasma Mass Spectrometry (ICP-MS). *Anal. Chem.* **2020**, *92*, 11664–11672. [CrossRef]

26. Primpke, S.; Christiansen, S.H.; Cowger, W.; De Frond, H.; Deshpande, A.; Fischer, M.; Holland, E.B.; Meyns, M.; O'Donnell, B.A.; Ossmann, B.E.; et al. Critical Assessment of Analytical Methods for the Harmonized and Cost-Efficient Analysis of Microplastics. *Appl. Spect.* **2020**, *74*, 1012–1047. [CrossRef] [PubMed]

27. Frias, J.P.; Sobral, P.; Ferreira, A.M. Organic pollutants in microplastics from two beaches of the Portuguese coast. *Mar. Pollut. Bull.* **2010**, *60*, 1988–1992. [CrossRef]

Article

Combined Effects of Polystyrene Nanoplastics and Enrofloxacin on the Life Histories and Gut Microbiota of *Daphnia magna*

Piotr Maszczyk [1,*], Bartosz Kiersztyn [1], Sebastiano Gozzo [2], Grzegorz Kowalczyk [1], Javier Jimenez-Lamana [2], Joanna Szpunar [2], Joanna Pijanowska [1], Cristina Jines-Muñoz [1], Marcin Lukasz Zebrowski [1] and Ewa Babkiewicz [1]

[1] Department of Hydrobiology, Institute of Functional Biology and Ecology, Faculty of Biology, University of Warsaw, 00-927 Warsaw, Poland
[2] IPREM, UMR 5254, E2S UPPA, CNRS, Universite de Pau et des Pays de l'Adour, 64053 Pau, France
* Correspondence: p.maszczyk@uw.edu.pl

Abstract: The effect of nanoplastics (NPs) has been shown to interact with the effect of pollutants, including antibiotics. However, little is known about studies performed on freshwater organisms. In this study, we aimed to test the hypothesis that both NPs and antibiotics affect the life history traits of freshwater planktonic *Daphnia magna*, a model organism in ecotoxicological research, as well as the metabolic and taxonomic fingerprint of their gut microbiota, and whether there is an interaction in the effect of both stressors. To assess this, we experimented with the effect of different spherical polystyrene nanoplastic concentrations and antibiotic enrofloxacin measured through (i) the *Daphnia* body size and their selected reproductive parameters (the clutch size, egg volume, and total reproductive investment), (ii) the metabolomic diversity of gut microbiota (the respiration rate and the relative use of different carbon sources), and (iii) the microbial taxonomic diversity in the *Daphnia* intestine. Our results supported the hypothesis as each of the stressors on its own significantly influenced most of the measured parameters, and because there was a significant interaction in the effect of both stressors on all of the measured parameters. Therefore, the results suggest an interactive negative effect of the stressors and a possible link between the observed effects at the different levels of a biological organisation.

Keywords: antibiotics; enrofloxacin; nanoplastics; polystyrene; stressors; *Daphnia magna*; gut microbiota; life history; Trojan horse effect

Citation: Maszczyk, P.; Kiersztyn, B.; Gozzo, S.; Kowalczyk, G.; Jimenez-Lamana, J.; Szpunar, J.; Pijanowska, J.; Jines-Muñoz, C.; Zebrowski, M.L.; Babkiewicz, E. Combined Effects of Polystyrene Nanoplastics and Enrofloxacin on the Life Histories and Gut Microbiota of *Daphnia magna*. *Water* **2022**, *14*, 3403. https://doi.org/10.3390/w14213403

Academic Editors: François Gagné, Stefano Magni and Valerio Matozzo

Received: 22 August 2022
Accepted: 24 October 2022
Published: 27 October 2022

Publisher's Note: MDPI stays neutral with regard to jurisdictional claims in published maps and institutional affiliations.

1. Introduction

Plastics comprise a wide range of synthetic or semi-synthetic organic compounds, usually polymers with a high molecular mass [1]. For many decades, their production and use has continued to increase worldwide, and many studies reported their increasing presence in natural environments [2–4]. Discarded plastic wastes can directly or indirectly enter the environment, and degrade due to solar radiation, mechanical forces, and the biological activity of organisms to smaller-sized pieces, including micro- (MPs, particles smaller than 5 mm) and nanoplastics (NPs, particles smaller than 1 μm, e.g., [5]; or smaller than 0.1 μm, [6]). In addition, primary nanoparticles originating from engineered materials (e.g., personal health care products) can be found. Among the MPs and NPs found in marine and freshwater environments, polystyrene (PS) is one of the most common types [7]. NPs pollution is an issue of special concern because of its unique features that include: (i) their colloidal properties, (ii) their tendency to aggregate, (iii) their high surface area to the volume ratio, resulting in a high ability to absorb and release chemicals, and (iv) the ability to penetrate cell membranes [8]. Because of their small size, adequate quantitative analytical techniques are not currently available to assess the NP concentrations in the environment [9]. While MPs' presence in marine and freshwater systems has been studied for several years, NPs have been detected in the North Atlantic Gyre only recently [10]. It is assumed that NP concentrations are even 10^{14} times higher than those currently measured for MPs [6].

Several studies have indicated that many organisms ingest NPs or absorb them on their surfaces, e.g., in [11–13], and that because of their small sizes and colloidal properties, they may cross biological barriers [14,15], negatively affecting the organisms. The negative effects depend on the particle type, size, density, charge, and origin (primary or secondary), and this may be related to mechanical (e.g., adhering to external surfaces hindering the mobility) and chemical effects [8]. In addition, chemical harmfulness results from the presence of additives that have the potential to leach into the environment, causing damage to organisms. Among the most common additives are plasticisers, which may affect life history and morphology in *Daphnia magna* [16], flame retardants that may cause induced significant sublethal chronic toxicity to *D. magna* [17], antioxidants that may reduce the hatching rates, increase the malformation rates and decrease the length of calcified vertebrae [18], and UV stabilisers that recently have been revealed to cause potential immune dysfunction [19]. Moreover, chemical harmfulness may be due to the particles that can act as a carrier for other co-occurring pollutants, resulting in organisms' accumulation of harmful hydrophobic substances from the surrounding water [13,15].

The ability of NPs, due to their high surface area to volume ratio, to adsorb, concentrate, and act as a vector of toxic pollutants can modify the environmental impact of the latter. In fact, it has been revealed that most of the combined toxic effects are not simply additive, but rather synergistic or antagonistic [20]. On the other hand, NPs may decrease the toxicity of other pollutants by absorbing and then agglomerating them to form larger particles, reducing the ease of uptake by organisms [21–23]. Additionally, their presence may cause an enhancement of toxicity occurring from the on-surface pre-concentration ("The Trojan horse effect") [22,24]. Although many ecotoxicological studies suggest that realistic environmental concentrations of micro- and nanoplastics may not induce significant detrimental effects on marine organisms nor threaten their survival [13,25], the co-exposure to NPs and other associated contaminants/stressors could exacerbate their effects [26–28].

A significant group of such pollutants are antibiotics, which are detectable in surface waters, including rivers, lakes, and seas [13], in the ng L^{-1} up to µg L^{-1} range, exceeding sometimes the predicted no-effect environmental concentration [29]. Their extensive and irregular use has induced multifaceted adverse impacts in recent years, such as the propagation of multi-drug-resistant bacteria, antibiotic-resistant bacteria (ARB), and antibiotic-resistant genes (ARGs) in the aquatic environment [30,31]. Their antibacterial impacts are not strain-specific; thus, while the pathogenic bacteria are killed, some bacteria which are beneficial for organisms' health are also targeted, which may cause several adverse effects, such as an intestinal flora imbalance [32,33]. In addition, different classes of antibiotics have been shown to be toxic to organisms at different trophic levels, such as algae, bacteria, crustaceans, and fish [34,35].

Among the most widely used antibiotics are fluoroquinolones (FQs), which are broad-spectrum synthetic antibiotics commonly used in human and veterinary medicine [36,37] and in agriculture and aquaculture [38]. Among FQs, enrofloxacin is used to prevent and treat a broad spectrum of gram-positive and -negative bacterial infections in livestock. Due to the spread of antibiotic resistance [39], it is listed among the compounds that can be considered to be of a high ecotoxicological concern [40]. It is usually detected in the effluents of municipal sewage plants and the related aquatic environments in the range of ng and µg L^{-1} [41,42] or even in extreme cases in mg L^{-1} [43]. Other examples concern their concentrations in surface waters (up to 248 ng L^{-1} [44]), in groundwater [45], and up to 7.7 mg kg^{-1} in sediments [46].

Some published studies have investigated the combined effect of NPs and antibiotics on cyanobacteria [47], algae [48], bivalvia [49], and fish [23,24,50,51]. However, there are no reviews yet on other aquatic organisms, including planktonic animals such as *D. magna*, a keystone species in the food webs of fishless ponds. The combined effect of the stressors may be different for different organisms. It is important to build experimental datasets using a range of different organisms to quantify and predict the factors and mechanisms responsible for the pattern under different contexts. The endpoint of many published studies has focused mainly on the effect at the molecular level: the integrated biomarkers response, antioxidant indexes, gene expression, and histological symptoms [24,52] rather than at the organismal level, e.g., as the combined effect on the life history traits and gut microbiota.

The present work aimed to test several hypotheses concerning the single and combined effects of polystyrene NPs and enrofloxacin on the selected life history traits of *D. magna* as well as the metabolic and taxonomic diversity of the bacterial community in their intestinal tracts. **First**, the presence of each of the stressors results in decreasing *Daphnia's* body size and reproductive parameters. The effect of enrofloxacin differs in the presence and absence of NPs. **Second**, as the NPs presence increases, the enrofloxacin presence decreases the metabolic rate of the gut microbiota of *Daphnia*. **Third**, the metabolic fingerprint measured as the relative use of various carbon sources is different in the presence of each of the stressors on its own and combined. **Finally**, those stressors affect *Daphnia's* taxonomic diversity in the gut microbiota. On the whole, there is an interaction in the effect of both stressors.

2. Materials and Methods

2.1. Experimental Animals

Three replicates of the experiments were performed. In order to assess the species—rather than the clone-specific effects—each replicate was performed using a different clone of *D. magna* (MB, MN, and MD of body size at first reproduction 1.86 ± 0.22, 1.82 ± 0.13, and 1.83 ± 0.31 mm, respectively) [53]. Clone MB was sampled from Lake Binnen (54°19′29″ N; 10°37′39″ E, Germany), clone MN from the Nový Rybník pond (50°13′27.8″ N; 14°4′3.1″ E, Czech Republic), and clone MD from the Domin pond (49°00′21.3″ N; 14°26′29.1″ E, Czech Republic). Daphnia was cultured in 5 L containers, with 25 individuals per container, at room temperature and with a natural photoperiod. The daily food supply was added ad libitum in the amount of 1.6 mg C × L^{-1} of unicellular green algae, *Chlamydomonas klinobasis* (strain SAG 56) from a stationary phase, a chemostat culture grown in a WC medium [54]. The algal concentration was assessed using a portable fluorometer (AquaFluor handheld fluorometer, Turner Designs®, San Jose, CA, USA).

2.2. Chemicals

Polystyrene (PS) NPs were synthesised in IPREM, Institut des Sciences Analytiques et de Physico-Chimie pour l'Environnement et les Matériaux, Pau, France, avoiding any additives, especially surfactants, bactericides (e.g., sodium azide), and the trace metals usually present in commercial standards [55,56]. The synthesis and characteristics of soap-free polystyrene models have been detailed elsewhere, e.g., in [55]. Spherical PS NPs were used (diameter of 420 ± 20 nm determined by scanning electron microscopy) with a surface functionalized by carboxylic groups, a low polydispersity (PDI of 0.009), and the zeta potential of −46 mV at a pH value of 7.

Enrofloxacin powder (purity ≥ 98%) was acquired from Sigma-Aldrich (St. Louis, MO, USA). The stock solution was prepared daily to get 100 µg ml^{-1} by dissolving the weighted amount in milliQ water in an ultrasonic bath. The stock solution of enrofloxacin was prepared without using organic solvents or buffer solutions to avoid changing the water parameters.

2.3. Experimental System

The system was installed in a room with a constant photoperiod (16 light:8 dark) and comprised of 12 glass containers (L = 25 cm, W = 25 cm, H = 40 cm, large enough to minimise the scale-effects) filled with a 9 L media placed in a water bath (L = 150 cm, W = 50 cm, H = 50 cm, and V = 200 L) with a submersible water-heater (Aquael Neoheather 150 W, Warsaw, Poland) and water pumps (Aquael Circulator 500, Warsaw, Poland) to maintain a stable temperature. The water bath had opaque walls with mounted warm white

(3000 K) LED lamps (5.76 Watts, manufacturer ID: FSLEDWW1200-EF, Green Lighting®, Worcester, UK) inside.

2.4. Experimental Design

2.4.1. Experimental Protocol

We performed the experiments at the Hydrobiological field station of the University of Warsaw in Pilchy (https://pilchy.biol.uw.edu.pl/, accessed on 1 June 2020). The samples obtained during the experiments were analysed at the station during and after the end of the experiments to assess the community-level metabolic fingerprinting and the life history parameters of *Daphnia*. Some analyses (e.g., the assessment of the taxonomic diversity of bacterial communities) were performed in the laboratories of the Department of Hydrobiology, Faculty of Biology, University of Warsaw. We completed the experimental part between May and July 2021 in 12 variants that represent the combination of 3 enrofloxacin concentrations (0, E_l = 10 and E_h = 100 ng × L^{-1}) and 4 densities of PS-NPs (0, N_l = 1 × 10^3, N_m = 1 × 10^6, and N_h = 1 × 10^9 particles × L^{-1}). The concentrations used in our study were within the environmental concentration range [41,42,44–46]. The concentration 1 × 10^9 × L^{-1} of NPs corresponds to the range of bacteria abundance in the lake samples. We fixed the temperature at 23 ± 0.3 °C, which is close to the thermal optimum of *D. magna* [57], and we supplied the media daily with the same amount of algal food (*Ch. klinobasis*) set close to the limiting concentration of 0.6 mg C_{org} × L^{-1}. We calculated the organic carbon content from the calibration curve relating the organic carbon concentration to the absorbance level at 800 nm. We chose a high temperature and meagre food for the experiment to increase the *Daphnia* filtration rate. The LED lamps inside the water bath provided homogeneity throughout the water column and a low light intensity (1.0 ± 0.4 μmol × m^{-2} × s^{-1}) measured by a Li-Cor 189 quantum sensor that measures the radiance (LiCor Biosciences), and it was used at a low light intensity. According to the literature, photodegradation is the main cause of the deactivation of fluoroquinolones, including enrofloxacin, in the environment [58].

At the beginning of each experiment, we added tap water to each of the 12 containers (one container per variant), filtered through 0.45 μm pore membrane filters and aerated for 24 h to reach an oxygen concentration up to 8.00 ± 0.08 mg × L^{-1} (pH = 7.4, μS cm^{-1} = 373 ± 0.7). The physicochemical parameters (the temperature, conductivity, and oxygen concentration) were determined using a multiparametric YSI 6000 probe (Yellow Spring, YSI Inc., Yellow Springs, OH, USA/Xylem Inc., Washington, DC, USA). We added NPs, enrofloxacin, and algal food in the following order. After, 90 newborn (0–24 h) *Daphnia* were collected and randomly distributed into all containers to a final density of 10 ind. × L^{-1} in each variant. Every six hours, the media was gently mixed. The new media was prepared and replaced every 24 h. During the media preparation, we removed the *Daphnia* from each container using a strainer with a plankton net and placed them temporarily in 250 mL glass containers with the respective media. We prepared the new media in the same order as the initial one. The experiments lasted five days when at least 50% of individuals produced eggs in each variant. At the end of each of the replicates of the experiments, we placed *Daphnia* temporarily in 250 mL glass containers with the respective media. Then, we photographed all of the individuals in order to determine the life history traits. The photographed individuals were transferred to sterile 100 mL plastic containers with milliQ water to remove non-symbiotic bacteria from their guts. Then, we used randomly selected individuals from each variant to determine the diversity of the bacterial community in *Daphnia* guts, both metabolic (25 individuals) and taxonomic (10 individuals).

2.4.2. Life History Parameters

We photographed all the individuals collected in each variant from the lateral side under a dissecting microscope connected with the camera and computer. The length and height of each *Daphnia* were measured in the photographs using the NIS program (Nikon Nis Elements). The length was measured from the top of the eye to the base of the tail spine and the height was measured along the body, starting from its greatest dimension. Based on these measurements, the body volume of each individual was calculated, assuming an ellipsoidal shape for each individual according to the formula $4/3\pi \times 1/2a \times 1/2b \times 1/2c$, where a stands for the length, b the height, and c the width (assuming that the body width is equal to the body height [59]. For each ovigerous female, we counted the eggs, and the volume of an egg (upon the mean dimensions for at least two eggs in the clutch) was calculated using the same formula as that which was employed for body volume evaluation. Finally, the clutch volume as the common currency of the reproductive investment was calculated as the number of eggs multiplied by the mean egg volume for each ovigerous female.

2.4.3. Metabolomic Diversity of Gut Microbiota

The extracted *Daphnia* guts were homogenised in 1 mL of milliQ water and transferred into 15 mL of their respective media, which were earlier filtered through 0.45 μm Nylon filters and autoclaved. The standard method of a Biolog EcoPlate [60,61] was used to analyse the metabolic diversity of the gut microbiota in these samples. An EcoPlate (Biolog, Hayward, CA, USA) was employed to measure the ability of the bacterial community to utilise the different carbon substrates. An EcoPlate is a 96-well microplate composed of the triplicates of control wells (containing no additional carbon source) and 31 wells containing various carbon sources (Table 1).

Table 1. Different carbon sources used by the Biolog EcoPlate method to measure the ability of the gut bacterial community of *Daphnia* to utilise carbon substrates.

Carbohydrates	D-cellobiose α-D-lactose β-methyl-D-glucoside D-xylose Erythritol D-mannitol N-acetyl-D-glucosamine D-galactonic acid γ-lactone
Phosphorylated carbons	glucose-1-phosphate D,L-α-glycerol phosphate
Amines	phenylethylamine putrescine
Carboxylic acids	D-glucosaminic acid D-galacturonic acid γ-hydroxybutyric acid itaconic acid α-ketobutyric acid D-malic acid pyruvic acid methyl ester 2-hydroxy benzoic acid 4-hydroxy benzoic acid
Complex carbon	Tween 40 Tween 80 α-cyclodextrin glycogen
Amino acids	L-arginine L-asparagine L-phenylalanine L-serine L-threonine glycyl-L-glutamic acid

Twelve plates were used, one for each variant. Each well of a single plate, except the control one (filled with milliQ water), was filled with 150 μL aliquots of homogenised and the diluted content of *Daphnia* guts from one variant. We incubated the plates in darkness at a temperature of 22 °C for 72 h. Because of the reduction in the tetrazolium chloride by the electrons which derived mainly from the oxidation chains, the colour in the wells increased the proportionally to the respiration rate. We measured the absorbance every four hours at 590 nm wavelength using a Biotek Synergy H1 plate reader (Biotek Corporation, Broadview, IL, USA). For the analysis, we used the maximal colour development rates (V_{max}). In calculating the Vmax values, we used Gen 5 software (Biotec Corporation, Broadview, IL, USA). Additionally, we identified the slope for every four consecutive reads. The Vmax was calculated using a linear regression model by determining the maximum slope during the 72 h incubation time. For the maximal rate of colour development, the V_{max} (mOD $\times$ min^{-1}) was treated as an indicator of the intensity of the respiration of a single carbon source by the *Daphnia* gut microbial communities. The difference in the V_{max} between the carbon-containing wells and the control wells was calculated to determine the influence of each additional carbon source on the respiration rate (delta). When the delta was lower than zero or equal to zero, we assumed the microorganism community did not use the source, and we set this value of V_{max} to zero. For the analyses of the community-level physiological fingerprinting, we used the relative values of the respiration rate for each carbon source, calculated as the percentage share of each carbon source in the sum of the V_{max} for the whole plate. The overall microbial activity in each microplate was expressed as the mean V_{max} for the plate, and was calculated as the average of all the wells, including the control.

2.4.4. Taxonomic Diversity of Gut Microbiota

The DNA was isolated from the microorganisms, inhabiting the intestinal tract of *Daphnia* from all replications of the experiment. The isolated DNA was then mixed v/v to obtain a representative sample and was then sequenced. In order to analyse the microbiota, we used a standard Miseq illumine sequencing method used in numerous previous studies, e.g., in [62–65]. The *D. magna*, the digestive tracts were collected in 1.5 mL Eppendorf and stored at −20 °C. Afterwards, the DNA extraction was performed in the spin-column-based method using the GeneMATRIX DNA Purification kit (EurX, Gdańsk, Poland), according to the manufacturer's procedure. The total DNA was assessed for its quality and quantity by the absorbance measurement in a Synergy H1 microplate reader (Gen5 software, BioTek, Broadview, IL, USA) equipped with a Take3 microvolume plate. We then stored the samples at −20 °C for further analysis. We performed the phylogenetic analysis of the bacterial community using Illumina sequencing method [66]. The 16S rRNA genes, V3–V4 hyper-variable regions (amplicons of approximately 459 bp), were selected. The PCR amplification was carried out using a Q5 Hot Start High-Fidelity 2X Master Mix, using reaction conditions as recommended by the manufacturer (95 °C for 3 min, 25 cycles of 95 °C for 30 s, 55 °C for 30 s, 72 °C for 30 s, and, after the last cycle, 72 °C for 5 min) with region-specific (341F and 785R) [67] primers that include the Illumina flow cell adapter sequences. The primer sequence was as follows: the forward primer was 5′ CGGGNGGCWGCAG 3′ and the reverse primer was 5′ GACTACHVGGGTATCTAATCC 3′. The Illumina overhang adapter sequences added to the locus-specific sequences were as follows: the forward overhang was 5′ TCGTCGGCAGCGTCAGATGTGTATAAGAGACAG (locus-specific sequence) and the reverse overhang was 5′ GTCTCGTGGGCTCGGAGATGTGTATAAGAGACAG (locus-specific sequence).

The amplicons were sequenced using a MiSeq (Illumina, San Diego, CA, USA) platform on a single run using the MiSeq Reagent Kit v2 (Illumina, San Diego, CA, USA) and the paired-end method (2 $\times$ 300 bp), according to the standard protocols by Genomed (Warsaw, Poland). A demultiplexing and trimming of the Illumina adapter sequences (cutadapt software [68]) were performed. FastQC (https://www.bioinformatics.babraham.ac.uk/projects/fastqc; acecssed on 4 October 2018) and MultiQC [69] were used to achieve the quality inspection, visualisation, and assessment of the raw FASTQ files. The sequences were processed using the DADA2 plugin within QIIME 2 [70]. We trimmed the sequence at 270 nt, while the first 8 nt were truncated. The alpha rarefaction plots confirmed that the number of the remaining sequences is sufficient for detecting the current microbial diversity. Taxonomies were assigned to the resulting amplicon sequence variants (ASV) with a q2-feature-classifier plug-in using a pre-trained Naive Bayes classifier based on a 16S rRNA silva 138 SILVA SSU gene database at 99% similarity. The core diversity metrics pipeline was the tool for calculating the phylogenetic and non-phylogenetic core diversity metrics. Data for this purpose were rarefied to a sampling-depth equal to the lowest frequency among the samples (23,500 reads).

2.5. Statistical Analysis

To assess the effect of the NPs and enrofloxacin on the life history parameters the (body length, body volume, clutch size, egg volume, and clutch volume) and on the respiration rate of the gut microbiota of *Daphnia*, we used an Aligned Rank Transform for a nonparametric factorial two-way ANOVA (ART ANOVA; ARTool package v.0.11.1 [71], with the "art" function (ARTool package 0.11.1 [72] Washington, USA), which allowed us to fit the model despite the non-normality and heteroscedasticity of the initial data distribution. To verify that the ART procedure was correctly applied and is appropriate for this dataset, the "summary()" function was used. To test the differences in the pairwise combinations of levels between the factors in the interactions, the ART-C method (multifactor contrast test) was conducted by using the "art.con" function (ARTool, [73]). The statistical analysis was performed using the R platform (v.4.2.0, R Team, Vienna, Austria) by setting the level of significance at $\alpha = 0.05$ for all of the statistics.

A Bray–Curtis-based NMDS (non-metric multidimensional scaling) was applied (PAST3 software, [74]), aiming to group the experimental variants according to the microorganism metabolic and phylogenetic differences. The Bray–Curtis dissimilarity was used because, unlike many other common statistical tools (e.g., Jaccard and unweighted UniFrac), it takes into account not only the number of observed ASVs, but also their relative abundance. For the analyses of community-level physiological fingerprinting, we used the relative values of the respiration rate for each carbon source, calculated as the percentage share of each carbon source in the sum of the Vmax for the whole plate. For taxonomic NMDS, we managed the relative abundance of ASV at the family level. The NMDS analysis and data visualisation were performed in Statistica 13 software (StatSoft, TIBCO Software Inc., Palo Alto, CA, USA).

Additionally, a Mantel correlation between the taxonomic and metabolic profiles was performed (PAST3 software) to reveal the correlation between two Bray–Curtis similarity-based matrices of the relative metabolic and relative phylogenetic data.

3. Results

3.1. The Effect of the Stressors on the Life History Parameters

None of the stressors (enrofloxacin and NPs) had a significant effect on the *Daphnia* body length (Table S1, Supplementary Materials). **The interaction between the stressors** was also not significant (at $p < 0.001$, Table S1), that is, the effect of one stressor has not been modified by the effect of another stressor. Neither the effect of a single stressor nor the effect of combined stressors was significant (Figures 1a and 2a), neither for the combined data from all the concentrations of NPs and enrofloxacin (Table S2, Figure 2a), nor for the data from each of the concentrations assessed, separately (Table S3, Figure 1a).

Both stressors significantly affected the *Daphnia* body volume (at $p \leq 0.001$, Table S1). The effect of the NPs was negative, which was apparent in the significant difference between the NE_{mean} and E_{mean} (Table S2, Figure 2b) and between $N_m E_l$ and E_l treatments (Table S4, Figure 1b). The effect of enrofloxacin was also negative, which was apparent in the significant difference between the NE_{mean} and N_{mean} treatments (Table S2, Figure 2b). **The interaction between the stressors** was also significant (at $p = 0.008$, Table S1); more specifically, the presence of one stressor resulted in increasing the negative effect of another one (Tables S2 and S4, Figures 1b and 2b). This was apparent: (1) in the significant difference between the NE_{mean} and E_{mean} treatments (at $p = 0.031$, Table S2, Figure 2b) in comparison to the non-significant difference between the N_{mean} and control treatments (Table S2, Figure 2b), (2) in the significant difference between the NE_{mean} and N_{mean} (at $p = 0.001$, Table S2, Figure 2b) in comparison to the non-significant difference between the E_{mean} and control treatments (Table S2, Figure 2b), and (3) in the significant difference between the $N_m E_l$ and E_l treatments (at $p \leq 0.017$, Table S4, Figure 1b) in comparison to the non-significant difference between the N_m and control treatments (Table S4, Figure 1b).

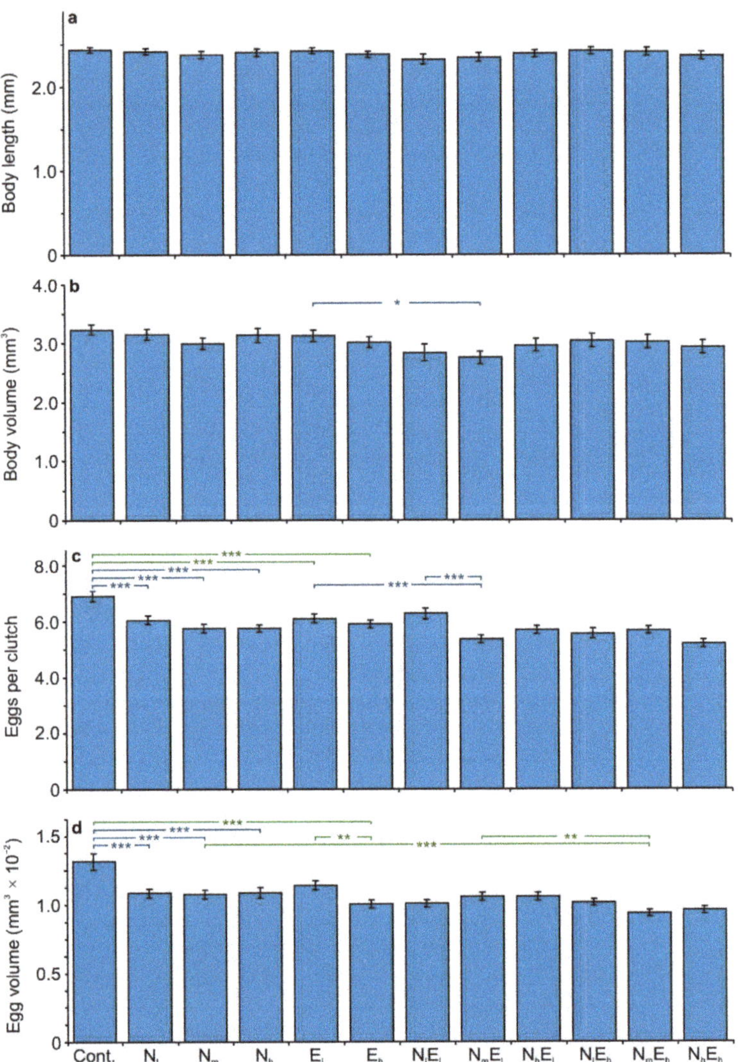

Figure 1. Mean values ($\pm$ 1 SE) of: (**a**) body length, (**b**) body volume, (**c**) clutch size (the number of eggs per ovigerous female), and (**d**) egg volume of 5-day-old *D. magna* from the control variant (Cont.) and from variants of a single or combined low, medium, and high density of polystyrene NPs ($N_l = 10^3$, $N_m = 10^6$, and $N_h = 10^9$ particles L^{-1}, respectively) and low and high concentration of enrofloxacin ($E_l = 10$ and $E_h = 100$ ng L^{-1}, respectively). Statistical significance is accepted at * $p < 0.05$, ** $p < 0.005$, or *** $p < 0.0005$. The NPs effect is marked on blue, and the enrofloxacin effect on green.

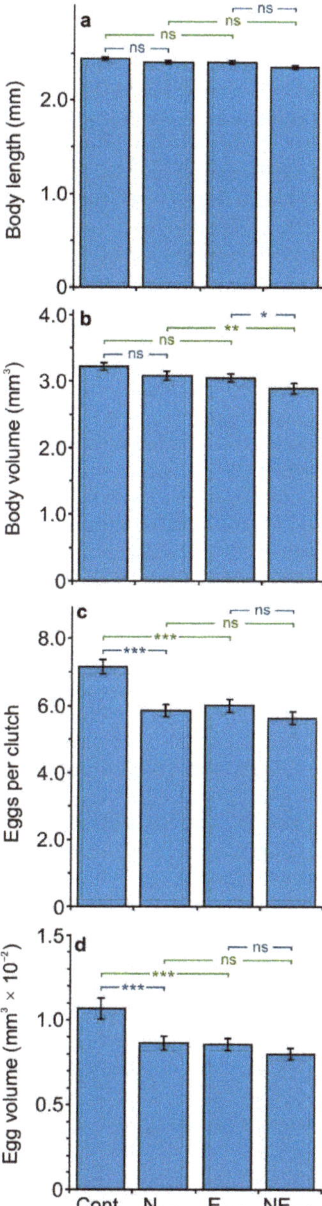

Figure 2. Mean values ($\pm$ 1 SE) of: (**a**) body length, (**b**) body volume, (**c**) clutch size (the number of eggs per ovigerous female), and (**d**) egg volume of 5-day-old *D. magna* from the control variant (Cont.) and from variants of a single or combined mean density of polystyrene NPs (N_{mean} for the combined data from $N_l = 10^3$, $Nm = 10^6$, and $N_h = 10^9$ particles L^{-1}) and the mean concentration of enrofloxacin (E_{mean} for the combined data from $E_l = 10$ and $E_h = 100$ ng L^{-1}). Statistical significance is accepted at * $p < 0.05$, ** $p < 0.005$, or *** $p < 0.0005$, ns stands for non-significant. The NPs effect is marked on blue, and the enrofloxacin effect on green.

The NPs and enrofloxacin significantly affected the clutch size (at $p < 0.001$, Table S1). The effect of the NPs was negative, which was apparent in the significant difference between: (1) the N_{mean} and control treatments (at $p < 0.001$, Table S2, Figure 2c), and (2) each of the concentrations of the NPs and the control, $N_m E_l$ and E_l, as well as the $N_m E_l$ and $N_l E_l$ treatments (at $p \leq 0.005$, Table S5, Figure 1c). The impact of enrofloxacin was also negative, which was apparent in the significant difference between: (1) the E_{mean} and control treatments (at $p < 0.001$, Table S2, Figure 2c), and (2) each of the two concentrations in relation to the control (at $p \leq 0.005$, Table S5, Figure 1c). **The interaction between the stressors** was also significant (at $p < 0.001$, Table S1); more specifically, the presence of one stressor resulted in decreasing the negative effect of another one (Tables S2 and S5, Figures 1c and 2c). This was apparent: (1) in the non-significant difference between the NE_{mean} and E_{mean} treatments in comparison to the significant difference between the N_{mean} and control treatments (at $p < 0.001$, Table S2, Figure 2c), (2) in the non-significant difference between the NE_{mean} and N_{mean} treatments in comparison to the significant difference between the E_{mean} and control treatments (at $p < 0.001$, Table S2, Figure 2c), and (3) in the non-significant differences in the majority of comparisons between the single and combined stressors in relation to the significant difference between the single stressors and control for the data from each of the concentrations of NPs and enrofloxacin, separately (Table S5, Figure 1c). The only exception was the significant differences between the $N_m E_l$ and E_l, and between the $N_m E_l$ and $N_l E_l$ treatments (Table S5, Figure 1c). The number of ovigerous females in relation to the females without eggs was the greatest in the control (84%), moderate in the presence of NPs on their own (81–83%), and in the presence of enrofloxacin on its own (79–80%), and was the lowest in the $N_m E_l$ (70%) and $N_h E_h$ (75%) treatments.

The NPs and enrofloxacin also significantly affected the egg volume (at $p < 0.001$, Table S1). The effect of the NPs was negative, which was apparent in the significant difference between: (1) the N_{mean} and control treatments (at $p < 0.001$, Table S2, Figure 2d), and (2) each of the concentrations of NPs and the control (at $p \leq 0.001$, Table S6, Figure 1d). The impact of enrofloxacin was also negative, which was apparent in the significant difference between: (1) the E_{mean} and control treatments (at $p < 0.001$, Table S2, Figure 2d), and (2) the E_h and the control, $N_m E_h$ and N_m, E_h and E_l, as well as the $N_m E_h$ and $N_m E_l$ treatments (at $p \leq 0.018$, Table S6, Figure 1d). **The interaction between the stressors** was also significant (at $p < 0.001$, Table S1); more specifically, the presence of one stressor resulted in decreasing the negative effect of another one (Tables S2 and S6, Figures 1d and 2d). This was apparent: (1) in the non-significant difference between the NE_{mean} and E_{mean} treatments in comparison to the significant difference between the N_{mean} and control treatments (at $p < 0.001$, Table S2, Figure 2d), (2) in the non-significant difference between the NE_{mean} and N_{mean} treatments in comparison to the significant difference between the E_{mean} and control treatments (at $p < 0.001$, Table S2, Figure 2d), and (3) in the non-significant differences in the majority of comparisons between the single and combined stressors in relation to the significant difference between the single stressors and the control for the data from each of the concentrations of NPs and enrofloxacin, separately (Table S6, Figure 1d).

Additionally, the NPs and enrofloxacin significantly affected the clutch volume (at $p \leq 0.001$, Table S1). The effect of the NPs was negative, which was apparent in the significant difference between: (1) the N_{mean} and control treatments (at $p < 0.001$, Table S2), and (2) each of the concentrations of the NPs and the control, $N_m E_l$ and E_l, $N_h E_l$ and E_l, as well as the $N_m E_h$ and E_h treatments (at $p \leq 0.040$, Table S7). The impact of enrofloxacin was also negative, which was apparent in the significant difference between: (1) the E_{mean} and control treatments (at $p < 0.001$, Table S2), and (2) each of the two concentrations in relation to the control, the $N_l E_h$ and N_l, $N_m E_h$ and N_m, and the $N_h E_h$ and N_h treatments (at $p \leq 0.031$, Table S7). **The interaction between the stressors** was also significant (at $p < 0.001$, Table S1); more specifically, the presence of one stressor resulted in decreasing the negative effect of another one (Tables S2 and S7). This was apparent: (1) in the non-significant difference between the NE_{mean} and E_{mean} treatments in comparison to the significant difference between the N_{mean} and control treatments (at $p < 0.001$, Table S2), (2) in the non-significant difference between the NE_{mean} and N_{mean} treatments in comparison to the significant difference between the E_{mean} and control treatments (at $p < 0.001$, Table S2), and (3) in the non-significant differences in the majority of comparisons between the single and combined stressors in relation to the significant difference between the single stressors and the control for the data from each of the concentrations of NPs and enrofloxacin, separately (Table S7).

3.2. Metabolomic Diversity of Gut Microbiota

The NPs and enrofloxacin significantly affected the mean respiration rate of the gut microbiota expressed as the V_{max} values (at $p < 0.001$, Table S1). The effect of the NPs was positive (Tables S2 and S8, Figures 3 and 4), which was apparent in the significant difference between: (1) the N_{mean} and control treatments (at $p < 0.001$, Table S2, Figure 4), and (2) the N_l and control, N_m and control, N_hE_l and E_l, as well as the N_hE_l and N_mE_l treatments (at $p \leq 0.002$, Table S8, Figure 3). The impact of enrofloxacin was negative, which was apparent in the significant difference between: (1) the NE_{mean} and N_{mean} treatments (at $p < 0.001$, Table S2, Figure 4), and (2) the N_lE_l and N_l, N_lE_h and N_l, N_mE_l and N_m, N_mE_h and N_m, N_hE_h and N_h, as well as the N_hE_h and N_hE_l treatments (at $p \leq 0.001$, Table S8, Figure 3). **The interaction between the stressors** was significant (at $p < 0.001$, Table S1), more specifically in the majority of comparisons, the presence of NPs increased the inhibitory effect of enrofloxacin, and the presence of enrofloxacin reduced the positive effect of the NPs (Tables S2 and S8, Figures 3 and 4). The former was apparent: (1) in the non-significant difference between the E_{mean} and control in comparison to the significant difference between the NE_{mean} and N_{mean} treatments (at $p < 0.001$, Table S2, Figure 4), and (2) in the non-significant difference between the E_l and control and the E_h and control in relation to the significant difference in the majority of comparisons (in six among nine) between the treatments in which enrofloxacin was combined with the NPs (at $p < 0.001$, Table S8, Figure 3). The latter was apparent (1) in the significant difference between the N_{mean} and the control (at $p < 0.001$, Table S2, Figure 4) in comparison to the non-significant difference between the NE_{mean} and E_{mean} treatments (Table S2, Figure 4), and (2) in the significant difference between the N_l and the control and the N_m and the control (at $p < 0.001$, Table S8, Figure 3) in relation to the non-significant difference in the majority of comparisons (in 10 among 12) between the treatments in which NPs were combined with enrofloxacin (Table S8, Figure 3).

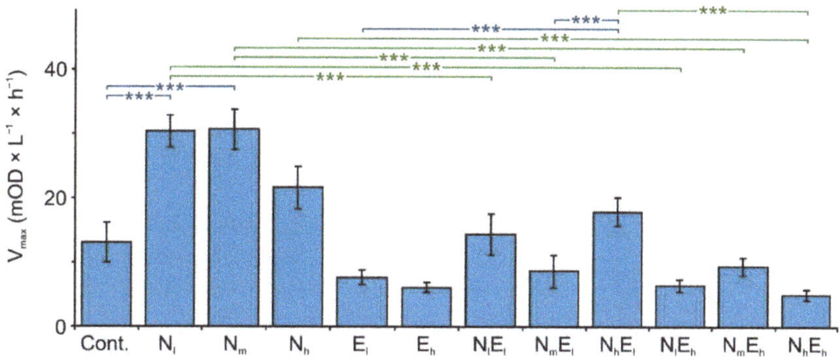

Figure 3. Mean values (± 1 SE) of respiration rate expressed as V_{max} values of 31 carbon sources by *Daphnia* gut microbiota from the control variant (Cont.) and from variants of a single or combined low, medium, and high density of polystyrene NPs ($N_l = 10^3$, $N_m = 10^6$, and $N_h = 10^9$ particles L^{-1}, respectively) and low and high concentration of enrofloxacin ($E_l = 10$ and $E_h = 100$ ng L^{-1}, respectively). Statistical significance is accepted at *** $p < 0.0005$. The NPs effect is marked on blue, and the enrofloxacin effect on green.

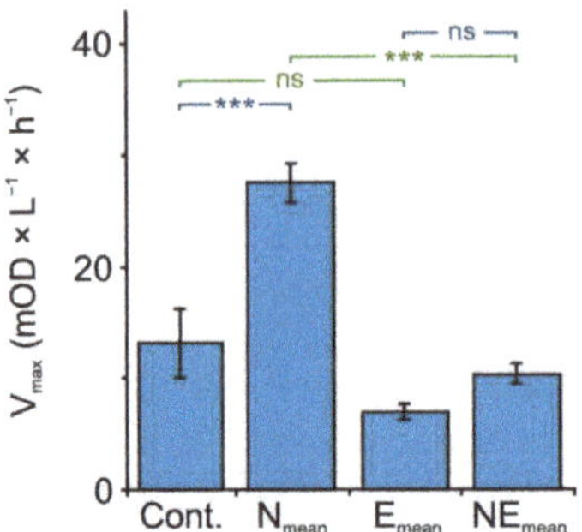

Figure 4. Mean values ($\pm$ 1 SE) of respiration rate expressed as V_{max} values of 31 carbon sources by *Daphnia* gut microbiota from the control variant (Cont.) and from variants of a single or combined mean density of polystyrene NPs (N_{mean} for the combined data from $N_l = 10^3$, $Nm = 10^6$, and $N_h = 10^9$ particles L^{-1}) and the mean concentration of enrofloxacin (E_{mean} for the combined data from $E_l = 10$ and $E_h = 100$ ng L^{-1}). Statistical significance is accepted at *** $p < 0.0005$, ns stands for non-significant. The NPs effect is marked on blue, and the enrofloxacin effect on green.

The analysis of the percentage share of the respiration rate of different carbon sources by the gut microbiota of *Daphnia* revealed that the presence of each of the stressors resulted in a relative increase in the usage of carboxylic acids, amino acids, and carbohydrates, as well as a relative decrease in the usage of phosphorylated carbons and complex carbon sources with the control (Figure 5a,b). The pattern was similar in the presence of single and combined stressors, which suggests a negative interaction between their effects.

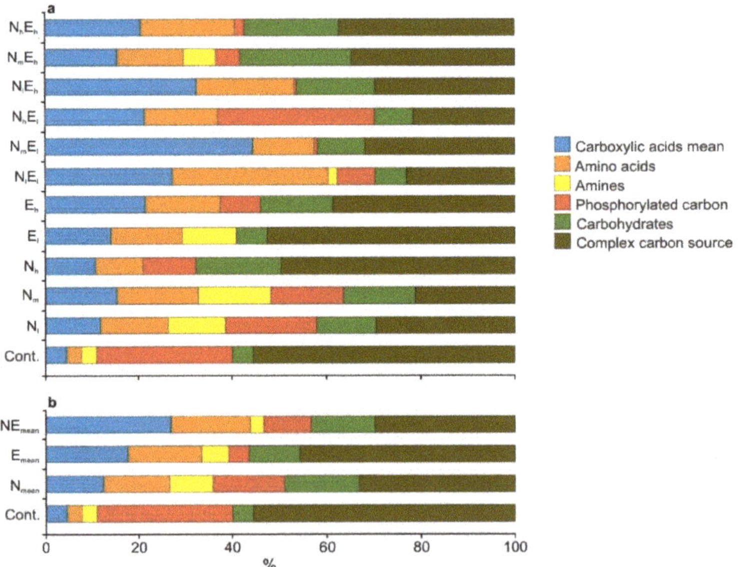

Figure 5. Percentage share of respiration rate (V_{max}) of different carbon sources by the gut microbiota of *Daphnia* (**a**) from the control variant (Cont.) and from variants of a single or combined low, medium and high density of polystyrene NPs ($N_l = 10^3$, $N_m = 10^6$, and $N_h = 10^9$ particles L^{-1}, respectively) and low and high concentration of enrofloxacin ($E_l = 10$ and $E_h = 100$ ng L^{-1}, respectively), and (**b**) from the control variant (Cont.) and from variants of a single or combined mean density of polystyrene NPs (N_{mean} for the combined data from N_l, N_m, and N_h) and the mean concentration of enrofloxacin (E_{mean} for the combined data from E_l and E_h).

A Bray–Curtis-based NMDS analysis revealed two distinct groups of variants: variants with a high concentration of enrofloxacin (in the presence and absence of NPs) and variants with a low and medium density of NPs, which suggests a different effect of each of the stressors on the metabolic profile of the gut microbial community (Figure 6). In the first group, a relatively low usage of phosphorylated carbon and amines compared to the control and the majority of the remaining variants was observed (Figure 5a). In the second group, there was relatively even usage of the different carbon sources with a relatively low usage of the complex carbon sources, as well as a relatively high usage of amines concerning the majority of the remaining variants (Figure 5a).

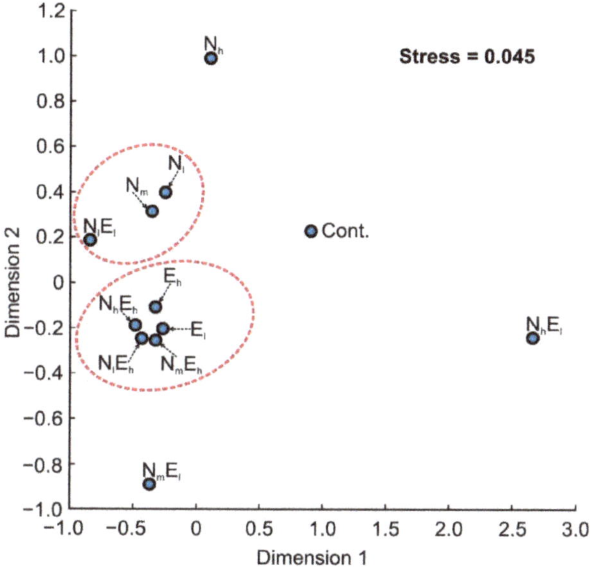

Figure 6. Bray–Curtis-based NMDS analysis of the relative respiration of 31 different carbon sources by *D. magna* gut microbiota from the control (Cont.), and from variants of the combination of two variables: (1) NPs in low, medium, and high concentrations (N_l, N_m, N_h, respectively) and (2) enrofloxacin in low and high concentrations (E_l and E_h, respectively).

3.3. Taxonomic Diversity of Gut Microbiota

We observed a low taxonomic diversity of the bacteria living in the digestive tract of *Daphnia* compared to typical diversity of the bacteria living in the lake waters and bacteria from the *Daphnia* digestive tracts described in other studies, e.g., in [61,75]. The gut microbiota was mainly represented by the bacteria belonging to three phyla—Actinobacteriota, Firmicutes, and Proteobacteria—with the predominance of Proteobacteria and Firmicutes (Figure 7a,b). The presence of each of the stressors, especially the high concentration of NPs, increased the Firmicutes' participation (Figure 7a,b). This effect was also apparent with the increasing concentrations of enrofloxacin (Figure 5). Despite the clear effect of both stressors on their own—the Firmicutes share the increase—the pattern was similar in the presence of single and combined stressors, which suggest a negative interaction between their effects (Figure 7a,b). The Bray–Curtis-based NMDS analyses at the family level did not show any apparent group of similar variants (Figure 8), which suggests the different taxonomic composition within phylum Firmicutes, whose relative abundance increased after adding both stressors. However, it revealed that the variants with the combined effect of both stressors and with high concentrations of each of the stressors on its own are further away from the control than the other variants (i.e., with low concentrations of each of the stressors on its own and combined).

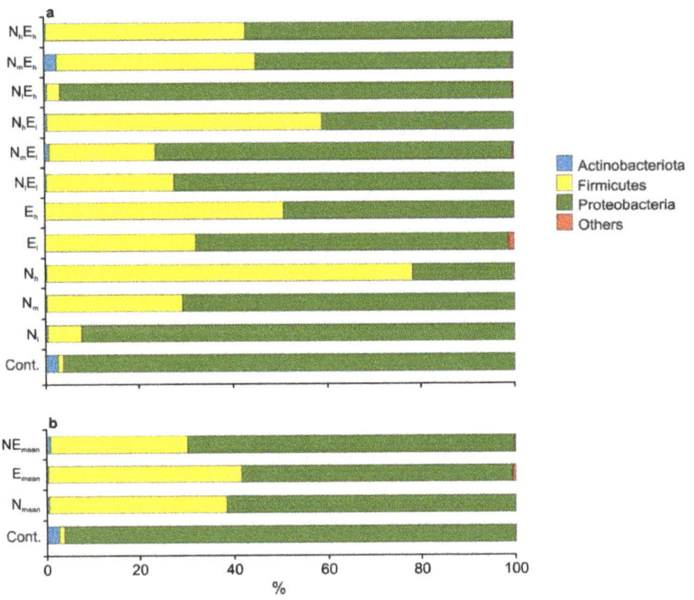

Figure 7. The relative abundances of the top three dominant bacteria phyla present in the gut of *D. magna* (**a**) from the control variant (Cont.) and from variants of a single or combined low, medium, and high density of polystyrene NPs ($N_l = 10^3$, $N_m = 10^6$, and $N_h = 10^9$ particles L^{-1}, respectively) and low and high concentration of enrofloxacin ($E_l = 10$ and $E_h = 100$ ng L^{-1}, respectively), as well as (**b**) from the control variant (Cont.) and from variants of a single or combined the mean density of polystyrene NPs (N_{mean} for the combined data from N_l, N_m, and N_h) and the mean concentration of enrofloxacin (E_{mean} for the combined data from E_l and E_h).

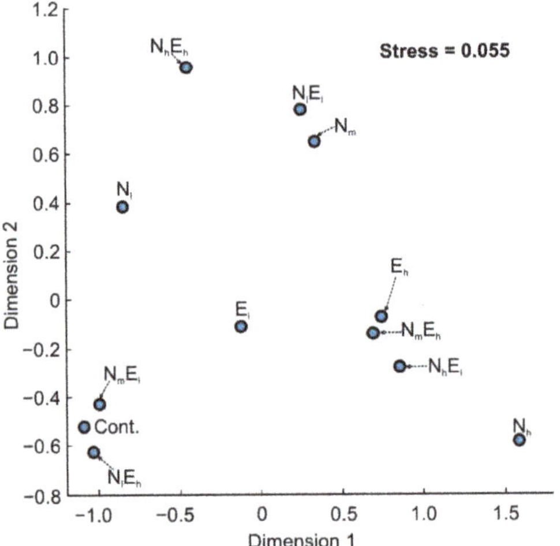

Figure 8. The graphical results of the Bray–Curtis-based NMDS analysis of sequence data, binned by taxonomic assignment to family. The figure shows the relative distances between the gut microbiota of *D. magna* from the control (Cont.) variant, and from variants of the combination of two variables: (1) NPs in low, medium, and high concentrations (N_l, N_m, N_h, respectively) and (2) enrofloxacin in low and high concentrations (E_l and E_h, respectively).

No significant correlation between the taxonomic profile (at the family level) and the Eco Plate-based metabolic profile was observed (Mantel Correlation, R = 0.153, p = 0.2052, permutation n = 9999).

4. Discussion

4.1. General Effect

The results of our study revealed that each of the stressors (the NPs and enrofloxacin) on their own influenced most of the measured life history parameters of *D. magna* (except the body length of 5-day-old individuals) and the metabolic and taxonomic diversity of the *Daphnia* gastrointestinal microbiota, and there was an interaction in the effect of both stressors on all of the measured parameters. On the one hand, since we used in our study temperature and food conditions very close to the optimal ones (23 °C and 0.6 mg $C_{org} \times L^{-1}$), it is rather unlikely that the stress caused by the suboptimal experimental conditions had any impact on the observed effects of NPs and enrofloxacin. On the other hand, it can be expected that the use of a lower temperature and a higher food concentration could reduce the *Daphnia* filtration rate and, consequently, could reduce the observed effects of the NPs and enrofloxacin, although they would not change the direction of the effects.

4.2. The Effect of Single and Combined Stressors on the Life History Parameters

The results confirmed our first hypothesis, as both stressors resulted in a decrease in the *Daphnia* body size and their reproductive parameters, including an average egg volume in the brood cavity and the number of eggs in the clutch of individuals during the first reproduction, which also resulted in decreasing the clutch volume being a common currency of the reproductive investment, which suggests that each of the stressors has a negative effect on the fitness of an individual. Although we did not assess the age of the first reproduction of *Daphnia* in the experiments, the greatest number of ovigerous females (in relation to the females without eggs) during the fifth day of the experiment in the control, moderate in the presence of each of the stressors on its own, and the lowest in the treatments with the combined stressors, suggests that each of the stressors delayed the start of reproduction and that there was an interaction in the effect of both stressors.

In the case of the NPs, these results may have been caused by the clogging of the filtration appendages and the gut which, in turn, results in decreasing the filtration and assimilation rate, as demonstrated in several earlier studies, e.g., in [76]. These findings are consistent with a great number of earlier studies in which the presence of NPs suspended in water alters the life history traits of various animals, and that these alterations are manifold [13], including a reduction in the body size of adult and juvenile individuals, reproduction (i.e., decreased numbers and body size of neonates), the individual growth rate and survival of freshwater, e.g., in [77], and saline lakes, e.g., in [78], planktonic animals, cnidarians [79], and fishes, e.g., in [23,80], although other studies did not find any effect of an acute exposure on the life history parameters, including negligible effects on the survival rate and development of *Danio rerio*, e.g., in [81], on the survival and individual growth rate of *Gammarus pulex* [82].

Several studies revealed the significance of antimicrobial drugs on aquatic organisms [83,84]. For example, the parental exposure of marine fish (*Oryzias melastigma*) to sulfamethazine (4.62 mg $\times$ g^{-1}) may negatively affect the growth performance in adults [14,85]. In the case of enrofloxacin, the decrease in all the measured life history parameters is consistent with the numerous earlier studies performed on fish [23,85] and *Daphnia* [75]. Another study revealed the negative effect of tetracycline (1 µg $\times$ L^{-1}) on the reproduction and survival of *D. magna* [75]. However, it should be pointed out that not all studies reported the effect of antibiotics on the life history parameters. For example, Nunes et al. [86] showed that ecologically relevant ciprofloxacin concentrations (0.005–0.195 mg $\times$ L^{-1}), did not cause significant impacts on the growth rate and reproductive parameters of *D. magna*, and Ma et al. [87] did not find the effect of tetracycline (10 mg $\times$ L^{-1}) on the growth rate of the soil annelid *Enchytraeus crypticus*. Moreover, in a single study, it was revealed that at very low ciprofloxacin concentrations (10 ng L^{-1}), that is, at a level of a lower concentration that the enrofloxacin used in our study, the growth and fecundity of *D. magna* were even higher than that in the control of animals [88].

We are aware of only two earlier studies in which the combined effect of the NPs and antibiotics on the life history parameters was determined [23,87]. While the first study revealed that the effect of tetracycline on the dry weight of *E. crypticus* after a seven-day exposure was stronger in the presence of polystyrene [87], the second study revealed that the adverse impact of the mixture of polystyrene NPs and sulfamethazine on the dry weight in the *O. melastigma* was weaker than the sole effect of the NPs [23]. In our study, the negative effect of each of the stressors (the NPs and enrofloxacin) on the body size was stronger, and on the reproductive parameters was lower in the presence of another factor, which may suggest that in the presence of cumulative stress, *Daphnia* redirects more resources to reproduction at the expense of somatic growth.

4.3. The Effect of Single and Combined Stressors on the Metabolomic Diversity of Gut Microbiota

The results also confirmed our second hypothesis, since NPs resulted in an increase and enrofloxacin resulted in the decrease in the overall carbon respiration rate of the *Daphnia* gut microbiota. This may be due to a changing bacteria abundance, their metabolic condition, and their ability of utilising different carbon sources. The decrease in the metabolic rate in the presence of enrofloxacin is consistent with the existing data on the negative effect of antibiotics on the bacteria metabolism, since there is a link between antibiotic-induced cellular respiration and bactericidal lethality. Antibiotics disturb the metabolic state of bacteria, which impacts the antibiotic efficacy [89,90]. The positive effect of NPs on the respiration of the bacteria in the *Daphnia* gastrointestinal tract may be due to the provision of an additional sorption surface, which not only facilitates the formation of bacterial biofilms, but may also affect the ability to degrade organic compounds that absorb on such surfaces. For example, the sorption of proteins on surfaces can change their conformation, making them more accessible to proteolytic enzymes and subject to faster hydrolysis [91]. Our results are the first to demonstrate the effect of NPs on the metabolic profile of the gut microbiota. Moreover, the results are also the first to demonstrate the interaction between the stressors on the metabolic profile of the gut microbiota, which was apparent in the increased inhibitory effect of enrofloxacin in the presence of NPs, and in the reduced positive effect of NPs in the presence of enrofloxacin.

The results also confirmed our third hypothesis, as the presence of NPs and enrofloxacin affected the metabolic fingerprints measured as the relative use of different carbon sources as compared to the control. The functional structure of the bacteria community had changed. The presence of each of the stressors resulted in a relative increase in the usage of carboxylic acids, amino acids, and carbohydrates and in the relative decrease in the usage of phosphorylated carbons and complex carbon sources with the control, which suggests that both the enrofloxacin and NPs forced the bacteria to exploit the easily digestible carbon sources [92]. In general, these results are in accordance with the existing literature. For instance, it has been shown that the presence of polystyrene NPs altered the carbohydrate metabolism of marine medaka fish gut microbiota [23]. However, Zhang et al. [51] did not show any significant effect of NPs on the predicted metabolic pathways of the gut microbiota in the same fish species. A picture of the negative impact of NPs on the metabolic fingerprint emerges from the research on the freshwater biofilm done with the Biolog EcoPlate method, in which the NPs' presence (1, 5 and 10 mg $\times$ L^{-1}) reduced the microbial metabolic functional diversity. The total carbon metabolism remained constant with growing NP concentrations, but the utilisation of some specific carbon sources (e.g., esters) had changed [93]. It was revealed that the presence of sulfamethazine affected the function of the gut microbiota of medaka fish, so the carbohydrate metabolism was significantly decreased [23]. Zhang et al. [51] presented similar results, showing a decrease in the carbohydrate metabolism in marine medaka females who had been exposed to low concentrations of sulfamethazine (0.5 mg $\times$ g^{-1} supplied in fish food); thus, also the lipid and amino acid metabolism was enhanced under these conditions. In the murine model, other antibiotics, like amoxicillin, have been shown to elevate the expression of genes responsible for starch utilisation by *Bacteroides thetaiotaomicron* [94].

Moreover, the NPs and enrofloxacin which were acting together had a different effect on the metabolic fingerprints than the NPs and enrofloxacin which were acting separately, suggesting the existence of the interaction between the two stressors. The simultaneous presence of sulfamethazine and NPs also significantly altered the carbohydrate metabolism, as He et al. [23] discovered. However, it should be pointed out that the results from He et al. [23] and Zhang et al. [51] concerning the effects of antibiotics and NPs on changes in the gut microbial metabolism fingerprint were obtained by the bioinformatics prediction. Our results stem from the direct measurement of the microbial metabolic fingerprint as the function of different respiration intensities of various carbon sources by the *Daphnia* gut microbial community.

4.4. The Effect of Single and Combined Stressors on the Taxonomic Diversity of Gut Microbiota

Finally, the results also confirmed the fourth hypothesis, as the taxonomic diversity of bacteria was affected by the presence of each of the two stressors on its own, which was apparent in the relative increase in the Firmicutes (mainly the Bacillaceae family, representatives of which occur in the digestive tracts of many aquatic animals, e.g., in [95]), in relation to Actinobacteria and Proteobacteria at the phylum level. Despite this, in most of the experimental variants, the dominant taxa at the phylum level were Proteobacteria, which is consistent with the previous studies on the microbiome of other animals, including the soil fauna [88,96,97]. On the one hand, the relative increase in Firmicutes may have a positive effect on *Daphnia*, as it increases the diversity of the microbial composition, which is often equated with the improvement of the host's health [98]. It has been disclosed that Firmicutes produce short-chain fatty acids, which could be used for the de novo synthesis of lipids or glucose and an additional energy source for the host [23,99,100]. Therefore, from the perspective of the host energy input, Firmicutes in the gut may play a positive role. On the other hand, it has been shown that an increased proportion of Firmicutes to other bacteria phyla is an indicator of metabolic disorders in animals [101,102], which may explain the reduction in the body size and reproductive potential of *Daphnia* by each of the stressors on its own in our study.

In the case of NPs, the relative increase in Firmicutes in the gut microbiota may be due to the fact that NPs can provide them with a better matrix for biofilm growth compared to other bacteria taxa. The results are consistent with several recent studies, which found that NPs may affect a community of free-living bacteria [103,104] and the microbiome of *E. crypticus* [67]. This also includes a marine mollusk *Mytilus galloprovincialis* [105] and various fish species, including *O. melastigma* [106] and *D. rerio*, e.g., [107,108], which can induce dysbiosis and inflammation in their intestine. For example, it has been discovered that the dietary NPs ($1 \text{ mg} \times \text{g}^{-1}$) affected the relative proportions of Microbacteriaceae, Streptococcaceae, Enterobacteriaceae, and Rhodocyclaceae in the whole body microbial community of *E. crypticus* [87]. It has been recognised that all of these taxa potentially negatively (with Enterobacteriaceae [109]) or positively (with the remaining taxa [110,111]) affect the host. However, our study contradicts the results of He et al. [23], who found a decrease in the relative abundance of Firmicutes in males exposed to ($3.45 \text{ mg} \times \text{g}^{-1}$) dietary polystyrene NPs. It is worth noting that, in our study, the taxonomic diversity of the bacteria increased in the presence of NPs, as the relative abundance of different taxa was more even compared to the controls, which is consistent with the earlier studies for the diversity of the microbial communities in the gut of *D. rerio*, e.g., [107].

The increase in the relative abundance of Firmicutes in the presence of enrofloxacin is consistent with several previous studies, which revealed that antibiotics might change the taxonomic diversity of the gut microbiota of animals, including mice [112] soil organisms [87,102], fish [23], and *Daphnia* [76]. For example, it has been revealed that an exposure to tetracycline ($1 \text{ μg} \times \text{L}^{-1}$) resulted in an increase in the relative abundance of

Pseudomonaceae in the intestinal microbial community of *D. magna* [75]. Other studies have shown that dietary tetracycline (0.01 mg $\times$ g^{-1}) affected the relative proportions of Microbacteriaceae, Streptococcaceae, Enterobacteriaceae, and Rhodocyclaceae in the whole body microbial community of *E. crypticus* [87]. In another study, Motiei et al. [88] showed that G + bacteria, mostly Actinobacteria and Firmicutes, was better equipped to withstand an exposure to ciprofloxacin (of a similar mechanism of action as enrofloxacin) as their relative abundance increased with the antibiotic concentration.

Moreover, the results also confirmed the fourth hypothesis as we observed an interaction between the two stressors on the bacteria community structure, which was apparent in the weaker effect of each of the stressors on its own than combined. To our knowledge, only two earlier studies investigated the combined effect of NPs and antibiotics [23,88]. The first study showed that the combined exposure of tetracycline and polystyrene NPs negatively affected the abundance of bacteria belonging to several families, including Microbacteriaceae, Streptococcaceae, and Enterobacteriaceae in the whole body microbiome of *E. crypticus*, and additionally, significantly higher ratios of Planococcaceae/Chitinophagaceae and Bacillaceae/Chitinophagaceae were observed after a tetracycline and polystyrene exposure, especially when the two pollutants were combined [87]. The second study revealed that a parental exposure to polystyrene NPs and sulfamethazine had a weaker effect on the gut microbial communities in the offspring of marine fish (*O. melastigma*) than each of the stressors on its own [23]. Moreover, it has been released that after terminating the exposure, the microbiome was not permanently changed but impacted reversibly [87].

Summing up, the NPs and enrofloxacin altered the taxonomic (and also metabolic) structure of the bacterial communities living in the gut of *Daphnia*. The interaction of the enrofloxacin and NPs resulted in changes in the taxonomic composition of the bacterial communities slightly different from those induced by each of the stressors acting separately. This fact may result from the system's complexity, where NPs constitute the sorption surface for the dissolved organic compounds and antibiotics, causing both their local concentration to increase and a change in the conformation of sorbed molecules that can change their physical properties. The formation of biofilms on aggregated plastic particles may also change the dominant bacteria's taxonomic structure. No clear relationship has been found between the bacterial taxonomic composition and metabolic fingerprints. Thus, this may indicate the occurrence of the phenomenon of redundancy and the high plasticity of microorganism communities which, when changing the taxonomic composition, can replace their environmental functions.

5. Conclusions

In conclusion, despite the growing interest in the effects of NPs and antibiotics on the biology of organisms, knowledge on the subject is still fragmentary and is often based on contradictory results. Our study seems to be the first one to investigate the combined effect of NPs and antibiotics on the life history parameters of freshwater organisms and the metabolomic and taxonomic diversity of their intestinal microbial community. Among the most important results that we obtained is that both stressors: (i) negatively affected most of the *Daphnia* life history parameters, (ii) modified the use of different carbon sources by the intestinal microbiota, (iii) increased the Firmicute phyla participation in the microbiota taxonomic composition, and (iv) interacted with each other, affecting all the measured parameters. Moreover, we also observed that the NPs increased and enrofloxacin decreased the metabolic rate of the gut microbiota. The results of our study suggest a possible link between the observed effects. Future studies concerning the issue should primarily focus on integrating the results from different levels of the biological organisation on the combined effect of both stressors on different taxa.

Supplementary Materials: The following are available online at https://www.mdpi.com/article/10.3390/w14213403/s1, Tables S1–S8: The results of the analysis of ART two-way ANOVA.

Author Contributions: Conceptualization: P.M. and B.K.; methodology: P.M. and B.K.; investigation: E.B., S.G., P.M., C.J.-M. and B.K.; resources: P.M.; data curation: S.G., B.K., P.M., G.K. and M.L.Z.; writing—original draft preparation: P.M.; writing—review and editing: J.P., J.S., J.J.-L., B.K., P.M., E.B. and C.J.-M.; visualization: B.K. and P.M.; supervision: B.K. and P.M.; project administration: P.M.; funding acquisition: P.M. and E.B. All authors have read and agreed to the published version of the manuscript.

Funding: The research described here was supported by the grants no. 2018/31/N/NZ8/03269 and 2019/35/B/NZ8/04523 from the National Science Center, Poland.

Institutional Review Board Statement: Not applicable.

Informed Consent Statement: Not applicable.

Data Availability Statement: Not applicable.

Conflicts of Interest: The authors declare no conflict of interest.

References

1. MacLeod, M.; Arp, H.P.H.; Tekman, M.B.; Jahnke, A. The global threat from plastic pollution. *Science* **2021**, *373*, 61–65. [CrossRef] [PubMed]
2. Barnes, D.K.; Galgani, F.; Thompson, R.C.; Barlaz, M. Accumulation and fragmentation of plastic debris in global environments. *Philos. Trans. R. Soc. B Biol. Sci.* **2009**, *364*, 1985–1998. [CrossRef] [PubMed]
3. Kedzierski, M.; Frčre, D.; Le Maguer, G.; Bruzaud, S. Why is there plastic packaging in the natural environment? Understanding the roots of our individual plastic waste management behaviours. *Sci. Total Environ.* **2020**, *740*, 139985. [CrossRef] [PubMed]
4. Geyer, R.; Jambeck, J.R.; Law, K.L. Production, use, and fate of all plastics ever made. *Sci. Adv.* **2017**, *3*, e1700782. [CrossRef] [PubMed]
5. Gigault, J.; Ter Halle, A.; Baudrimont, M.; Pascal, P.Y.; Gauffre, F.; Phi, T.L.; Reynaud, S. Current opinion: What is a nanoplastic? *Environ. Pollut.* **2018**, *235*, 1030–1034. [CrossRef]
6. Besseling, E.; Redondo-Hasselerharm, P.; Foekema, E.M.; Koelmans, A.A. Quantifying ecological risks of aquatic micro-and nanoplastic. *Crit. Rev. Environ. Sci. Technol.* **2019**, *49*, 32–80. [CrossRef]
7. Egessa, R.; Nankabirwa, A.; Ocaya, H.; Pabire, W.G. Microplastic pollution in surface water of Lake Victoria. *Sci. Total Environ.* **2020**, *741*, 140201. [CrossRef]
8. Huang, D.; Chen, H.; Shen, M.; Tao, J.; Chen, S.; Yin, L.; Li, R. Recent advances on the transport of microplastics/nanoplastics in abiotic and biotic compartments. *J. Hazard. Mater.* **2022**, *438*, 129515. [CrossRef]
9. Koelmans, A.A.; Besseling, E.; Shim, W.J. Nanoplastics in the aquatic environment. Critical review. *Mar. Anthropog. Litter* **2015**, 325–340.
10. Ter Halle, A.; Jeanneau, L.; Martignac, M.; Jardé, E.; Pedrono, B.; Brach, L.; Gigault, J. Nanoplastic in the North Atlantic subtropical gyre. *Environ. Sci. Technol.* **2017**, *51*, 13689–13697. [CrossRef]
11. Bhattacharya, P.; Lin, S.; Turner, J.P.; Ke, P.C. Physical adsorption of charged plastic nanoparticles affects algal photosynthesis. *J. Phys. Chem. C* **2010**, *114*, 16556–16561. [CrossRef]
12. Booth, A.M.; Hansen, B.H.; Frenzel, M.; Johnsen, H.; Altin, D. Uptake and toxicity of methylmethacrylate-based nanoplastic particles in aquatic organisms. *Environ. Toxicol. Chem.* **2016**, *35*, 1641–1649. [CrossRef] [PubMed]
13. Maszczyk, P.; Pijanowska, J.; Mrówka, P.; Babkiewicz, E. Effects of nanoplastics on aquatic organisms. In *Environmental Nanopollutants: Sources, Occurrence, Analysis and Fate*; CRC Press: Boca Raton, FL, USA, 2022.
14. El Hadri, H.; Gigault, J.; Maxit, B.; Grassl, B.; Reynaud, S. Nanoplastic from mechanically degraded primary and secondary microplastics for environmental assessments. *NanoImpact* **2020**, *17*, 100206. [CrossRef]
15. Gaylarde, C.C.; Neto, J.A.B.; da Fonseca, E.M. Nanoplastics in aquatic systems-are they more hazardous than microplastics? *Environ. Pollut.* **2021**, *272*, 115950. [CrossRef] [PubMed]
16. Schrank, I.; Trotter, B.; Dummert, J.; Scholz-Böttcher, B.M.; Löder, M.G.; Laforsch, C. Effects of microplastic particles and leaching additive on the life history and morphology of *Daphnia magna*. *Environ. Pollut.* **2019**, *255*, 113233. [CrossRef] [PubMed]
17. Song, J.; Na, J.; An, D.; Jung, J. Role of benzophenone-3 additive in chronic toxicity of polyethylene microplastic fragments to *Daphnia magna*. *Sci. Total Environ.* **2021**, *800*, 149638. [CrossRef]
18. Zhao, H.J.; Xu, J.K.; Yan, Z.H.; Ren, H.Q.; Zhang, Y. Microplastics enhance the developmental toxicity of synthetic phenolic antioxidants by disturbing the thyroid function and metabolism in developing zebrafish. *Environ. Int.* **2020**, *140*, 105750. [CrossRef]
19. Li, Z.; Li, W.; Zha, J.; Chen, H.; Martyniuk, C.J.; Liang, X. Transcriptome analysis reveals benzotriazole ultraviolet stabilizers regulate networks related to inflammation in juvenile zebrafish (*Danio rerio*) brain. *Environ. Toxicol.* **2019**, *34*, 112–122. [CrossRef]

20. Gauthier, P.T.; Norwood, W.P.; Prepas, E.E.; Pyle, G.G. Metal–PAH mixtures in the aquatic environment: A review of co-toxic mechanisms leading to more-than-additive outcomes. *Aquat. Toxicol.* **2014**, *154*, 253–269. [CrossRef]

21. Trevisan, R.; Voy, C.; Chen, S.; Di Giulio, R.T. Nanoplastics decrease the toxicity of a complex PAH mixture but impair mitochondrial energy production in developing zebrafish. *Environ. Sci. Technol.* **2019**, *53*, 8405–8415. [CrossRef]

22. Menéndez-Pedriza, A.; Jaumot, J. Interaction of environmental pollutants with microplastics: A critical review of sorption factors, bioaccumulation and ecotoxicological effects. *Toxics* **2020**, *8*, 40. [CrossRef] [PubMed]

23. He, S.; Li, D.; Wang, F.; Zhang, C.; Yue, C.; HuangF, Y.; Mu, J. Parental exposure to sulfamethazine and nanoplastics alters the gut microbial communities in the offspring of marine madaka (*Oryzias melastigma*). *J. Hazard. Mater.* **2022**, *423*, 127003. [CrossRef]

24. Liu, S.; Yan, L.; Zhang, Y.; Junaid, M.; Wang, J. Polystyrene nanoplastics exacerbated the ecotoxicological and potential carcinogenic effects of tetracycline in juvenile grass carp (*Ctenopharyngodon idella*). *Sci. Total Environ.* **2022**, *803*, 150027. [CrossRef] [PubMed]

25. Canniff, P.M.; Hoang, T.C. Microplastic ingestion by *Daphnia magna* and its enhancement on algal growth. *Sci. Total Environ.* **2018**, *633*, 500–507. [CrossRef] [PubMed]

26. Trevisan, R.; Ranasinghe, P.; Jayasundara, N.; Di Giulio, R.T. Nanoplastics in Aquatic Environments: Impacts on Aquatic Species and Interactions with Environmental Factors and Pollutants. *Toxics* **2022**, *10*, 326. [CrossRef] [PubMed]

27. Chen, Q.; Yin, D.; Jia, Y.; Schiwy, S.; Legradi, J.; Yang, S.; Hollert, H. Enhanced uptake of BPA in the presence of nanoplastics can lead to neurotoxic effects in adult zebrafish. *Sci. Total Environ.* **2017**, *609*, 1312–1321. [CrossRef]

28. Mendes, L.A.; Barreto, A.; Santos, J.; Amorim, M.J.; Maria, V.L. Co-Exposure of Nanopolystyrene and Other Environmental Contaminants—Their Toxic Effects on the Survival and Reproduction of *Enchytraeus crypticus*. *Toxics* **2022**, *10*, 193. [CrossRef]

29. Booth, A.; Aga, D.S.; Wester, A.L. Retrospective analysis of the global antibiotic residues that exceed the predicted no effect concentration for antimicrobial resistance in various environmental matrices. *Environ. Int.* **2020**, *141*, 105796. [CrossRef]

30. Fair, R.J.; Tor, Y. Antibiotics and bacterial resistance in the 21st century. *Perspect. Med. Chem.* **2014**, *6*, S14459. [CrossRef]

31. Zainab, S.M.; Junaid, M.; Xu, N.; Malik, R.N. Antibiotics and antibiotic resistant genes (ARGs) in groundwater: A global review on dissemination, sources, interactions, environmental and human health risks. *Water Res.* **2020**, *187*, 116455. [CrossRef]

32. Hill, D.A.; Hoffmann, C.; Abt, M.C.; Du, Y.; Kobuley, D.; Kirn, T.J.; Artis, D. Metagenomic analyses reveal antibiotic-induced temporal and spatial changes in intestinal microbiota with associated alterations in immune cell homeostasis. *Mucosal Immunol.* **2010**, *3*, 148–158. [CrossRef] [PubMed]

33. Zhou, L.; Limbu, S.M.; Shen, M.; Zhai, W.; Qiao, F.; He, A.; Zhang, M. Environmental concentrations of antibiotics impair zebrafish gut health. *Environ. Pollut.* **2018**, *235*, 245–254. [CrossRef] [PubMed]

34. Kovalakova, P.; Cizmas, L.; McDonald, T.J.; Marsalek, B.; Feng, M.; Sharma, V.K. Occurrence and toxicity of antibiotics in the aquatic environment: A review. *Chemosphere* **2020**, *251*, 126351. [CrossRef] [PubMed]

35. Duan, W.; Cui, H.; Jia, X.; Huang, X. Occurrence and ecotoxicity of sulfonamides in the aquatic environment: A review. *Sci. Total Environ.* **2022**, *820*, 153178. [CrossRef]

36. Baggio, D.; Ananda-Rajah, M.R. Fluoroquinolone antibiotics and adverse events. *Aust. Prescr.* **2021**, *44*, 161. [CrossRef]

37. Brown, S.A. Fluoroquinolones in animal health. *J. Vet. Pharmacol. Ther.* **1996**, *19*, 1–14. [CrossRef]

38. Migliore, L.; Cozzolino, S.; Fiori, M. Phytotoxicity to and uptake of flumequine used in intensive aquaculture on the aquatic weed, *Lythrum salicaria* L. *Chemosphere* **2000**, *40*, 741–750. [CrossRef]

39. Jacoby, G.A. Mechanisms of resistance to quinolones. *Clin. Infect. Dis.* **2005**, *41* (Suppl. 2), S120–S126. [CrossRef]

40. Välitalo, P.; Kruglova, A.; Mikola, A.; Vahala, R. Toxicological impacts of antibiotics on aquatic micro-organisms: A mini-review. *Int. J. Hyg. Environ. Health* **2017**, *220*, 558–569. [CrossRef]

41. Kim, Y.; Jung, J.; Kim, M.; Park, J.; Boxall, A.B.; Choi, K. Prioritizing veterinary pharmaceuticals for aquatic environment in Korea. *Environ. Toxicol. Pharmacol.* **2008**, *26*, 167–176. [CrossRef]

42. Kern, W.V. New plasmid-borne quinolone-resistance determinant in *Escherichia coli*. *Future Microbiol.* **2007**, *2*, 473–475. [CrossRef]

43. Ding, G.; Chen, G.; Liu, Y.; Li, M.; Liu, X. Occurrence and risk assessment of fluoroquinolone antibiotics in reclaimed water and receiving groundwater with different replenishment pathways. *Sci. Total Environ.* **2020**, *738*, 139802. [CrossRef] [PubMed]

44. Wagil, M.; Kumirska, J.; Stolte, S.; Puckowski, A.; Maszkowska, J.; Stepnowski, P.; Białk-Bielińska, A. Development of sensitive and reliable LC-MS/MS methods for the determination of three fluoroquinolones in water and fish tissue samples and preliminary environmental risk assessment of their presence in two rivers in northern Poland. *Sci. Total Environ.* **2014**, *493*, 1006–1013. [CrossRef]

45. Ma, Y.; Li, M.; Wu, M.; Li, Z.; Liu, X. Occurrences and regional distributions of 20 antibiotics in water bodies during groundwater recharge. *Sci. Total Environ.* **2015**, *518*, 498–506. [CrossRef] [PubMed]

46. Hu, Y.; Yan, X.; Shen, Y.; Di, M.; Wang, J. Antibiotics in surface water and sediments from Hanjiang River, Central China: Occurrence, behavior and risk assessment. *Ecotoxicol. Environ. Saf.* **2018**, *157*, 150–158. [CrossRef] [PubMed]

47. You, X.; Cao, X.; Zhang, X.; Guo, J.; Sun, W. Unraveling individual and combined toxicity of nano/microplastics and ciprofloxacin to *Synechocystis* sp. at the cellular and molecular levels. *Environ. Int.* **2021**, *157*, 106842. [CrossRef]

48. Feng, L.J.; Shi, Y.; Li, X.Y.; Sun, X.D.; Xiao, F.; Sun, J.W.; Yuan, X.Z. Behavior of tetracycline and polystyrene nanoparticles in estuaries and their joint toxicity on marine microalgae *Skeletonema costatum*. *Environ. Pollut.* **2020**, *263*, 114453. [CrossRef] [PubMed]

49. Guo, X.; Cai, Y.; Ma, C.; Han, L.; Yang, Z. Combined toxicity of micro/nano scale polystyrene plastics and ciprofloxacin to *Corbicula fluminea* in freshwater sediments. *Sci. Total Environ.* **2020**, *789*, 147887. [CrossRef]

50. Almeida, M.; Martins, M.A.; Soares, A.M.; Cuesta, A.; Oliveira, M. Polystyrene nanoplastics alter the cytotoxicity of human pharmaceuticals on marine fish cell lines. *Environ. Toxicol. Pharmacol.* **2019**, *69*, 57–65. [CrossRef]
51. Zhang, Y.T.; Chen, H.; He, S.; Wang, F.; Liu, Y.; Chen, M.; Mu, J. Subchronic toxicity of dietary sulfamethazine and nanoplastics in marine medaka (*Oryzias melastigma*): Insights from the gut microbiota and intestinal oxidative status. *Ecotoxicol. Environ. Saf.* **2021**, *226*, 112820. [CrossRef]
52. Baysal, A.; Saygin, H. Co-occurence of antibiotics and micro (nano) plastics: A systematic review between 2016–2021. *J. Environ. Sci. Health Part A* **2022**, *57*, 519–539. [CrossRef] [PubMed]
53. Zebrowski, M.; Babkiewicz, E.; Błażejewska, A.; Pukos, S.; Wawrzeńczak, J.; Wilczynski, W.; Zebrowski, J.; Slusarczyk, M.; Maszczyk, P. The effect of microplastics on the interspecific competition of *Daphnia*. *Environ. Pollut.* **2022**, *313*, 120121. [CrossRef] [PubMed]
54. Guillard, R.R.L. Culture of phytoplankton for feeding marine invertebrates. In *Culture of Marine Invertebrate Animals*; Smith, W.L., Chanley, M.H., Eds.; Plenum Press: New York, NY, USA, 1975; pp. 29–60.
55. Pessoni, L.; Veclin, C.; El Hadri, H.; Cugnet, C.; Davranche, M.; Pierson-Wickmann, A.C.; Reynaud, S. Soap-and metal-free polystyrene latex particles as a nanoplastic model. *Environ. Sci. Nano* **2019**, *6*, 2253–2258. [CrossRef]
56. Jiménez-Lamana, J.; Marigliano, L.; Allouche, J.; Grassl, B.; Szpunar, J.; Reynaud, S. A novel strategy for the detection and quantification of nanoplastics by single particle inductively coupled plasma mass spectrometry (ICP-MS). *Anal. Chem.* **2020**, *92*, 11664–11672. [CrossRef] [PubMed]
57. Bruijning, M.; ten Berge, A.C.; Jongejans, E. Population-level responses to temperature, density and clonal differences in *Daphnia magna* as revealed by integral projection modelling. *Funct. Ecol.* **2018**, *32*, 2407–2422. [CrossRef]
58. Ge, L.; Na, G.; Zhang, S.; Li, K.; Zhang, P.; Ren, H.; Yao, Z. New insights into the aquatic photochemistry of fluoroquinolone antibiotics: Direct photodegradation, hydroxyl-radical oxidation, and antibacterial activity changes. *Sci. Total Environ.* **2015**, *527*, 12–17. [CrossRef] [PubMed]
59. Bartosiewicz, M.; Jabłoński, J.; Kozłowski, J.; Maszczyk, P. Brood space limitation of reproduction may explain growth after maturity in differently sized *Daphnia* species. *J. Plankton Res.* **2015**, *37*, 417–428. [CrossRef]
60. Dickerson, T.L.; Williams, H.N. Functional diversity of bacterioplankton in three North Florida freshwater lakes over an Annual Cycle. *Microb. Ecol.* **2011**, *67*, 34–44. [CrossRef]
61. Kiersztyn, B.; Chróst, R.; Kaliński, T.; Siuda, W.; Bukowska, A.; Kowalczyk, G.; Grabowska, K. Structural and functional microbial diversity along a eutrophication gradient of interconnected lakes undergoing anthropopressure. *Sci. Rep.* **2019**, *9*, 11144. [CrossRef]
62. Macke, E.; Callens, M.; De Meester, L.; Decaestecker, E. Host-genotype dependent gut microbiota drives zooplankton tolerance to toxic cyanobacteria. *Nat. Commun.* **2017**, *8*, 1608. [CrossRef]
63. Callens, M.; Watanabe, H.; Kato, Y.; Miura, J.; Decaestecker, E. Microbiota inoculum composition affects holobiont assembly and host growth in *Daphnia*. *Microbiome* **2018**, *6*, 56. [CrossRef] [PubMed]
64. Hegg, A.; Radersma, R.; Uller, T. A field experiment reveals seasonal variation in the *Daphnia* gut microbiome. *Oikos* **2021**, *130*, 2191–2201. [CrossRef]
65. Bulteel, L.; Houwenhuyse, S.; Declerck, S.A.; Decaestecker, E. The role of microbiome and genotype in *Daphnia* magna upon parasite re-exposure. *Genes* **2021**, *12*, 70. [CrossRef] [PubMed]
66. Caporaso, J.G.; Kuczynski, J.; Stombaugh, J.; Bittinger, K.; Bushman, F.D.; Costello, E.K.; Knight, R. QIIME allows analysis of high-throughput community sequencing data. *Nat. Methods* **2010**, *7*, 335–336. [CrossRef] [PubMed]
67. Beckers, B.; Op De Beeck, M.; Thijs, S.; Truyens, S.; Weyens, N.; Boerjan, W.; Vangronsveld, J. Performance of 16s rDNA primer pairs in the study of rhizosphere and endosphere bacterial microbiomes in metabarcoding studies. *Front. Microbiol.* **2016**, *7*, 650. [CrossRef]
68. Martin, M. Cutadapt removes adapter sequences from high-throughput sequencing reads. *EMBnet. J.* **2011**, *17*, 10–12. [CrossRef]
69. Ewels, P.; Magnusson, M.; Lundin, S.; Käller, M. MultiQC: Summarize analysis results for multiple tools and samples in a single report. *Bioinformatics* **2016**, *32*, 3047–3048. [CrossRef]
70. Bolyen, E.; Rideout, J.R.; Dillon, M.R.; Bokulich, N.A.; Abnet, C.C.; Al-Ghalith, G.A.; Caporaso, J.G. Reproducible, interactive, scalable and extensible microbiome data science using QIIME 2. *Nat. Biotechnol.* **2019**, *37*, 852–857. [CrossRef]
71. Kay, M.; Elkin, L.A.; Higgins, J.J.; Wobbrock, J.O. *ARTool: Aligned Rank Transform for Nonparametric Factorial ANOVAs*, R package version 0.11.1; Association for Computing Machinery: New York, NY, USA, 2021; Available online: https://github.com/mjskay/ARTool (accessed on 27 April 2021). [CrossRef]
72. Wobbrock, J.O.; Findlater, L.; Gergle, D.; Higgins, J.J. The aligned rank transform for nonparametric factorial analyses using only ANOVA procedures. In Proceedings of the ACM Conference on Human Factors in Computing Systems (CHI 2011), Vancouver, BC, Canada, 7–12 May 2011; ACM Press: New York, NY, USA, 2011; pp. 143–146. Available online: https://depts.washington.edu/acelab/proj/art/ (accessed on 7 May 2011). [CrossRef]
73. Elkin, L.A.; Kay, M.; Higgins, J.J.; Wobbrock, J.O. An aligned rank transform procedure for multifactor contrast tests. In Proceedings of the ACM Symposium on User Interface Software and Technology (UIST 2021), Virtual Event, 10–14 October 2021; ACM Press: New York, NY, USA, 2021; pp. 754–768. [CrossRef]
74. Hammer, Ř.; Harper, D.A.; Ryan, D.D. Past: Paleontological statistics software package for education and data analysis. *Palaeontol. Electron.* **2001**, *4*, 9.

75. Akbar, S.; Gu, L.; Sun, Y.; Zhou, Q.; Zhang, L.; Lyu, K.; Yang, Z. Changes in the life history traits of *Daphnia magna* are associated with the gut microbiota composition shaped by diet and antibiotics. *Sci. Total Environ.* **2020**, *705*, 135827. [CrossRef]

76. Rist, S.; Baun, A.; Hartmann, N.B. Ingestion of micro- and nanoplastics in *Daphnia magna*—Quantification of body burdens and assessment of feeding rates and reproduction. *Environ. Pollut.* **2017**, *228*, 398–407. [CrossRef] [PubMed]

77. Besseling, E.; Wang, B.; Lurling, M.; Koelmans, A.A. Nanoplastic affects growth of *S. obliquus* and reproduction of *D. magna*. *Environ. Sci. Technol.* **2014**, *48*, 12336–12343. [CrossRef] [PubMed]

78. Bergami, E.; Pugnalini, S.; Vannuccini, M.L.; Manfra, L.; Faleri, C.; Savorelli, F.; Corsi, I. Long-term toxicity of surface-charged polystyrene nanoplastics to marine planktonic species *Dunaliella tertiolecta* and *Artemia franciscana*. *Aquat. Toxicol.* **2017**, *189*, 159–169. [CrossRef] [PubMed]

79. Venâncio, C.; Ferreira, I.; Martins, M.A.; Soares, A.M.; Lopes, I.; Oliveira, M. The effects of nanoplastics on marine plankton: A case study with polymethylmethacrylate. *Ecotoxicol. Environ. Saf.* **2019**, *184*, 109632. [CrossRef] [PubMed]

80. Sökmen, T.Ö.; Sulukan, E.; Türkoğlu, M.; Baran, A.; Özkaraca, M.; Ceyhun, S.B. Polystyrene nanoplastics (20 nm) are able to bioaccumulate and cause oxidative DNA damages in the brain tissue of zebrafish embryo (*Danio rerio*). *Neurotoxicology* **2020**, *77*, 51–59. [CrossRef]

81. Brun, N.R.; Koch, B.E.; Varela, M.; Peijnenburg, W.J.; Spaink, H.P.; Vijver, M.G. Nanoparticles induce dermal and intestinal innate immune system responses in zebrafish embryos. *Environ. Sci. Nano* **2018**, *5*, 904–916. [CrossRef]

82. Redondo-Hasselerharm, P.E.; Vink, G.; Mitrano, D.M.; Koelmans, A.A. Metal-doping of nanoplastics enables accurate assessment of uptake and effects on *Gammarus pulex*. *Environ. Sci. Nano* **2021**, *8*, 1761–1770. [CrossRef]

83. Fernandez, R.; Colás-Ruiz, N.R.; Bolívar-Anillo, H.J.; Anfuso, G.; Hampel, M. Occurrence and Effects of Antimicrobials Drugs in Aquatic Ecosystems. *Sustainability* **2021**, *13*, 13428. [CrossRef]

84. Kostich, M.S.; Lazorchak, J.M. Risks to aquatic organisms posed by human pharmaceutical use. *Sci. Total Environm.* **2008**, *389*, 329–339. [CrossRef]

85. Qiu, W.; Fang, M.; Magnuson, J.T.; Greer, J.B.; Chen, Q.; Zheng, Y.; Xiong, Y.; Luo, S.; Zheng, C.; Schlenk, D. Maternal exposure to environmental antibiotic mixture during gravid period predicts gastrointestinal effects in zebrafish offspring. *J. Hazard. Mater.* **2020**, *399*, 123009. [CrossRef]

86. Nunes, B.; Leal, C.; Rodrigues, S.; Antunes, S.C. Assessment of ecotoxicological effects of ciprofloxacin in *Daphnia magna*: Life-history traits, biochemical and genotoxic effects. *Water Sci. Technol.* **2018**, *2017*, 835–844. [CrossRef] [PubMed]

87. Ma, J.; Sheng, G.D.; Chen, Q.L.; O'Connor, P. Do combined nanoscale polystyrene and tetracycline impact on the incidence of resistance genes and microbial community disturbance in *Enchytraeus crypticus*? *J. Hazard. Mater.* **2020**, *387*, 122012. [CrossRef] [PubMed]

88. Motiei, A.; Brindefalk, B.; Ogonowski, M.; El-Shehawy, R.; Pastuszek, P.; Ek, K.; Gorokhova, E. Disparate effects of antibiotic-induced microbiome change and enhanced fitness in *Daphnia magna*. *PLoS ONE* **2020**, *15*, e0214833. [CrossRef] [PubMed]

89. Lobritz, M.A.; Belenky, P.; Porter, C.B.; Gutierrez, A.; Yang, J.H.; Schwarz, E.G.; Collins, J.J. Antibiotic efficacy is linked to bacterial cellular respiration. *Proc. Natl. Acad. Sci. USA* **2015**, *112*, 8173–8180. [CrossRef]

90. Stokes, J.M.; Lopatkin, A.J.; Lobritz, M.A.; Collins, J.J. Bacterial metabolism and antibiotic efficacy. *Cell Metab.* **2019**, *30*, 251–259. [CrossRef]

91. Siuda, W.; Kiersztyn, B.; Chrost, R.J. The dynamics of protein decomposition in lakes of different trophic status-reflections on the assessment of the real proteolytic activity in situ. *J. Microbiol. Biotechnol.* **2007**, *17*, 897–904.

92. Kataria, R.; Ruhal, R. Microbiological metabolism under chemical stress. In *Microbial Biodegradation and Bioremediation*; Elsevier: Amsterdam, The Netherlands, 2014; pp. 497–509.

93. Miao, L.; Guo, S.; Liu, Z.; Liu, S.; You, G.; Qu, H.; Hou, J. Effects of nanoplastics on freshwater biofilm microbial metabolic functions as determined by BIOLOG ECO microplates. *Int. J. Environ. Res. Public Health* **2019**, *16*, 4639. [CrossRef]

94. Cabral, D.J.; Penumutchu, S.; Reinhart, E.M.; Zhang, C.; Korry, B.J.; Wurster, J.I.; Belenky, P. Microbial metabolism modulates antibiotic susceptibility within the murine gut microbiome. *Cell Metab.* **2019**, *30*, 800–823. [CrossRef]

95. Zuo, Y.; Xie, W.; Pang, Y.; Li, T.; Li, Q.; Li, Y. Bacterial community composition in the gut content of *Lampetra japonica* revealed by 16S rRNA gene pyrosequencing. *PLoS ONE* **2017**, *12*, e0188919. [CrossRef]

96. Zheng, F.; Zhu, D.; Giles, M.; Daniell, T.; Neilson, R.; Zhu, Y.G.; Yang, X.R. Mineral and organic fertilization alters the microbiome of a soil nematode *Dorylaimus stagnalis* and its resistome. *Sci. Total Environ.* **2019**, *680*, 70–78. [CrossRef]

97. Berg, M.; Stenuit, B.; Ho, J.; Wang, A.; Parke, C.; Knight, M.; Alvarez-Cohen, L.; Shapira, M. Assembly of the *Caenorhabditis elegans* gut microbiota from diverse soil microbial environments. *ISME J.* **2016**, *10*, 1998–2009. [CrossRef] [PubMed]

98. Xiong, J.B.; Nie, L.; Chen, J. Current understanding on the roles of gut microbiota in fish disease and immunity. *Zool. Res.* **2019**, *40*, 70–76. [PubMed]

99. Wolever, T.; Brighenti, F.; Royall, D.; Jenkins, A.L.; Jenkins, D.J. Effect of rectal infusion of short chain fatty acids in human subjects. *Am. J. Gastroenterol.* **1989**, *84*, 1027–1033. [PubMed]

100. Schwiertz, A.; Taras, D.; Schäfer, K.; Beijer, S.; Bos, N.A.; Donus, C.; Hardt, P.D. Microbiota and SCFA in lean and overweight healthy subjects. *Obesity* **2010**, *18*, 190–195. [CrossRef]

101. Shin, N.R.; Whon, T.W.; Bae, J.W. Proteobacteria: Microbial signature of dysbiosis in gut microbiota. *Trends Biotechnol.* **2015**, *33*, 496–503. [CrossRef]

102. Zhu, D.; An, X.L.; Chen, Q.L.; Yang, X.R.; Christie, P.; Ke, X.; Wu, L.H.; Zhu, Y.G. Antibiotics disturb the microbiome and increase the incidence of resistance genes in the gut of a common soil collembolan. *Environ. Sci. Technol.* **2018**, *52*, 3081–3090. [CrossRef]
103. Saygin, H.; Baysal, A. Similarities and discrepancies between bio-based and conventional submicron-sized plastics: In relation to clinically important bacteria. *Bull. Environ. Contam. Toxicol.* **2020**, *105*, 26–35. [CrossRef]
104. Zhu, D.; Li, G.; Wang, H.T.; Duan, G.L. Effects of nano-or microplastic exposure combined with arsenic on soil bacterial, fungal, and protistan communities. *Chemosphere* **2021**, *281*, 130998. [CrossRef]
105. Auguste, M.; Lasa, A.; Balbi, T.; Pallavicini, A.; Vezzulli, L.; Canesi, L. Impact of nanoplastics on hemolymph immune parameters and microbiota composition in *Mytilus galloprovincialis*. *Mar. Environ. Res.* **2020**, *159*, 105017. [CrossRef]
106. Kang, H.M.; Byeon, E.; Jeong, H.; Kim, M.S.; Chen, Q.; Lee, J.S. Different effects of nano-and microplastics on oxidative status and gut microbiota in the marine medaka *Oryzias melastigma*. *J. Hazard. Mater.* **2021**, *405*, 124207. [CrossRef]
107. Gu, W.; Liu, S.; Chen, L.; Liu, Y.; Gu, C.; Ren, H.Q.; Wu, B. Single-cell RNA sequencing reveals size-dependent effects of polystyrene microplastics on immune and secretory cell populations from zebrafish intestines. *Environ. Sci. Technol.* **2020**, *54*, 3417–3427. [CrossRef] [PubMed]
108. Xie, S.; Zhou, A.; Wei, T.; Li, S.; Yang, B.; Xu, G.; Zou, J. Nanoplastics induce more serious microbiota dysbiosis and inflammation in the gut of adult zebrafish than microplastics. *Bull. Environ. Contam. Toxicol.* **2021**, *107*, 640–650. [CrossRef]
109. Yoon, J.H.; Kang, S.J.; Schumann, P.; Oh, T.K. Yonghaparkia alkaliphila gen. nov., sp nov., a novel member of the family Microbacteriaceae isolated from an alkaline soil. *Int. J. Syst. Evol. Microbiol.* **2006**, *56*, 2415–2420. [CrossRef] [PubMed]
110. Pishgar, R.; Dominic, J.A.; Sheng, Z.; Tay, J.H. Denitrification performance and microbial versatility in response to different selection pressures. *Bioresour. Technol.* **2019**, *281*, 72–83. [CrossRef]
111. Oh, S.; Choi, D. Microbial community enhances biodegradation of bisphenol A through selection of Sphingomonadaceae. *Microb. Ecol.* **2019**, *77*, 631–639. [CrossRef] [PubMed]
112. Yin, J.B.; Zhang, X.X.; Wu, B.; Xian, Q.M. Metagenomic insights into tetracycline effects on microbial community and antibiotic resistance of mouse gut. *Ecotoxicology* **2015**, *24*, 2125–2132. [CrossRef]

Article

Investigation of Global Trends of Pollutants in Marine Ecosystems around Barrang Caddi Island, Spermonde Archipelago Cluster: An Ecological Approach

Ismail Marzuki [1,*], Early Septiningsih [2], Ernawati Syahruddin Kaseng [2], Herlinah Herlinah [2], Andi Sahrijanna [2], Sahabuddin Sahabuddin [2], Ruzkiah Asaf [2], Admi Athirah [2], Bambang Heri Isnawan [3], Gatot Supangkat Samidjo [3], Faizal Rumagia [4], Emmy Hamidah [5], Idum Satia Santi [6] and Khairun Nisaa [7]

[1] Department of Chemical Engineering, Fajar University, Makassar 90231, South Sulawesi, Indonesia
[2] Research Institute for Coastal Aquaculture and Fisheries Extension, Maros 90512, South Sulawesi, Indonesia; earlyseptiningsih@gmail.com (E.S.); ernawatisyahruddin71@gmail.com (E.S.K.); hjompa@yahoo.com (H.H.); idsarsompa@gmail.com (A.S.); s.abud_din@yahoo.co.id (S.S.); qiaasaf@gmail.com (R.A.); m.athirah@gmail.com (A.A.)
[3] Department of Agrotechnology, Universitas Muhammadiyah Yogyakarta, Bantul 55183, DI Yogyakarta, Indonesia; bambanghi@umy.ac.id (B.H.I.); supangkat@umy.ac.id (G.S.S.)
[4] Study Program of Fisheries Resource Utilization, Faculty of Fisheries and Marine, Khairun University, Ternate 97719, North Maluku, Indonesia; faizal.rumagia@unkhair.ac.id
[5] Department of Agrotechnology, Universitas Islam Darul 'Ulum, Lamongan 62253, Jawa Timur, Indonesia; emmyhamidah@unisda.ac.id
[6] Department of Agrotechnology, Institut Pertanian Stiper, Yogyakarta 55283, DI Yogyakarta, Indonesia; idum@instiperjogja.ac.id
[7] National Research and Innovation Agency (BRIN), Jakarta 10340, DKI, Indonesia; nisauicha27@gmail.com
* Correspondence: ismailmz@unifa.ac.id

Citation: Marzuki, I.; Septiningsih, E.; Kaseng, E.S.; Herlinah, H.; Sahrijanna, A.; Sahabuddin, S.; Asaf, R.; Athirah, A.; Isnawan, B.H.; Samidjo, G.S.; et al. Investigation of Global Trends of Pollutants in Marine Ecosystems around Barrang Caddi Island, Spermonde Archipelago Cluster: An Ecological Approach. *Toxics* **2022**, *10*, 301. https://doi.org/10.3390/toxics10060301

Academic Editors: François Gagné, Stefano Magni and Valerio Matozzo

Received: 2 May 2022
Accepted: 27 May 2022
Published: 1 June 2022

Publisher's Note: MDPI stays neutral with regard to jurisdictional claims in published maps and institutional affiliations.

Abstract: High-quality marine ecosystems are free from global trending pollutants' (GTP) contaminants. Accuracy and caution are needed during the exploitation of marine resources during marine tourism to prevent future ecological hazards that cause chain effects on aquatic ecosystems and humans. This article identifies exposure to GTP: microplastic (MP); polycyclic aromatic hydrocarbons (PAH); pesticide residue (PR); heavy metal (HM); and medical waste (MW), in marine ecosystems in the marine tourism area (MTA) area and Barrang Caddi Island (BCI) waters. A combination of qualitative and quantitative analysis methods were used with analytical instruments and mathematical formulas. The search results show the average total abundance of MPs in seawater (5.47 units/m^3) and fish samples (7.03 units/m^3), as well as in the sediment and sponge samples (8.18 units/m^3) and (8.32 units/m^3). Based on an analysis of the polymer structure, it was identified that the dominant light group was MPs: polyethylene (PE); polypropylene (PP); polystyrene (PS); followed by polyamide-nylon (PA); and polycarbonate (PC). Several PAH pollutants were identified in the samples. In particular, naphthalene (NL) types were the most common pollutants in all of the samples, followed by pyrene (PN), and azulene (AZ). Pb^{+2} and Cu^{+2} pollutants around BCI were successfully calculated, showing average concentrations in seawater of 0.164 ± 0.0002 mg/L and 0.293 ± 0.0007 mg/L, respectively, while in fish, the concentrations were 1.811 ± 0.0002 µg/g and 4.372 ± 0.0003 µg/g, respectively. Based on these findings, the BCI area is not recommended as a marine tourism destination.

Keywords: pollutants; microplastics; heavy metals; polycyclic aromatic hydrocarbons; pesticide residues; medical waste

1. Introduction

Marine ecosystems are formed by two main components: biotic (organic) and abiotic (inorganic) elements [1–3]. The seascape is vast and is influenced by many factors, causing the characteristics of the sea area to vary [4,5]. Each marine area has different characteristics and ecosystems, such as deep-sea, poles, beaches, coasts, coral reefs, and tides [6]. The types and amounts of materials in marine ecosystems are not limited, and comprise both those that are naturally occurring and those that are produced due to human activities and dynamics [7,8]. Most human waste material can occur in all of the regions of the earth. Due to the influences of ecology and gravity, garbage has the potential to end up in the sea, either intentionally or due to negligence in managing the marine environment [9–11].

Human waste materials that enter marine waters are classified as marine debris and generally contain hazardous and toxic components, such as heavy metals (HM), polycyclic aromatic hydrocarbon components (PAH), microplastics (MP), pesticide residues (PR), medical waste (MW), and radioactive waste. They can be categorized as global trending pollutants (GTP) [12–20]. The term global trending pollutants was introduced to the world community to describe at least five types of pollutants (MP, PAH, PR, MW, and HM), which are pollutant materials that have become global environmental issues to date, especially for marine ecosystems [14,18,21–24].

The handling of GTP as toxic components is a problem that has occurred for a long time, with impacts including contributions to the triggering of global warming and climate change; thus, they have become a severe problem for many countries in terms of handling and managing them [25–28]. Many countries have difficulty handling these GTP, especially developing countries with relatively low economic levels. The concern of the world's population about the massive quantity of GTP is indicated by the issuance of G20 recommendations for developed countries to reduce global carbon production [2,29]. Whether we realize it or not, these global trending pollutants' inputs enter the ocean continuously and are difficult to prevent, and it is feared that the toxic components in the waste have the potential to cause anthropogenic hazards in the oceans [3,5,30].

The interactions among materials in the sea occur continuously and produce various types of new compounds, due to the natural ability of the sea to maintain the sustainability of life in the ocean [31,32]. Photodegradation due to sunlight, interaction by ocean currents, mobilization of materials due to differences in the waters, effects of salinity, consequences arising from temperature differences, and collisions between materials due to tidal movements in coastal areas are natural processes of the sea that occur continuously [17]. The occurrence of these events in marine ecosystems enriches the production of components with diverse marine biological and non-biological materials [33–37].

The effect of applying polymer technology to the production of various types of industrial equipment, household appliances, and plastic-based packaging was previously predicted to produce components that are harmful to human health and the environment, such as microplastics. However, it cannot be avoided due to the rapid flow of human needs [18,19,38,39]. Pesticide residue pollutants, the birth of synthetic products for activities to accelerate and increase agricultural production, and fulfilment of the need for treatment with various types of synthetic drugs and medical equipment were studied before. These factors have the potential to cause negative impacts, but cannot be suppressed due to the swift pace of demand [40–42]. Estimates of the emergence of PAH-type components in petroleum processing and fossil combustion, heavy metal contaminants in mineral processing, and the introduction of various metal-based equipment were all previously predicted. However, this cannot counter the solid demand of meeting human needs, particularly in terms of preserving life [43,44].

The distribution of GTP is almost even among many countries, and their presence is especially significant in marine and coastal areas. GTP have toxic and carcinogenic properties, are generally difficult to decompose, have accumulative properties, and can combine and even become assimilated into the processes of the objects, especially their metabolic processes [45,46]. The particle size of MP is minimal (<5 nm), allowing it to contribute to the cell division of living things, not least in marine ecosystems, such as fish and other marine biota and various marine biological components, which, at the end of the biogeochemical cycle, have the potential to become materials that accumulate in human foodstuffs [47,48].

GTP are pollutants that are very difficult to decompose through biodegradation, bio-adsorption, and bioreduction methods, including destruction methods. The treatment approach using bioremediation methods for global trending pollutants is challenging to conduct on a large scale with a technological approach [49,50]. Bioremediation microorganisms can be used, where the natural aid of biological materials, such as bio-adsorbent plants, is employed. However, they certainly cannot balance the speed of increase in the volume of pollutants entering the environment, when operating at a limited rate, and especially in the most vulnerable marine ecosystems [51]. This level of pollution could become a severe threat to the Earth's population, especially in marine areas and ecosystems, which could act as giant containers for almost all types of waste that are left over from human activities and dynamic natural processes [52,53].

The phenomenon of tourism activities that make marine ecosystem objects into marine tourism areas is occurring in many countries, including Indonesia [54]. Exploiting marine areas as tourist areas by presenting a variety of marine beauty characteristics, both on the surface and deep in the sea, is tempting because many people are interested in them [55]. Barrang Caddi Island (BCI) is one of the five islands included in the Marine Tourism Area (MTA), in addition to Samalona Island, Kodingareng Keke, Barrang Lompo, and Langkai. These islands are designated as an MTA that are managed by the local government of Makassar City, South Sulawesi [56]. The islands in the MTA, especially BCI, are included in a national marine conservation park and are part of the Spermonde Archipelago Cluster, which is world famous among marine sponge observers, because it is home to tens of thousands of species and marine sponge populations. The Spermonde Archipelago Cluster is known worldwide as a sponge study and research laboratory [57,58].

Barrang Caddi Island (BCI) in the south is directly adjacent to Makassar City. Along the coastline, there are the Soekarno Hatta seaports and fish landing ports. This coastal area operates two regional and two industrial hospitals and several large and medium sized hotels [54,55]. The beach is also full of marine tourism objects, which are visited by many domestic and foreign tourists. All forms of operations and services around BCI have the potential to produce waste containing GTP group contaminants. This contradicts the island's status as a marine tourism area and the viability of the various types of sponge living around it [59,60].

Research in this area is exciting and essential. We present an update on the data and information related to GTP contaminants in a review of the seawater, sediments, fish, and sponges, as well as the dual role of BCI as part of an MTA and the living area of various types of sponge [14–17]. The data presented are expected to act as a reference for the management of BCI as part of the MTA to prevent marine tourism activities from adding to the burden of island sustainability for marine life by ensuring that the balance of marine ecosystems is maintained and, most importantly, that the sustainability of the sponge population is maintained.

2. Materials and Methods

2.1. Materials and Equipment

Sediment samples, seawater, sponges, and demersal fish were the main materials. Other materials included H_2O_2 p.a., 30%, Fe(II) 0.05 M, N-hexane p.a., ethanol p.a., standard analysis of PAH (anthracene) 1000 mg/L (Supelco®), Na_2SO_4, H_2SO_4 p.a., NaCl p.a., HCl p.a., NaOH p.a., HNO_3 p.a., $HClO_4$ p.a., $Pb(NO_3)_2$ p.a.(standard solution of Pb analysis 1.000 mg/L), and $CuSO_4$ (standard solution of Cu 1000 mg/L), which were all obtained from Sigma-Aldrich, Saint Louis, MO, USA. Equipment used included the Fourier Transform Infra-red (FTIR) Shimadzu IR Prestige-21, Shimadzu Europa GmbH, Duisburg, Germany, a Gas Chromatography/Mass Spectrometer (GC/MS) from Agilent Technologies 7890A, Santa Clara, CA, USA (the operating conditions for GC/MS max. temperature 350 °C, the increase in temperature of 10 °C, every 5 min, pressure 18.406 psi), a Helium gas carrier, a speed of 150 mL/min, capillary column (Agilent 19019S-436HP-5 ms, Santa Clara, CA, USA), (dimensions of 60 m × 250 μm × 0.25 μm, the pressure 18,406 psi, separation 26,128 cm/sec, a retention time of max. 30 min) [12,44], an Atomic absorption spectroscope (AAS) (variant type AA240FS), a Muffle Furnace (type Thermolyne F6010, Shanghai, China), Portable Water (Quality AZ 8361, AZ Instrument Corp., Taichung, Taiwan), a set of glassware (Pyrex), a plankton net (mesh size 0.4 mm), stainless-steel mesh filters (5 mm and 0.3 mm), a sieve shaker, a density separator, and a vacuum pump [33,36,48].

2.2. Sampling

The total sample was 36 packages, each consisting of three sample packages of each type (sediment, seawater, sponges, and fish). The samples were obtained around the BCI waters at three different sampling points (Figure 1) [3,21,61–63]. Physical characteristics were observed at each station, such as the sampling coordinates, pH, salinity, electrical conductivity (EC), total dissolved solids (TDS), and others (Table 1). For the fish and sponge samples, a morphological analysis was also carried out to determine the species in each sample. Sampling was carried out following standard sampling methods. Each sample was put in a box in the laboratory for immediate preparation until it was ready to be analyzed using the appropriate instrument [64,65].

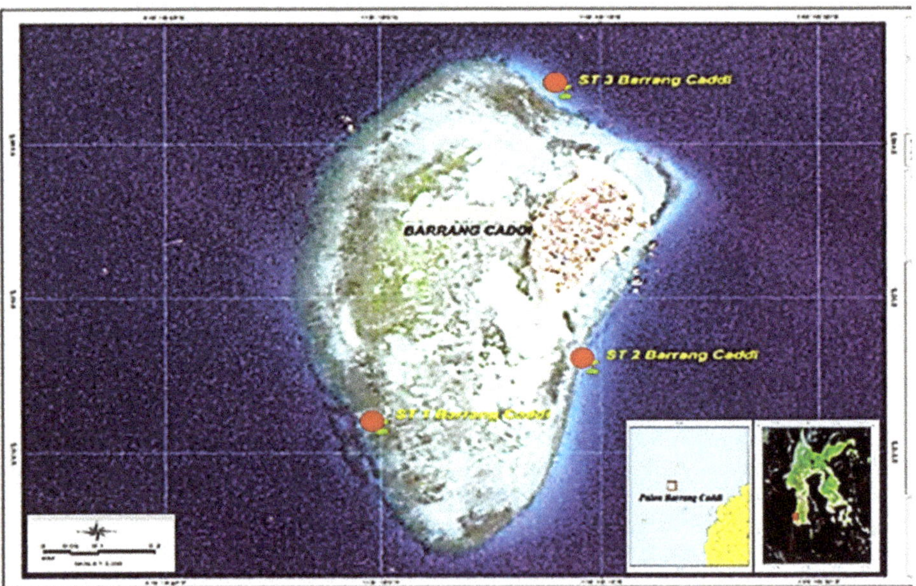

Figure 1. Map of Barrang Caddi Island (BCI). The red dots indicate the sampling stations (STs).

Sediment, sponge, seawater, and fish samples were obtained from three different sampling stations around BCI. This island is included in the Marine Tourism Area of Makassar City [1,5] The distance between BCI and the coast of Makassar City is ±7.5 km. The small red circles (Figure 1) are non-sampling areas, referred to as stations, which are coded ST 1, ST 2, and ST 3 [22,24].

According to the data (Table 1), the physical characteristics of the sampling stations are similar to those of other marine areas around them, especially in the Spermonde Islands cluster [24,33]. The data show that the seawater shows no physical signs that the area is exposed to the contaminants that fall into the GTP category.

Table 1. Physical characteristics of the seawater sampling stations in the waters around BCI.

Station Sampling	Coordinate	Depth MSL (m)	Distance from the Beach (m)	pH	Temperature (°C)	EC (ds/m)	TDS (mg/L)
ST 1	5°5′12.48216″ S 119°19′0.16536″ E	5	±300	7.32	29.8	14.91	7.56
ST 2	5°5′70.1664″ S 119°19′14.20716″ E	6	±350	7.30	30.4	16.69	8.12
ST 3	4°46′01664″ S 119°19′12.58932″ E	8	±370	7.31	29.9	16.33	8.22

2.3. Sample Preparation

Samples were prepared according to the type of analysis conducted, based on the target data. Sample preparation was carried out to determine the abundance of microplastics by filtering using a plankton net (adjusted for the samples of fish, sponges, and sediments). The samples were dissolved and the organic matter and the microplastic particles were separated [66]. The total abundance of microplastics was calculated in each sample [19,28]. The microplastic polymer structure was analyzed using microplastic particles collected based on differences in density, put in a cuvette, and then run using FTIR [67,68]. This instrument can produce a chromatogram as a spectrum in the form of a wavenumber that can be converted into the chemical structure with functional groups. Based on the functional groups, the type of MP can be predicted. Seven types of MP pollutants may be present in a sample, among which the most common are polypropylene, polyethylene, polystyrene, and polycarbonate [69,70].

Sample preparation for the analysis of the abundance and types of PAH was completed using an extraction ethanol, where the ethanol was used to extract all chemical components in the sample. Then extraction with N-hexane as a solvent was executed to separate the nonpolar components (PAH) [71]. The N-hexane extraction was run using GC/MS [72]. The visible chromatograms were analyzed to determine the abundance and types of PAH components present. Sample preparation for the analysis of the concentrations of Pb^{+2} and Cu^{+2} contaminants in the sample was carried out with the dry destruction method for the sediment, fish, and sponge samples, while the alkaline digestion method was used for seawater samples [73–75]. Each sample was dissolved with a concentrated acid solvent. Samples were injected into the AAS [44,45,55,76].

2.4. Sample Measurement

There were five types of pollutant in the GTP category. However, in the analysis of each sample, the identification of microplastics, PAH, and heavy metal contaminants was prioritized, while the identification of pesticide residues and medical waste was not prioritized to limit the publication's scope. However, these two types of contaminants can be viewed, together with PAH, based on GC/MS chromatogram data:

- The analysis of microplastic components, especially the total abundance of microplastics was completed using 30% H_2O_2 solvent. Measurement of the total abundance of microplastics was carried out in triplicate. The polymer structure was analyzed as the main characteristic of each microplastic, using FTIR combined with GC/MS chromatogram data [26,76,77];
- The types and abundance of the PAH contaminants were analyzed based on GC/MS chromatogram data. Information on the pesticide residue contaminants and media waste can also be obtained from the GC/MS chromatogram [24,76,77]. However, the determination of components was more specific, because in general, pesticide residues and several types of medical waste have chemical structures with reactive groups that should be identified using a combination of pyrolysis-GC/MS [54,78,79];
- Determination of the concentration of heavy metal contaminants, particularly exposure to Lead (Pb) and Copper (Cu) ions, was conducted for each sample according to

AAS absorption at the maximum wavelength (Pb: λmax. 228.9 nm and Cu: λmax. 324.7 nm) [33,34,80]. The method for determining the pollutant concentrations of Pb^{+2} and Cu^{+2} first makes a calibration curve to make ten series of Pb^{+2} and Cu^{+2} standard solutions whose estimated sample concentrations fall within the range of the standard solution, then the absorption of each concentration is measured. Then, the standard deviation and slope determination calculations are carried out. The concentration of each sample (Pb^{+2} and Cu^{+2}) was calculated based on the absorption obtained from AAS after being plotted into the regression equation [22,33,80]. Determination of pollutant concentrations of Pb^{+2} and Cu^{+2} carried out measurements of three replications for each type of sample obtained at three different sampling points. The data from the measurement results were calculated on average and summarized in a table. The quality of seawater according to the quality standard for pollutants Pb^{+2} and Cu^{+2} is a maximum of 0.05 mg/L. The quality standard for fish and other non-spongy biota is a maximum of 0.008 mg/L [55,76,80–82]. The maximum limit is not specified for sponges because they are included in the category of biota that are not eaten. The maximum limit of Pb^{+2} and Cu^{+2} for sediment is 0.10 μg/g. The quality of seawater and fish in BCI is determined by comparing the average pollutant concentration calculated compared to the standard for seawater and fish [4,10,59,83].

2.5. Data Presentation and Analysis

The presentation of analytical data as quantitative data was completed, using the appropriate equation. Data on the total abundance of MP, Pb, and Cu ion contaminant concentrations are displayed. The qualitative analysis data are presented based on the chromatogram results [26,28,84]. The polymer structure types of microplastic contaminants were analyzed using the FTIR chromatogram reading data, while data on PAH exposure types, including the pesticide residues and medical waste contaminants in each sample, were analyzed using GC/MS chromatography [85–87].

3. Results

Five types of pollutant fall into the GTP group, and the target of the GTP contaminant analysis was samples of sediment, sponge, seawater, and fish obtained from BCI (Makassar City MTA). The priority for the analysis was MP-type pollutants, of which the main hydrocarbon components are PAHs and heavy metal pollutants (HM). Pesticide residue pollutants and medical waste are not detailed, because of concerns that the data obtained are not significant due to the limitations of the analytical instrument as, ideally, a combined pyrolysis-GC/MS analysis tool should be used to analyze such substances [86,87].

3.1. Microplastic Pollutant Analysis

A specific morphological analysis of fish and sponge samples is essential because each type of marine biota has a different lifestyle and habitat, so the response to GTP contaminants or contaminant species (MP, PAH, PR, MW, and HM) can vary. The results of the morphological analysis of fish and sponge samples obtained at each station are presented in full in Table 2.

Table 2. Fish and sponge morphology based on the sampling point.

Station Sampling	Sponges	Fishes
ST 1	*Cribrochalina olemda*	*Chrysiptera Unimaculata*
ST 2	*Clathria Reinwardtii*	*Ambhygphidodon Curacao*
ST 3	*Clathria* sp.	*Scolopsis Brenatus*

The grouping of samples by pairing sediment and sponge, and seawater and fish to present the analysis results was based on the similarity of the media. The growth media and the habitat of sponges are more dominant in sediment, while the habitat and growth of fish are highly dependent on water. Under these conditions, the type of MP pollutant in fish bodies is estimated to be similar to the types of MP in water, in the bodies of sponges, and in sediments [21,22].

Two parameters were used to analyze the MP pollution in the sample: the total MP abundance (Figures 2 and 3), and the MP type (Figures 4 and 5).

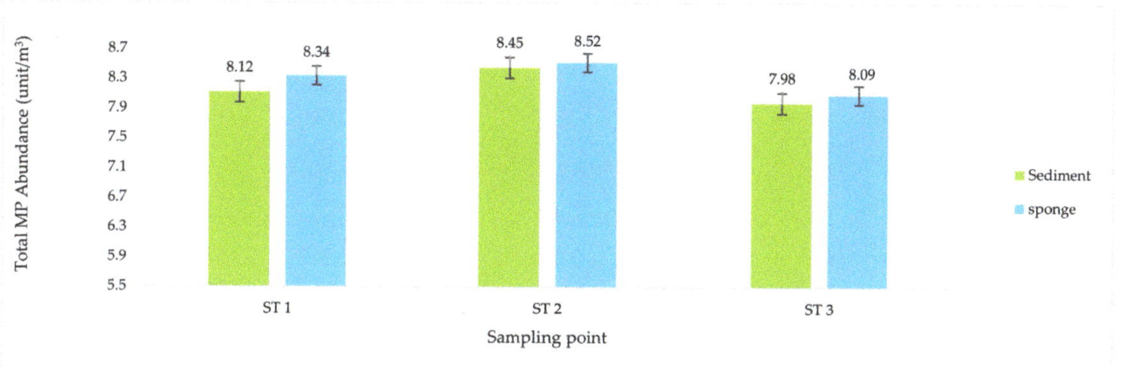

Figure 2. Total abundance of microplastics (MP) in the sediment and sponge samples based on the sampling points.

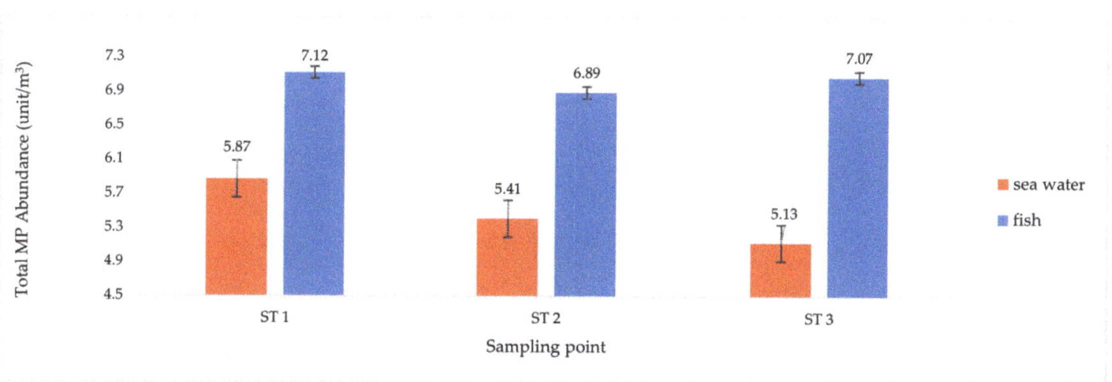

Figure 3. Total abundance of microplastics (MP) in seawater and fish samples based on the sampling points.

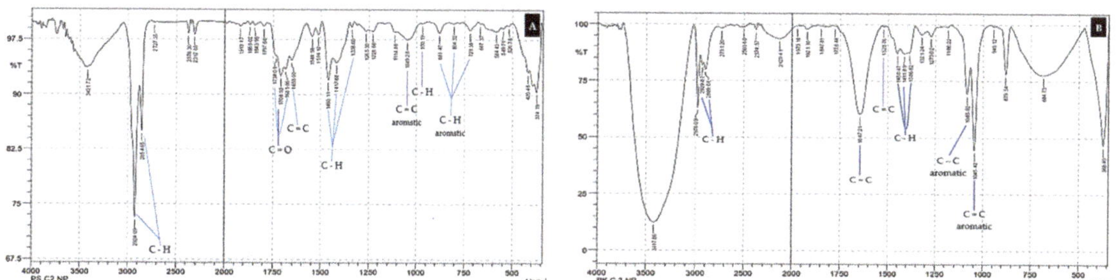

Figure 4. Identification of the type of MP pollutant in the sample based on the FTIR chromatogram. (**A**) Sediment samples; (**B**) sponge sample.

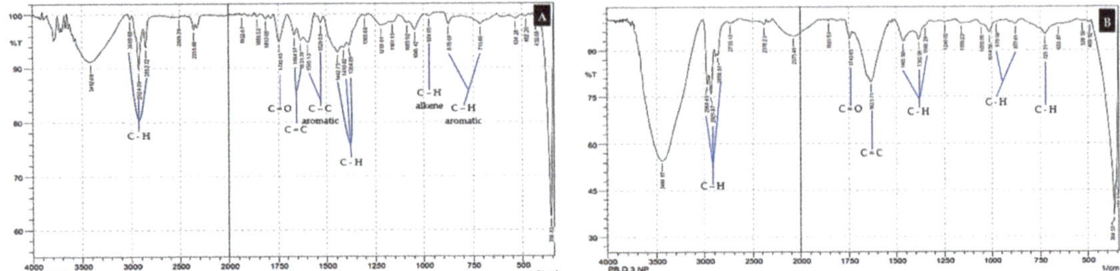

Figure 5. Identification of the type of MP pollutant in each sample based on the FTIR chromatogram. (**A**) Marine water samples; (**B**) fish samples.

The total MP abundance in the sediment samples was similar to the total MP abundance in the sponge samples (Figure 2). These results illustrate that MP exposure in sediment samples is relatively consistent with that in the sponge samples and is generally thought to be dominated by fiber and fragment types [18,26]. This is based on the difference in density between the two types of MP, which have higher specific gravity than water and film-type MP, so the two types of MP accumulate in greater concentrations on the bottom of the water [27,64]. The relatively settled lifestyle of sponges and their method of obtaining food nutrients by sucking particles from the mud and removing the dregs (filter feeder) are solid arguments for why the total abundance of MP in sediments is similar to that in sponges [26,27,39].

Total MP abundance in fish samples was higher than the total MP abundance in seawater samples. This occurred at all three sampling points (Figure 3). This result is entirely rational, as fish mobility is wider, so the presence of fish is not expected to always be in the zone of seawater exposed to MP [38,66]. The type of MP that fish samples are exposed to is also thought to be dominated by film-type MP, which have a lower density than water, meaning that they float on the surface zone, and are prone to being swallowed by fish, as the fish consider them to be food. The difference in the total MP abundance of fish samples between stations 1, 2, and 3 (Figure 3) is thought to be influenced by the different fish samples at each station (Table 2). The average total abundance of MP in the samples was 8.18 units/m^3 in sediment, 8.32 units/m^3 in sponge, 5.47 units/m^3 in seawater, and 7.03 units/m^3 in fish [19,22,28].

The type of MP pollutant was based on the polymer structure thought to be present in each analyzed sample. This was determined by analyzing the spectrum read by the FTIR chromatogram. The FTIR spectrum (Figures 4 and 5) shown for each type of sample is that obtained from station 1 (ST 1), which was considered representative of the samples collected at ST 2 and ST 3 [28,39].

The analysis of the MP pollutants was completed with an FTIR chromatogram and was based on the unique monomer structures of functional groups in each type of MP. Seven types of MP are commonly found in marine ecosystems [18,70]. Their respective monomers are as follows: polypropylene (PP) with the molecular formula (C_3H_6)n; polyethylene (PE), a monomer of ethylene (C_2H_4)n; polystyrene (PS) (C_8H_8)n; polycarbonate (PC), which has a carbonyl functional group (O-(C=O)-O)n; polyamide-nylon (PA), with the monomer formula $NH_2(CH_2)_6.NHCO(CH_2)_4COOH$; polyvinyl chloride (PVC) with unique components, such as chloride (C_2H_3Cl)n; and acrylonitrile butadiene styrene (ABS), which is composed of three monomers with the formula (C_8H_8)x(C_4H_6)y(C_3H_3N)z [18,19,26,27,38,64]. The results of the MP-type analysis based on the FTIR spectrum are displayed for each sample obtained at ST 1 (Figure 4). The MP, PP, and PS types existed in all of the samples. The ABS type was found in sediment, sponge, and seawater samples, while PVC was identified only in sea-water samples [19,38].

The types of MP identified in the seawater and fish samples did not differ significantly from the types of MP identified in the sediment and sponge samples (Figure 5). All results for the identification of MP species, based on the polymer structures found in the samples obtained at the three stations (ST 1, 2 and 3) and shown on the FTIR chromatogram, are presented more fully in Table 3.

Table 3. Type and distribution of MP pollutants in each sample.

Station Sampling	Sample Type			
	Sediment	Sponges	Marine Water	Fishes
ST 1	PA, PC, PE, PP, PS, ABS	PA, PE, PP, PS, ABS	PC, PE, PP, PS, PVC, ABS	PA, PE, PC, PP, PS
ST 2	PA, PC, PE, PP, PS	PA, PC, PE, PP, PS	PC, PE, PP, PS, ABS	PA, PC, PE, PP, PS
ST 3	PA, PC, PE, PP, PS, PVC	PA, PC, PE, PS, PVC	PA, PC, PP, PS, PVC	PA, PC, PE, PS, PVC

Note: PP = polypropylene; PE = polyethylene; PS = polystyrene; PC = polycarbonate; PA = polyamide nylon; PVC = polyvinyl chloride; ABS = acrylonitrile butadiene styrene.

In terms of the types of MP identified in each sample at the sampling points (ST 1–3) (Table 3), the pollutant MP type PA was identified in all of the samples, except for the seawater samples (ST 1 and 2). The PC type was seen in most of the samples, except for the sponge and seawater samples (ST 1) [64,68]. The PE pollutants were also recorded in most of the samples, except for the seawater samples (ST 3). PP contaminants were not found in the sponge or fish samples (ST 3), while PS was found in all of the samples at each station. PVC-type pollutants were identified in all of the sample types obtained from ST 3, and ABS-type pollutants were identified in sediment, sponge, and seawater samples from ST 1 only [66,68].

3.2. PAH Pollutant Analysis

The identification of hydrocarbon pollutants, especially the PAH type, and other hydrocarbon components in the samples was completed using the GC/MS chromatogram data. The parameters used to determine the presence of PAH-type pollutants in the samples were based on several recorded chromatogram data units, namely, the peak number, retention time, peak height as a manifestation of the component abundance, the content in percentage units in relation to the total PAH, and the component name.

According to the source (ST 1–3), each sample's complete details regarding the PAH-type hydrocarbon pollutants are presented in tabular form (Tables 4–6). As shown in Tables 4–6, the peak numbers are not sequential, and it is known that each peak is identical to one component [88,89]. Some peak numbers are not included in the table, due to several different reasons: first, the peak that appears is not a PAH component. Second, it is a PAH component, but the level of similarity with the comparison component (library instrument) does not reach the minimum standard of 85% [44,54,73]. Third, the components that appear are PAH derivatives. The PAH data from the four samples obtained from ST 1 (Table 4) indicate that the component is naphthalene (NL). The abundance of PAH components in the samples occurred in the following order: NL >> BZ >AZ >PR >PH. The peak numbers in the recorded GS/MS chromatograms are sequential, according to the number of components [9,74]. However, some peak numbers are not included in the data (Table 1), indicating that the peak number is not considered to be a PAH component or that it may be a derivative of a type of PAH, but the level of similarity with the standard library on the instrument does not reach 85% [88,90,91].

Table 4. Types and abundance of PAH in samples collected from ST 1 based on the GC/MS chromatogram.

Sample Type	Peak Number	Retention Time	Peak Height	Quality (%)	Abundance (%)	Compound Name
Sediment	4	9.167	2,027,937	91	66,386	NL
	6	15.550	68,138	90	2.324	AZ
	9	17.788	26,924	85	0.786	PH
	12	19.023	31,894	87	1.278	PN
Sponge	2	9.168	2,565,156	91	76.539	NL
	6	19.023	28,386	93	2.410	PN
	7	26.736	62,698	87	11.672	BZ
Sea water	1	9.168	1,581,333	91	78.456	NL
	3	15.550	87,843	91	3.045	AZ
Fish	2	9.167	2,134,672	91	74.456	NL
	6	26.736	54,789	88	8.192	BZ

Note: the compounds included are those with a level of similarity (quality) reaching $\geq 85\%$. NL = Naphthalene group; AZ = Azulene class; PH = Phenyl group; PN = Pyrene group; BZ = Benzene group.

The results of the identification of the PAH pollutant samples obtained from ST 2 were similar to those collected from ST 1. The PAH pollutant types, NL and PN, were found in all of the samples. The abundance of components in each sample was in the following order: NL > > PN >PD. The AZ, BZ, and PH were only identified in the sediment samples, and PT was only found in the sponge samples (Table 5).

Table 5. Types and abundance of compound hydrocarbon and PAH in samples at ST 2 station based on GC/MS chromatogram.

Sample Type	Peak Number	Retention Time	Peak Height	Quality (%)	Abundance (%)	Compound Name
Sediment	1	9.168	1,125,323	91	84.221	NL
	2	13.205	20,184	96	1.186	PD
Sediment	3	15.549	54,212	85	3.643	AZ
	4	17.788	18,406	85	1.369	PH
	6	19.023	62,311	89	3.409	PN
	8	26.735	71,925	86	5.357	BZ
Sponge	3	9.167	3,219,575	91	86.046	NL
	4	13.206	44,603	97	0.978	PD
	6	16.283	14,355	86	0.696	PT
	8	19.023	31,674	87	1.312	PN
Sea water	3	9.168	860,498	91	63.945	NL
	4	13.205	20,327	96	1.204	PD
	9	19.023	138,767	96	10.165	PN
Fish	2	9.168	4,322,267	87	89.859	NL
	4	19.023	12,458	92	3.146	PN

Note: the compounds included are those that have a level of similarity (quality) reaching $\geq 85\%$. NL = naphthalene group; AZ = Azulene class; PT = phenanthrene; PN = Pyrene group; BZ = Benzene group; PH = Phenyl group; PD = Pentadecane.

The results of the analysis of PAH pollutants obtained from ST 3 (Table 6) are similar to those for the samples from ST 1 and 2. The NL and PR were found in the four sample types, while the PT types were identified in the sediment, sponge, and water samples. AZ PAH pollutants were identified in the sediment and sponge samples, while BZ was only found in the sediment samples (Table 6) [54,79]. The abundance of components was, in order, NL > > PN >PT. Comparing the variation of PAH components identified at all sampling stations (Tables 4–6), the types of PAH in the sediment samples were more varied, followed by the sponge samples, and then the seawater and fish samples [14,86,87]. The types of PAH pollutants identified in each sample showed that the PAH contained in the sediment samples were similar to those found in the sponge samples, while the seawater samples showed similar results to the fish samples [92,93]. This result is quite rational if we consider that sponges tend to settle at the bottom of the water (sediment), and the dominant habitat of fish is the water.

Table 6. Types and abundance of PAH in samples obtained from ST 3 based on the GC/MS chromatogram.

Sample Type	Peak Number	Retention Time	Peak Height	Quality (%)	Abundance (%)	Compound Name
	2	9.168	1,222,751	91	82.356	NL
	4	15.549	88,566	90	3.087	AZ
Sediment	5	16.283	22,123	86	1.126	PT
	7	19.023	99,864	88	4.106	PN
	9	26.734	67,854	86	2.123	BZ
	2	9.167	2,423,789	91	78.127	NL
Sponge	4	15.550	65,832	85	3.983	AZ
	6	16.283	16,732	86	0.883	PT
	8	19.023	34,376	87	1.515	PN
	2	9.168	1,222,751	91	82.356	NL
Sea water	4	16.283	21,874	86	1.083	PT
	6	19.023	64,357	88	2.982	PN
Fish	3	9.168	1,111,244	91	64.362	NL
	5	19.023	13,318	87	0.734	PN

Note: the compounds included are those that have a level of similarity (quality) reaching ≥ 85%. NL = Naphthalene group; AZ = Azulene class; PT = phenanthrene; PN = Pyrene group; BZ = Benzene group.

3.3. Pesticide Residue Pollutant Analysis and Medical Waste

It is suspected that pesticide residue and medical waste are contaminating the marine waters around BCI. This conclusion is based on the GC/MS chromatogram, which identified several types of organic compounds containing acidic groups (phenol, arsenous acid, sulfurous acid), carboxylic acid (hexadecenoic, benzoic acid), halogen groups (chloridanthene, fluoranthene, blomidanthene), phosphate, cyclopentasiloxane, thiophene, silicic acid, benzo[h]quinolone, and cyclotrisiloxane [16,40,42]. These compounds are thought to be derived from pesticide residues and medical waste and then change during their stay in the aquatic environment. This assumption was confirmed by the GC/MS chromatogram [77,94,95].

The components of pesticide residue and medical waste are not presented significantly in tabular form. The abundance of chromatogram peaks is due to the similarity or suitability of components with references in the tool library, which is less than 85% [12,26,96]. Determination of a component based on the GS/MS measurement is considered valid if it has a similarity level or quality of 80%, as was completed for the hydrocarbon and PAH components [14,17,42,48,97]. Ideally, for the analysis of pesticide residue and medical waste in samples obtained from marine ecosystems, such as from fish, sponge, seawater, and sediment, the preparation of these samples should be carried out using a combination of pyrolysis-GC/MS instruments [46,48,98,99].

3.4. Heavy Metal Pollutant Analysis

Many heavy metal pollutants are thought to enter the marine ecosystem. However, only Lead (Pb) and Copper (Cu) ions were identified in this research. Observations made in the Makassar City MTA area concluded that these two types of pollutants are the most significant components of pollution in marine areas, especially around BCI in the Spermonde Islands group [11,13,36,100].

Based on the quality standard according to KLH RI Num. 51/2004, the average concentration of the heavy metals Pb^{+2} and Cu^{+2} in seawater should be a maximum of 0.05 mg/L [4,10,33,62]. Comparing the concentrations identified in the results (Table 7) with the seawater quality standards showed that the pollutant concentrations of Pb^{+2} and Cu^{+2} around BCI were 20 times and 25 times higher than the required quality standards, respectively [44,48,62,100]. According to the Director General of BPOM RI No. 03725/B/SK/VII/89, the maximum threshold value of heavy metal contamination in fish for pollutants Pb^{+2} and Cu^{+2} = 0.008 µg/g [33,55,82,101]

Table 7. Heavy metal pollutant (HM) concentration analysis.

Type of Pollutant	Sampling Station	Average Concentration of Pollutants Replication (n = 3)			
		Sediment (µg/g)	Sponge (µg/g)	Sea Water (mg/L)	Fish (µg/kg)
Lead ion (Pb^{+2})	ST 1	4.041 ± 0.0004	3.871 ± 0.0003	0.104 ± 0.0002	2.452 ± 0.0003
	ST 2	4.676 ± 0.0003	3.725 ± 0.0002	0.251 ± 0.0002	1.656 ± 0.0001
	ST 3	4.643 ± 0.0005	3.813 ± 0.0004	0.137 ± 0.0005	1.326 ± 0.0004
Average:		4.453 ± 0.0003	3.803 ± 0.0003	0.164 ± 0.0001	1.811 ± 0.0002
Copper ion (Cu^{+2})	ST 1	9.279 ± 0.0001	6.166 ± 0.0002	0.319 ± 0.0002	4.822 ± 0.0001
	ST 2	8.843 ± 0.0001	5.567 ± 0.0003	0.286 ± 0.0003	4.474 ± 0.0002
	ST 3	7.920 ± 0.0002	5.474 ± 0.0002	0.275 ± 0.0004	3.821 ± 0.0004
Average:		8.681 ± 0.0004	5.735 ± 0.0007	0.293 ± 0.0007	4.372 ± 0.0003

The data on the average concentrations of the two heavy metal pollutants (Table 7) show that exposure to Pb^{+2} and Cu^{+2} is still below the threshold value. However, using the Australian and New Zealand Environment and Conservation Council (ANZECC) standard, with a smaller tolerance value than the maximum threshold, it was concluded that the two pollutants found in fish around BCI exceeded the standard [10,55,76,102,103]. Based on the data, Table 7 shows that all types of samples (fish, sponges, seawater, and sediments) obtained in the waters around BCI have exceeded the maximum threshold value specified [4,10,48,104]. This is understandable and quite rational because consumption patterns, activities, and dynamics in each type of marine fish vary greatly. Fish samples were obtained for analysis from different stations (Table 2). The type of fish sampled at ST 1 (Chrysiptera unimaculata) differed from the fish samples from ST 2 and ST 3 [62,105,106].

4. Discussion

Physical observation of the waters around BCI did not find plastic waste, media waste, or oil spills. The physical characteristics of Barrang Caddi waters (EC, TDS, pH, and temperature) and other parameters (Table 1) were similar to those of surrounding marine waters [107]. It is known that the contaminants in the GTP group are invisible. Demographically, the distance between BCI and Makassar City is ±11.5 km, the land area is ±45,000 m², and the population is ±3000 people (612 families) (Figure 1). The residents generally make their livelihoods through the fishing industry, and some have vegetable gardens and undergo traditional household processing, generally burning [108,109]. Such conditions have ecological relationships and the potential to contribute to exposure to GTP contaminants [110].

The morphology of the fish and sponges differed at each station (Table 2). In marine ecosystems, fish interact with seawater as a growth medium throughout their lives, and the dominant sponges interact more with sediment (Figures 2–5). These factors cause differences in life patterns, the dynamics that occur between types of fish and sponges, and differences in the nutritional patterns of the two types of marine biota (fish and sponges) in terms of their growth media, their level of preference for types of nutrients, and their limit of tolerance to GTP contaminants [111,112].

The results for the total abundance of MP for each type of sample analyzed (Figures 2 and 3) showed that the waters around BCI are exposed to MP pollutants [18]. The type of MP contaminant contained in each sample type was analyzed further in terms of the density (film, fiber, fragment). The suspected types of MP based on polymer structures were distinguished by measurement and analysis, using the difference spectrum seen on the FTIR chromatogram (Figures 4 and 5) [38,77]. These observations show that the marine ecosystem around BCI waters is exposed to almost all types of MP contaminants [34,113]. The dominant MP found in all samples was the mild MP group (PE, PP, PS), followed by the heavy group (PA and PC), while few of the PVC and ABS types were found (Table 3) [26,42]. The data illustrate that the dominant source of MP contaminants is community activities, including industry, services, and household activities [87]. This type of MP is widely used in packaging applications, such as in the manufacture of plastic bags, clothes, ropes, coloring, and recycling containers, as well as various types of plastic caps, bottles, disposable tableware, and optical discs (CDs). Almost all of the uses of this type of MP are part of the daily activities of society and industry [114,115].

The other GTP contaminants analyzed were hydrocarbon components, especially PAHs. Research was carried out on all types of samples from three different stations (Tables 4–6), and eight types of PAH were identified. In terms of types of low-toxic PAH, the NL type was identified in all samples. NL is a simple aromatic hydrocarbon [9,38,116]. Naphthalene-based products, such as camphor, are thought to come from household activities, for example, in products used to eradicate insects and absorb odors in bathroom toilets. Another source can be the distillation of coal tar. The nature of NL is volatile and carcinogenic [14,117].

Pyrene is a second order PAH that was identified in several samples. PN is a group of highly toxic and carcinogenic PAHs. The presence of PR in the sample is thought to have come from the transportation activities of petroleum carriers. This assumption was reinforced by the identification of the PAH type AZ, as the PAH type is a component of petroleum as are some of its derivative products, including those from coal processing [9,84,118]. This allegation is quite reasonable, because BCI operates three ports: a container port, a passenger port, and a fish-landing port. Other PAHs, such as PT, PH, and BZ, are derivative products resulting from chemical reactions, such as oxidation reactions that produce these derivatives or community waste burning activities [71,119].

Pesticide residues and medical waste are strongly suspected to be pollutants in the marine ecosystem around BCI. This assumption is based on GC/MS chromatogram peaks of varying abundances and the recorded GC/MS chromatogram peaks identified with PR and MW components and observed in the PAH chromatogram data (Tables 4–6), where the peak numbers not listed in the table are the peaks of the PR component or residual MW [73,109,116]. This assumption is based on the fact that two hospitals operate around the coast of Makassar City, which is adjacent to BCI [51,120]. The presence of the pollutant types PR and MW in samples of fish, seawater, sponges, and sediments around the marine waters of BCI could also have come from the activities of the people who inhabit BCI or be due to biological activities, where these pollutants undergo dynamic activities and eventually empty into the sea [24,55,96,110].

Heavy metal (HM) pollutants, especially Pb^{+2} and Cu^{+2}, were identified in all of the samples obtained from the three different stations (Table 7). These results indicate that the concentrations are within the tolerable concentration limits for marine biota and are not dangerous according to Indonesian KLH standards, but they are in the worrying category based on ANZECC standards [12,62]. Pb^{+2} and Cu^{+2} pollutants in marine ecosystems around BCI are also thought to be due to natural and anthropogenic activities [20,121,122].

The concentration of all types of GTP identified in the waters around BCI is at alert status, not only in terms of the quality and sustainability of the marine ecosystem around BCI but also for the people who inhabit the island and the tourists visiting the island. In addition, the survival of sponges is important, as they are known to have primary and secondary metabolic contents that are used as raw materials for medicinal products, food concentrates, and beauty products [62,83,111,123].

It is estimated that exposure to contaminants from the GTP group around BCI will increase in the future due to the island's status as part of the Marine Tourism Area (MTA), meaning that it is visited by many tourists, which is seen as a potential factor contributing to the pollutant component, especially the pollutant types MP and PAH [65,69,80]. BCI is adjacent to the city of Makassar, which has a beach length of ± 7.5 km. Anthropogenic sources of pollutants are thought to heavily contribute to this sector. Along the coast, there are hotels, hospitals, industrial operations, culinary areas, and residential housing [1,11,40]. The area is very densely populated, and community activities occur every day for almost 24 h per day [7,10,80]. This situation is thought to be the main factor causing exposure to MP pollutants of the MW type. The vicinity of BCI is also adjacent to the Soekarno-Hatta Port, a fish-landing port. Thus, there are many transportation activities for various types of ships. This significantly contributes to exposure to pollutants, especially pollutants of the PAH, HM, and MP types [13,46,101].

We recommend that the Makassar City MTA manager should use this data as material for consideration in MTA operations, so that the managed marine tourism does not cause increases in pollutants from the GTP category [3,47]. It is expected that the tourist area could be improved by increasing the population of marine biota, especially sponges, through the transplant method, as sponges are known as biota with functions, such as biodegradation, bio-absorption, and the bioremediation of several types of pollutants including PAHs, microplastics, and heavy metals [20,121]. Sponges can be used as bioindicators and for the biomonitoring of PAH and HM contamination [44,73,122]. Efforts should be made along the shoreline of BCI, and it is recommended that plants that have a biofilter function against pollutants, such as mangroves, should be planted [124]. Makassar City MTA managers are also advised to carry out socialization, education, and advocacy with the community to make them aware of the effects of waste, so that their household waste can be appropriately managed and not enter sea waters.

5. Conclusions

This article reviewed microplastic pollutants, PAHs, pesticide residue, heavy metals, and medical waste, which are known as global trending pollutants (GTP) and are present in marine ecosystems around BCI's MTA waters. The average total MP abundance was found to be 5.47 units/m^3 in seawater samples, 7.03 units/m^3 in fish, 8.18 units/m^3 in sediment samples, and 8.32 units/m^3 in sponges. Seven types of MP were identified in the BCI aquatic ecosystem, dominated by the mild categories of MP, PE, PP, and PS, followed by PA and PC. The marine ecosystem around BCI was found to be contaminated with PAH-type pollutants, especially the NL type, which was found in all samples, followed by PN and AZ. BCI waters are also suspected of being exposed to pesticide residue and medical waste. The heavy metal pollutants Pb^{+2} and Cu^{+2} were also identified in the marine waters around BCI, especially in fish and seawater samples. The average concentrations of Pb^{+2} and Cu^{+2} in seawater were 0.164 ± 0.0001 mg/L and 0.293 ± 0.0007 mg/L, while the average concentrations of these two types of heavy metal pollutants were 1.811 ± 0.0002 µg/g and 4.372 ± 0.0003 µg/g, respectively.

The five types of pollutants identified in the marine ecosystem around BCI illustrate that this MTA is not conducive to being a recommended marine tourism destination. It is estimated that the concentrations of these pollutants will increase, given the status of BCI as one of the islands in the MTA area. If the BCI is maintained as a marine tourism area, it is feared that its ecosystem will enter an ecological hazard state, which will have a chain effect, not only on the aquatic ecosystem around the BCI but also on the sponge populations and fish, leaving them unfit for consumption and subsequently causing health problems for the community.

Author Contributions: Conceptualization, I.M., E.S., E.S.K., A.A. and K.N.; methodology, I.M., H.H., R.A. and A.A.; software, I.M. and A.S.; validation, I.M., S.S. and B.H.I.; formal analysis, I.M., F.R. and E.H.; investigations, I.M., R.A. and I.S.S.; resources, I.M., G.S.S. and R.A.; data curation, E.S.; writing—original draft preparation, I.M., I.S.S. and K.N.; writing—review and editing, K.N., E.S.K. and F.R.; visualization, H.H. and A.S.; supervision, S.S., I.S.S. and E.H.; project administration, I.M., A.A.; funding acquisition, I.M., B.H.I. and G.S.S. All authors have read and agreed to the published version of the manuscript.

Funding: This research received external funding from the "Ministry of Research, Technology and Higher Education/National Research and Technology Agency, who granted funds in 2021 under the applied research scheme, according to Master Contract numbers 315/E4.1/AK.04.PT/2021" and 1868/E4/AK.04/2021, dated 7 June 2021". Support was also received from Fajar University under the Work Implementation Order, contract Number: 29/AI/LPPM-UNIFA/IV/2021, dated 19 May 2021.

Institutional Review Board Statement: Not applicable.

Informed Consent Statement: Not applicable.

Data Availability Statement: Data sharing is not applicable to this article.

Acknowledgments: We thank Hasanuddin Biochemistry Laboratory, Department of Chemistry for their support through the use of their facilities. Makassar Plantation Product Industry Center (BBIHP) is a partner for this research. We acknowledge the research collaboration with the Research Institute for Coastal Aquaculture and Fisheries Extension, Maros, and BRIN as parties, including several individual partners from several campuses.

Conflicts of Interest: The authors declare no conflict of interest.

References

1. Bergmann, M.; Gutow, L.; Klages, M. Marine anthropogenic litter. In *Marine Anthropogenic Litter*; Springer: Cham, Switzerland; Berlin/Heidelberg, Germany; New York, NY, USA, 2015; pp. 1–447. [CrossRef]
2. Del-Mondo, G.; Peng, P.; Gensel, J.; Claramunt, C.; Lu, F. Leveraging spatio-temporal graphs and knowledge graphs: Perspectives in the field of maritime transportation. *ISPRS Int. J. Geo-Inf.* **2021**, *10*, 541. [CrossRef]
3. Fauzi, A.I.; Sakti, A.D.; Robbani, B.F.; Ristiyani, M.; Agustin, R.T.; Yati, E.; Nuha, M.U.; Anika, N.; Putra, R.; Siregar, D.I.; et al. Assessing potential climatic and human pressures in Indonesian coastal ecosystems using a spatial data-driven approach. *ISPRS Int. J. Geo-Inf.* **2021**, *10*, 778. [CrossRef]
4. Garba, F.; Ogidiaka, E.; Akamagwuna, F.C.; Nwaka, K.H.; Edegbene, A.O. Deteriorating water quality state on the structural assemblage of aquatic insects in a North-Western NigerianRiver. *Water Sci.* **2022**, *36*, 22–31. [CrossRef]
5. Mustafa, A.; Paena, M.; Athirah, A.; Ratnawati, E.; Asaf, R.; Suwoyo, H.S.; Sahabuddin, S.; Hendrajat, E.A.; Kamaruddin, K.; Septiningsih, E.; et al. Temporal and Spatial Analysis of Coastal Water Quality to Support Application of Whiteleg Shrimp Litopenaeus vannamei Intensive Pond Technology. *Sustainability* **2022**, *14*, 2659. [CrossRef]
6. Rakib, M.R.J.; Jolly, Y.N.; Dioses-Salinas, D.C.; Pizarro-Ortega, C.I.; De-la-Torre, G.E.; Khandaker, M.U.; Alsubaie, A.; Almalki, A.S.A.; Bradley, D.A. Macroalgae in biomonitoring of metal pollution in the Bay of Bengal coastal waters of Cox's Bazar and surrounding areas. *Sci. Rep.* **2021**, *11*, 20999. [CrossRef]
7. Busuricu, F.; Schroder, V.; Margaritti, D.; Anghel, A.H.; Tomos, S. Nutritional quality of some non-alcoholic beverages from the Romanian market. *Tech. Biochem.* **2022**, *3*, 1–6. [CrossRef]
8. Das, A.K.; Islam, M.N.; Billah, M.M.; Sarker, A. COVID-19 pandemic and healthcare solid waste management strategy–A mini-review. *Sci. Total Environ.* **2021**, *778*, 146220. [CrossRef]
9. Freeman, C.J.; Easson, C.G.; Fiore, C.L.; Thacker, R.W. Sponge–Microbe Interactions on Coral Reefs: Multiple Evolutionary Solutions to a Complex Environment. *Front. Mar. Sci.* **2021**, *8*, 705053. [CrossRef]
10. Marzuki, I.; Gusty, S.; Armus, R.; Sapar, A.; Asaf, R.; Athirah, A.; Jaya, J. Secondary Metabolite Analysis and Anti-Bacteria and Fungal Activities of Marine Sponge Methanol Extract Based on Coral Cover. In Proceedings of the 6th International Conference on Basic Sciences, Maluku, Indonesia, 4–5 November 2020; Volume 2360, pp. 1–9. [CrossRef]
11. Selvin, J.; Shanmugha, P.S.; Seghal, K.G.; Thangavelu, T.; Sapna, B.N. Sponge-associated marine bacteria as indicators of heavy metal pollution. *Microbiol. Res.* **2009**, *164*, 352–363. [CrossRef]
12. Kui, M.; Wu, L.; Fam, H. Heavy metal ions affecting the removal of polycyclic aromatic hydrocarbons by fungi with heavy-metal resistance. *Appl. Microbiol. Biotechnol.* **2014**, *98*, 9817–9827. [CrossRef]
13. Marzuki, I. The Bio-adsorption Pattern Bacteria Symbiont Sponge Marine Against Contaminants Chromium and Manganese in The Waste Modification of Laboratory Scale. *Indones. Chim. Acta* **2020**, *13*, 1–9. Available online: http://journal.unhas.ac.id/index.php/ica/article/view/9972 (accessed on 1 January 2022). [CrossRef]
14. Fu, W.; Xu, M.; Sun, K.; Cao, L.H.; Dai, C.; Jia, Y. Biodegradation of phenanthrene by endophytic fungus Phomopsis liquidambari in vitro and in vivo. *Chemosphere* **2018**, *203*, 160–169. [CrossRef] [PubMed]

15. Imachi, H.; Aoi, K.; Tasum, E.; Saito, Y.; Yamanaka, Y.; Saito, Y.; Takashi, Y.; Tomaru, H.; Takeuchi, R.; Morono, Y.; et al. Cultivation of methanogenic community from subseafloor sediments using a continuous-flow bioreactor. *ISME J.* **2011**, *5*, 1913–1925. [CrossRef] [PubMed]

16. Gao, J.; Luo, Y.; Li, Q.; Zhang, H.; Wu, L.; Song, J.; Qian, W.; Christie, P.; Chen, S. Distribution patterns of polychlorinated biphenyls in soils collected from Zhejiang province, east China. *Environ. Geochem. Health.* **2006**, *28*, 79–87. [CrossRef] [PubMed]

17. Essumang, D.K.; Togoh, G.K.; Chokky, L. Pesticide residues in the water and Fish (lagoon tilapia) samples from lagoons in Ghana. *Bull. Chem. Soc. Ethiop.* **2009**, *23*, 19–27. [CrossRef]

18. Alava, J.J.; Tirapé, A.; Mc-Mullen, K.; Uyaguari, M.; Domínguez, G.A. Microplastics and Macroplastic Debris as Potential Physical Vectors of SARS-CoV-2: A Hypothetical Overview with Implications for Public Health. *Microplastics* **2022**, *1*, 10. [CrossRef]

19. Baalkhuyur, F.M.; Bin- Dohaish, E.-J.A.; Elhalwagy, M.E.A.; Alikunhi, N.M.; AlSuwailem, A.M.; Rostad, A.; Coker, D.J.; Berumen, M.L.; Duarte, C.M. Microplastic in the gastrointestinal tract of fishes along the Saudi Arabian Red Sea coast. *Mar. Pol. Bull.* **2018**, *131*, 407–415. [CrossRef]

20. Marzuki, I.; Ahmad, R.; Kamaruddin, M.; Asaf, R.; Armus, R.; Siswanty, I. Performance of cultured marine sponges-symbiotic bacteria as a heavy metal bio-adsorption. *Biodiversitas* **2021**, *22*, 5536–5543. [CrossRef]

21. Abass, O.K.; Zhuo, M.; Zhang, K. Concomitant degradation of complex organics and metals recovery from fracking wastewater: Roles of nano zerovalent iron. *Chem. Eng. J.* **2017**, *328*, 159–171. [CrossRef]

22. Sobota, M.; Swi, M. Marine Waste—Sources, Fate, Risks, Challenges and Research Needs. *Int. J. Environ. Res. Public Heal.* **2021**, *18*, 433. [CrossRef]

23. Chulalaksananukul, S.; Gadd, G.M.; Sangvanich, P.; Sihanonth, P.; Piapukiew, J.; Vangnai, A.S. Biodegradation of benzo(a)pyrene by a newly isolated *Fusarium* sp. *FEMS Microbiol. Lett.* **2006**, *262*, 99–106. [CrossRef] [PubMed]

24. Akoto, O.; Azuure, A.A.; Adotey, K.D. Pesticide residues in water, sediment and fish from Tono Reservoir and their health risk implications. *Springer Plus* **2016**, *5*, 1849. [CrossRef]

25. Alghazeer, R.; Azwai, S.; Garbaj, A.M.; Amr, A.; Elghmasi, S.; Sidati, M.; Yudiati, E.; Kubbat, M.G.; Eskandrani, A.A.; Shamlan, G.; et al. Alkaloids Rich Extracts from Brown Algae Against Multidrug-Resistant Bacteria by Distinctive Mode of Action. *Arab. J. Sci. Eng.* **2022**, *47*, 179–188. [CrossRef]

26. Castelluccio, S.; Alvim, C.B.; Bes-Piá, M.A.; Mendoza-Roca, J.A.; Fiore, S. Assessment of Microplastics Distribution in a Biological Wastewater Treatment. *Microplastics* **2022**, *1*, 9. [CrossRef]

27. Fitri, S.; Patria, M.P. Microplastic contamination on Anadara granosa Linnaeus 1758 in Pangkal Babu mangrove forest area, Tanjung Jabung Barat district, Jambi. *J. Phys. Conf. Ser.* **2019**, *1282*, 012109. [CrossRef]

28. Gomiero, A.; Øysæd, K.B.; Agustsson, T.; van Hoytema, N.; van Thiel, T.; Grati, F. First record of characterization, concentration and distribution of microplastics in coastal sediments of an urban fjord in south west Norway using a thermal degradation method. *Chemosphere* **2019**, *227*, 705–714. [CrossRef]

29. El-Naggar, N.A.; Moawad, M.N.; Ahmed, E.F. Toxic phenolic compounds in the Egyptian coastal waters of Alexandria: Spatial distribution, source identification, and ecological risk assessment. *Water Sci.* **2022**, *36*, 32–40. [CrossRef]

30. Medaura, M.C.; Guivernau, M.; Moreno-Ventas, X.; Prenafeta-Boldú, F.X.; Viñas, M. Bioaugmentation of Native Fungi, an Efficient Strategy for the Bioremediation of an Aged Industrially Polluted Soil With Heavy Hydrocarbons. *Front. Microbiol.* **2021**, *12*, 1–18. [CrossRef]

31. Cui, X.; Tian, J.; Yu, Y.; Chand, A.; Zhang, S.; Meng, Q.; Li, X.; Wang, S. Multifunctional graphene-based composite sponge. *Sensors* **2020**, *20*, 329. [CrossRef]

32. Keller, C.T.; Jousset, A.; Van, O.L.; Van, E.J.D.; Costa, R. The freshwater sponge Ephydatia fluviatilis harbours diverse Pseudomonas species (Gammaproteobacteria, Pseudomonadales) with broad-spectrum antimicrobial activity. *PLoS ONE* **2014**, *9*, e88429. [CrossRef]

33. Dural, M.; Göksu, M.Z.L.; Özak, A.A. Investigation of heavy metal levels in economically important fish species captured from the Tuzla lagoon. *Food Chem.* **2007**, *102*, 415–421. [CrossRef]

34. Pratiwi, N.T.M.; Wulandari, D.Y.; Iswantari, A. Horizontal Distribution of Zooplankton in Tangerang Coastal Waters, Indonesia. *Procedia Environ. Sci.* **2016**, *33*, 470–477. [CrossRef]

35. Bart, M.C.; Hudspith, M.; Rapp, H.T.; Verdonschot, P.F.M.; de Goeij, M.J. A Deep-Sea Sponge Loop? Sponges Transfer Dissolved and Particulate Organic Carbon and Nitrogen to Associated Fauna. *Front. Mar. Sci.* **2021**, *8*, 229. [CrossRef]

36. Tepe, Y.; Türkmen, M.; Türkmen, A. Assessment of heavy metals in two commercial fish species of four Turkish seas. *Environ. Monit. Assess.* **2008**, *146*, 277–284. [CrossRef]

37. Ye, J.; Song, Y.; Liu, Y.; Zhong, Y. Assessment of medical waste generation, associated environmental impact, and management issues after the outbreak of COVID-19: A case study of the Hubei Province in China. *PLoS ONE* **2022**, *17*, e0259207. [CrossRef] [PubMed]

38. Anbumani, S.; Kakkar, P. Ecotoxicological effects of microplastics on biota: A review. *Environ. Sci. Pollut. Res.* **2018**, *25*, 14373–14396. [CrossRef]

39. Gad, A.K.; Midway, S.R. Relationship of Microplastics to Body Size for Two Estuarine Fishes. *Microplastics* **2022**, *1*, 14. [CrossRef]

40. Brodie, J.; Landos, M. Pesticides in Queensland and Great Barrier Reef Waterways-Potential Impacts on Aquatic Ecosystems and The Failure of National Management. *Estuar. Coast. Shelf Sci.* **2019**, *230*, 106447. [CrossRef]

41. Kotinagu, K.; Krishnaiah, N. Organochlorine and Organophosphorus Pesticide Residues in Todder and Milk Samples Along Musi River Belt, India. *Vet. World* **2015**, *8*, 545–550. [CrossRef]

42. Polat, A.; Polat, S.; Simsek, A.; Kurt, T.T.; Ozyurt, G. Pesticide residues in muscles of some marine fish species and seaweeds of Iskenderun Bay (Northeastern Mediterranean), Turkey. *Environ. Sci. Pollut. Res.* **2018**, *25*, 3756–3764. [CrossRef]

43. Cao, H.; Wang, C.; Liu, H.; Jia, W.; Sun, H. Enzyme activities during Benzo[a]pyrene degradation by the fungus Lasiodiplodia theobromae isolated from a polluted soil. *Sci. Rep.* **2020**, *10*, 865. [CrossRef]

44. Gran, S.A.; Ramos, Z.J.; Fuentes, E.; Bravo, D.; Pérez, D.J.M. Effect of co-contamination by PAHs and heavy metals on bacterial communities of diesel contaminated soils of south shetland islands, antarctica. *Microorganisms* **2020**, *8*, 1749. [CrossRef] [PubMed]

45. Zaimee, M.Z.A.; Sarjadi, M.S.; Rahman, M.L. Heavy metals removal from water by efficient adsorbents. *Water* **2021**, *13*, 2659. [CrossRef]

46. Marzuki, I.; Sinardi, S.; Pratama, I.; Chaerul, M.; Paserangi, I.; Kamaruddin, M.; Asaf, R. Performance of sea sponges micro symbionts as a biomaterial in biodegradation naphthalene waste of modified. In Proceedings of the 5th International Seminar on Sustainable Urban Development, Jakarta, Indonesia, 10–11 October 2021; p. 012016. [CrossRef]

47. Curchod, L.; Christelle, O.; Marion, J.; Christian, S.; Mohamed, A.D.; Martin, R.; Samuel, F. Temporal Variation of Pesticide Mixtures in Rivers of Three Agricultural Watersheds During a Major Drought in The Western Cape, South Africa. *Water Res.* **2020**, *10*, 100039. [CrossRef] [PubMed]

48. Topal, T.; Onac, C. Determination of Heavy Metals and Pesticides in Different Types of Fish Samples Collected from Four Different Locations of Aegean and Marmara Sea. *J. Food Qual.* **2020**, *2020*, 12. [CrossRef]

49. Alaboudi, A.A.; Ahmed, B.; Brodie, G. Annals of Agricultural Sciences Phytoremediation of Pb and Cd contaminated soils by using sun fl ower (Helianthus annuus) plant. *Ann. Agric. Sci.* **2018**, *63*, 123–127. [CrossRef]

50. Marzuki, I.; Alwi, R.S.; Erniati; Mudyawati; Sinardi; Iryani, A.S. *Chitosan Performance of Shrimp Shells in The Biosorption Ion Metal of Cadmium, Lead and Nickel Based on Variations pH Interaction*; Proceeding in Advances in Engineering Research; Atlantik Press: Amsterdam, The Netherlands, 2019; pp. 6–11. [CrossRef]

51. Liu, X.; Hu, X.; Cao, Y.; Jing, W.P.; Yu, H.J.; Guo, P.; Huang, L. Biodegradation of Phenanthrene and Heavy Metal Removal by Acid-Tolerant Burkholderia fungorum FM-2. *Front. Microbiol.* **2019**, *10*, 408. [CrossRef]

52. Baburam, C.; Feto, N.A. Mining of two novel aldehyde dehydrogenases (DHY-SC-VUT5 and DHY-G-VUT7) from metagenome of hydrocarbon contaminated soils. *BMC Biotechnol.* **2021**, *21*, 18. [CrossRef]

53. Kamaruddin, M.; Marzuki, I.; Burhan, A.; Ahmad, R. Screening acetylcholinesterase inhibitors from marine-derived actinomycetes by simple chromatography. In Proceedings of the 1st International Conference on Biotechnology and Food Sciences, Surabaya, Indonesia, 11 September 2020; p. 012011. [CrossRef]

54. Marzuki, I.; Ali, M.Y.; Syarif, H.U.; Gusty, S.; Daris, L.; Nisaa, K. Investigation of Biodegradable Bacteria as Bio indicators of the Presence of PAHs Contaminants in Marine Waters in the Marine Tourism Area of Makassar City. In Proceedings of the 6th International Conference on Tropical Coastal Region Eco-Development 2020, Indonesia, 27–28 October 2020; p. 012006. [CrossRef]

55. Marzuki, I.; Daris, L.; Nisaa, K.; Emelda, A. The power of biodegradation and bio-adsorption of bacteria symbiont sponges sea on waste contaminated of polycyclic aromatic hydrocarbons and heavy metals. *IOP Conf. Ser. Earth Env. Sci.* **2020**, *584*, 012013. [CrossRef]

56. Su, G.; Greaves, A.K.; Teclechiel, D.; Letcher, R.J. In Vitro Metabolism of Photolytic Breakdown Products of Tetradecabromo-1,4-diphenoxybenzene Flame Retardant in Herring Gull and Rat Liver Microsomal Assays. *Environ. Sci. Technol.* **2016**, *50*, 8335–8343. [CrossRef]

57. Marzuki, I.; Kamaruddin, M.; Ahmad, R. Identification of marine sponges-symbiotic bacteria and their application in degrading polycyclic aromatic hydrocarbons. *Biodiversitas* **2021**, *22*, 1481–1488. [CrossRef]

58. Omoni, V.T.; Lag-Brotons, A.J.; Ibeto, C.N.; Semple, K.T. Effects of biological pre-treatment of lignocellulosic waste with white-rot fungi on the stimulation of 14C-phenanthrene catabolism in soils. *Int. Biodeterior. Biodegrad.* **2021**, *165*, 105324. [CrossRef]

59. Orani, A.M.; Barats, A.; Vassileva, E.; Thomas, O.P. Marine sponges as a powerful tool for trace elements biomonitoring studies in coastal environment. *Mar. Pollut. Bull.* **2018**, *131*, 633–645. [CrossRef]

60. Rua, C.P.J.; Oliveira, L.S.; De-Froes, A.; Tschoeke, D.A.; Soares, A.C.; Leomil, L.; Gregoracci, G.B.; Coutinho, R.; Hajdu, E.; Thompson, C.C.; et al. Microbial and Functional Biodiversity Patterns in Sponges that Accumulate Bromopyrrole Alkaloids Suggest Horizontal Gene Transfer of Halogenase Genes. *Microb. Ecol. J.* **2018**, *76*, 825–838. [CrossRef] [PubMed]

61. Marzuki, I.; Daris, L.; Yunus, S.; Riana, A.D. Selection and characterization of potential bacteria for polycyclic aromatic biodegradation of hydrocarbons in sea sponges from Spermonde Islands, Indonesia. *AACL Bioflux.* **2020**, *13*, 3493–3506.

62. Ziarati, P.; Ziarati, N.N.; Nazeri, S.; Saber, G.M. Phytoextraction of heavy metals by two sorghum spices in treated soil using black tea residue for cleaning-uo the contaminated soil. Orient. *J. Chem.* **2015**, *31*, 317–326. [CrossRef]

63. Bibi, F.; Faheem, M.; Azhar, E.I.; Yasir, M.; Alvi, S.A.; Kamal, M.A.; Ikram, U.; Naseer, M.I. Bacteria From Marine Sponges: A Source of New Drugs. *Curr. Drug Metab.* **2016**, *17*, 1–6. [CrossRef]

64. Ibrahim, Y.S.; Rathnam, R.; Anuar, S.T.; Khalik, W.M.A.W.M. Separation and characterization of microplastics in the Lates calcarifer from the soils of setiu Malaysia. *Malays. J. Anal. Sci.* **2017**, *21*, 1054–1064. [CrossRef]

65. Marzuki, I.; Asaf, R.; Paena, M.; Athirah, A.; Nisaa, K.; Ahmad, R.; Kamaruddin, M. Anthracene and Pyrene Biodegradation Performance of Marine Sponge Symbiont Bacteria Consortium. *Molecules* **2021**, *26*, 6851. [CrossRef]

66. Jung, S.; Cho, S.-H.; Kim, K.-H.; Kwon, E.E. Progress in quantitative analysis of microplastics in the environment: A review. *Chem. Eng. J.* **2021**, *422*, 130154. [CrossRef]

67. Löder, M.G.J.; Kuczera, M.; Mintenig, S.; Lorenz, C.; Gerdts, G. Focal plane array detector-based micro-Fourier-transform infrared imaging for the analysis of microplastics in environmental samples. *Environ. Chem.* **2015**, *12*, 563–581. [CrossRef]

68. Roebroek, C.T.J.; Harrigan, S.; Van-Emmerik, T.H.M.; Baugh, C.; Eilander, D.; Prudhomme, C.; Pappenberger, F. Plastic in global rivers: Are floods making it worse? *Environ. Res. Lett.* **2021**, *16*, 025003. [CrossRef]

69. Schwinghammer, L.; Krause, S.; Schaum, C. Determination of large microplastics: Wet-sieving of dewatered digested sludge, co-substrates, and compost. *Water Sci. Technol.* **2021**, *84*, 384–392. [CrossRef] [PubMed]

70. Onyena, A.P.; Aniche, D.C.; Ogbolu, B.O.; Rakib, M.R.J.; Uddin, J.; Walker, T.R. Governance Strategies for Mitigating Microplastic Pollution in the Marine Environment: A Review. *Microplastics* **2021**, *1*, 3. [CrossRef]

71. Agrawal, N.; Verma, P.; Shahi, S.K. Degradation of polycyclic aromatic hydrocarbons (phenanthrene and pyrene) by the ligninolytic fungi Ganoderma lucidum isolated from the hardwood stump. *Bioresour. Bioprocess.* **2018**, *5*, 11. [CrossRef]

72. Ahmad, M.; Wang, P.; Li, J.L.; Wang, R.; Duan, L.; Luo, X.; Irfan, M.; Peng, Z.; Yin, L.; Li, W.-J. Impacts of bio-stimulants on pyrene degradation, prokaryotic community compositions, and functions. *Environ. Pollut.* **2021**, *289*, 117863. [CrossRef]

73. Atagana, H.I. Biodegradation of PAHs by fungi in contaminated-soil containing cadmium and nickel ions. *Afr. J. Biotechnol.* **2009**, *8*, 5780–5789. [CrossRef]

74. Bisht, S.; Pandey, P.; Bhargava, B.; Sharma, S.; Kumar, V.; Krishan, D. Bioremediation of polyaromatic hydrocarbons (PAHs) using rhizosphere technology. *Braz. J. Microbiol.* **2015**, *46*, 7–21. [CrossRef]

75. Dotsenko, I.V.; Mikhailenko, A.V. Phytoplankton and Its Role in Accumulation of Microelements in Bottom Deposits of Azov Sea. *Sci. World J.* **2019**, *2019*, 7. [CrossRef]

76. Selvam, S.; Jesuraja, K.; Venkatramanan, S.; Roy, P.D.; Kumari, P. J. Hazardous microplastic characteristics and its role as a vector of heavy metal in groundwater and surface water of coastal south India. *J. Hazard. Mater.* **2021**, *402*, 123786. [CrossRef]

77. Dimitrov, N.; Kratofil Krehula, L.; Ptiček Siročić, A.; Hrnjak-Murgić, Z. Analysis of recycled PET bottles products by pyrolysis-gas chromatography. *Polym. Degrad. Stab.* **2013**, *98*, 972–979. [CrossRef]

78. Leila, M.; Moscariello, A. Depositional and petrophysical controls on the volumes of hydrocarbons trapped in the Messinian reservoirs, onshore Nile Delta. *Egypt Petroleum.* **2018**, *4*, 250–267. [CrossRef]

79. Liu, S.H.; Zeng, G.M.; Niu, Q.Y.; Liu, Y.; Zhou, L.; Jiang, L.H.; Tan, X.-F.; Xu, P.; Zhang, C.; Cheng, M. Bioremediation mechanisms of combined pollution of PAHs and heavy metals by bacteria and fungi: A mini review. *Bioresour. Technol.* **2017**, *224*, 25–33. [CrossRef] [PubMed]

80. Skoog, D.A.; Holler, F.J.; Crouch, S.R. *Principles of Instrumental Analysis*, 7th ed.; Brooks/Cole, Ed.; Cengage Learning: Pacific Grove, CA, USA, 2021.

81. Bell, J.J.; Davy, S.K.; Jones, T.; Taylor, M.W.; Webster, N.S. Could some coral reefs become sponge reefs as our climate changes? *Glob. Chang. Biol.* **2013**, *19*, 2613–2624. [CrossRef] [PubMed]

82. Gawad, A.S.S. Acute toxicity of some heavy metals to the fresh water snail, Theodoxus niloticus (Reeve, 1856). *Egypt J. Aquat. Res.* **2018**, *44*, 83–87. [CrossRef]

83. Igiri, B.E.; Okoduwa, S.I.R.; Idoko, G.O.; Akabuogu, E.P.; Adeyi, A.O.; Ejiogu, I.K. Toxicity and Bioremediation of Heavy Metals Contaminated Ecosystem from Tannery Wastewater: A Review. *J. Toxicol.* **2018**, *2018*, 2568038. [CrossRef]

84. Armus, R.; Selry, C.; Marzuki, I.; Hasan, H.; Syamsia; Sapar, A. Investigation of Potential Marine Bacterial Isolates in Biodegradation Methods on Hydrocarbon Contamination. In Proceedings of the 2nd Workshop on Engineering, Makassar, Indonesia, 5 October 2020. [CrossRef]

85. Melawaty, L.; Noor, A.; Harlim, H.; de Voogd, N. Essential metal Zn in sponge Callyspongia aerizusa from Spermonde Archipelago. *Adv. Biol. Chem.* **2014**, *4*, 86–90. [CrossRef]

86. Çoban-Yıldız, Y.; Chiavari, G.; Fabbri, D.; Gaines, A.F.; Galletti, G.; Tuğrul, S. The chemical composition of Black Sea suspended particulate organic matter: Pyrolysis-GC/MS as a complementary tool to traditional oceanographic analyses. *Mar. Chem.* **2000**, *69*, 55–67. [CrossRef]

87. Hermabessiere, L.; Himber, C.; Boricaud, B.; Kazour, M.; Amara, R.; Cassone, A.-L.; Laurentie, M.; Paul-Pont, I.; Soudant, P.; Dehaut, A.; et al. Optimization, performance, and application of a pyrolysis-GC/MS method for the identification of microplastics. *Anal. Bioanal. Chem.* **2018**, *410*, 6663–6676. [CrossRef]

88. Mao, J.; Guan, W. Fungal degradation of polycyclic aromatic hydrocarbons (PAHs) by Scopulariopsis brevicaulis and its application in bioremediation of PAH-contaminated soil. *Acta Agric. Scand Sect. B Soil Plant Sci.* **2016**, *66*, 399–405. [CrossRef]

89. Parhamfar, M.; Abtahi, H.; Godini, K.; Saeedi, R.; Sartaj, M.; Villaseñor, J.; Coulon, F.; Kumar, V.; Soltanighias, T.; Ghaznavi-Rad, E.; et al. Biodegradation of heavy oily sludge by a two-step inoculation composting process using synergistic effect of indigenous isolated bacteria. *Process Biochem.* **2020**, *91*, 223–230. [CrossRef]

90. Roy, A.; Sar, P.; Sarkar, J.; Dutta, A.; Gupta, A.; Mohapatra, B.; Pal, S.; Kazy, S.K. Petroleum hydrocarbon rich oil refinery sludge of North-East India harbours anaerobic, fermentative, sulfate-reducing, syntrophic and methanogenic microbial populations. *BMC Microbiol.* **2018**, *18*, 151. [CrossRef] [PubMed]

91. Saraswath, A.; Hallberg, R. Degradation of pyrene by indigenous fungi from a former gasworks site. *FEMS Microbiol. Lett.* **2002**, *210*, 227–232. [CrossRef] [PubMed]

92. Shehu, U.; Ahmad, F.A.; Yusuf, F.; Muhammad, F.; Yakasai, H.M. Isolation and Identification of Anthracene Utilizing Proteus vulgaris from Oil Spill Contaminated Soil at NNPC Depot Kano State Nigeria. *J. Adv. Biol. Biotechnol.* **2021**, *24*, 46–53. [CrossRef]

93. Souza, H.M.; de L-Barreto, L.R.; da Mota, A.J.; de Oliveira, L.A.; Barroso, H.; Zanotto, S.S.P. Tolerance to polycyclic aromatic hydrocarbons (PAHs) by filamentous fungi isolated from contaminated sediment in the Amazon region. *Acta. Sci.-Biol. Sci.* **2017**, *39*, 481–488. [CrossRef]

94. Govarthanan, M.; Fuzisawa, S.; Hosogai, T.; Chang, Y.C. Biodegradation of aliphatic and aromatic hydrocarbons using the filamentous fungus Penicillium sp. CHY-2 and characterization of its manganese peroxidase activity. *RSC Adv.* **2017**, *7*, 20716–20723. [CrossRef]

95. Irfa'I, M.; Arifin, A.; Kriswandana, F.; Thohari, I. The Design of Medical Waste Treatment in Public Health Center (MWT-P) for Reducing Total Bacteria Count in Banjarbaru. *J. Kesehat. Lingkung* **2020**, *12*, 254. [CrossRef]

96. Mormede, S.; Davies, I.M. Polychlorobiphenyl and pesticide residues in monkfish Lophius piscatorius and black scabbard Aphanopus carbo from the Rockall Trough. *ICES J. Mar. Sci.* **2001**, *58*, 725–736. [CrossRef]

97. Busch, K.; Wurz, E.; Rapp, H.T.; Bayer, K.; Franke, A.; Hentschel, U. Chloroflexi Dominate the Deep-Sea Golf Ball Sponges Craniella zetlandica and Craniella infrequens Throughout Different Life Stages. *Front. Mar. Sci.* **2020**, *7*, 674. [CrossRef]

98. Costa, G.; Violi, B.; Bavestrello, G.; Pansini, M.; Bertolino, M. Aplysina aerophoba (Nardo, 1833) (Porifera, Demospongiae): An unexpected miniaturised growth form from the tidal zone of Mediterranean caves: Morphology and DNA barcoding. *Eur. Zool. J.* **2020**, *87*, 73–81. [CrossRef]

99. Landrigan, P.J.; Stegeman, J.J.; Fleming, L.E.; Allemand, D.; Anderson, D.M.; Backer, L.C.; Brucker-Davis, F.; Chevalier, N.; Corra, L.; Czerucka, D.; et al. Human health and ocean pollution. *Ann. Glob. Health* **2020**, *86*, 1–64. [CrossRef] [PubMed]

100. Mostafidi, M.; Shirkhan, F.; Zahedi, M.T.; Ziarati, P.; Hochwimmer, B. Risk Assessment of the Heavy Metals Contents in Ready-to-eat Vegetables in Distributed Brands at the Supply Level of Sales Centers in Tehran, Iran. *Res. Line (IJCSREM)* **2019**, *2*, 1–12. [CrossRef]

101. Ziss, E.; Friesl-Hanl, W.; Noller, C.; Watzinger, A.; Hood-Nowotny, R. Heavy Metal City-Zen. Exploring the potential risk of heavy metal contamination of food crop plants in urban gardening contexts using a citizen science approach. *EGU Gen. Assem. Conf. Abstr.* **2020**, *15*, 8626. [CrossRef]

102. Zhou, Y.; Xia, S.; Zhang, Z.; Zhang, J.; Hermanowicz, S.W. Associated Adsorption Characteristics of Pb (II) and Zn (II) by a Novel Biosorbent Extracted from Waste-Activated Sludge. *Environ. Eng.* **2016**, *142*, 04016032. [CrossRef]

103. Wu, C.; Liu, M.; Liu, X.; Wang, T.; Wang, L. Developing a new spectral index for detecting cadmium-induced stress in rice on a regional scale. *Int. J. Environ. Res. Public. Health* **2019**, *16*, 4811. [CrossRef]

104. Wei, Y.; Chen, W.; Liu, C.; Wang, H. Facial synthesis of adsorbent from hemicelluloses for Cr(VI) adsorption. *Molecules* **2021**, *26*, 1443. [CrossRef]

105. Tenea, A.G.; Vasile, G.G.; Dinu, C.; Gheorgh, S.; Pascu, L.F.; Mureseanu, M.; Ene, C. Behavior of Cd accumulation in sinapis alba L. In the presence of essential elements (Ca, Mg, Fe, Zn, Mn, Cu, Ni). *Rev. Chim.* **2020**, *71*, 378–389. [CrossRef]

106. Qu, R.; Han, G.; Liu, M.; Li, X. The mercury behavior and contamination in soil profiles in mun river basin, Northeast Thailand. *Int. J. Environ. Res. Public Health* **2019**, *16*, 4131. [CrossRef]

107. Zou, X.; Liu, X.; Liu, M.; Liu, M.; Zhang, B. A framework for rice heavy metal stress monitoring based on phenological phase space and temporal profile analysis. *Int. J. Environ. Res. Public Health* **2019**, *16*, 350. [CrossRef]

108. Korytár, P.; Covaci, A.; De Boer, J.; Gelbin, A.; Brinkman, U.A.T. Retention-time database of 126 polybrominated diphenyl ether congeners and two Bromkal technical mixtures on seven capillary gas chromatographic columns. *J. Chromatogr. A* **2005**, *1065*, 239–249. [CrossRef]

109. Gomiero, A.; Øysæd, K.B.; Palmas, L.; Skogerbø, G. Application of GC/MS-pyrolysis to estimate the levels of microplastics in a drinking water supply system. *J. Hazard. Mater.* **2021**, *416*, 125708. [CrossRef]

110. Ping, N.-C. Organochlorine Pesticide Residues in Plankton, Rangsit Agricultural Area, Central Thailand W. *Bull. Env. Contam. Toxicol.* **2008**, *81*, 608–612. [CrossRef]

111. Zhang, Q.; Han, G.; Liu, M.; Li, X.; Wang, L.; Liang, B. Distribution and contamination assessment of soil heavy metals in the jiulongjiang river catchment, southeast China. *Int. J. Environ. Res. Public Health* **2019**, *16*, 4674. [CrossRef]

112. Fomina, M.; Charnock, J.M.; Hillier, S.; Alvarez, R.; Livens, F.; Gadd, G.M. Role of fungi in the biogeochemical fate of depleted uranium. *Curr. Biol.* **2008**, *18*, 375–377. [CrossRef] [PubMed]

113. Ohowa, B.; Kiteresi, L.I.; Wanjeri, V.W.; Mwamburi, S.M.; Tunje, S.L. Sponges as simple biomonitoring tools for trace element pollution in marine environments: Insights from a Kenyan study focused on the leaf sponge Phyllospongia foliascens. *Afr. J. Mar. Sci.* **2021**, *43*, 533–538. [CrossRef]

114. De-Vries, I.; Schreiber, S.; Boßmann, D.; Hellmann, Z.; Kopatz, J.; Neumann, H.; Beutel, S. Single-use membrane adsorbers for endotoxin removal and purification of endogenous polysialic acid from Escherichia coli K1. *Biotechnol. Rep.* **2018**, *17*, 110–116. [CrossRef] [PubMed]

115. Yang, Q.; Franco, C.M.M.; Lin, H.W.; Zhang, W. Untapped sponge microbiomes: Structure specificity at host order and family levels. *FEMS Microbiol. Ecol.* **2019**, *95*, fiz136. [CrossRef]

116. Käppler, A.; Fischer, M.; Scholz-Böttcher, B.M.; Oberbeckmann, S.; Labrenz, M.; Fischer, D.; Eichhorn, K.-J.; Voit, B. Comparison of μ-ATR-FTIR spectroscopy and py-GCMS as identification tools for microplastic particles and fibers isolated from river sediments. *Anal. Bioanal. Chem.* **2018**, *410*, 5313–5327. [CrossRef]

117. Tsaboula, A.; Papadakis, E.; Vryzas, Z.; Kotopoulou, A.; Kintzikoglou, K.; Papadopoulou, M.E. Assessment and Management of Pesticide Pollution at a River Basin Level Part II: Optimization of Pesticide Monitoring Networks on Surface Aquatic Ecosystems by Data Analysis Methods. *Sci. Total Environ.* **2019**, *653*, 1597–1611. [CrossRef]

118. Smith, M.; Love, D.C.; Rochman, C.M.; Neff, R.A. Microplastics in Seafood and the Implications for Human Health. *Curr. Environ. Heal Rep.* **2018**, *5*, 375–386. [CrossRef]

119. Li, Q.; Liu, J.; Gadd, G.M. Fungal bioremediation of soil co-contaminated with petroleum hydrocarbons and toxic metals. *Appl. Microbiol. Biotechnol.* **2020**, *104*, 8999–9008. [CrossRef] [PubMed]

120. Galitskaya, P.; Biktasheva, L.; Blagodatsky, S.; Selivanovskaya, S. Response of bacterial and fungal communities to high petroleum pollution in different soils. *Sci. Rep.* **2021**, *11*, 164. [CrossRef]

121. Marzuki, I.; Pratama, I.; Ismail, H.E.; Paserangi, I.; Kamaruddin, M.; Chaerul, M.; Ahmad, R. The Identification and Distribution Components of Polycyclic Aromatic Hydrocarbon Contaminants at the Port of Paotere, Makassar, South Sulawesi. In Proceedings of the 1st International Conference on Biotech. and Food Sciences, Surabaya, Indonesia, 11 September 2020; p. 012017. [CrossRef]

122. Gawad, A.S.S. Concentrations of heavy metals in water, sediment and mollusk gastropod, Lanistes carinatus from Lake Manzala, Egypt. *Egypt J. Aquat. Res.* **2018**, *44*, 77–82. [CrossRef]

123. Shimoda, T.; Suryati, E.; Ahmad, T. Evaluation in a Shrimp Aquaculture System Using Mangroves, Oysters, and Seaweed as Biofilters Based on the Concentrations of Nutrients and Chlorophyll. *JARQ* **2006**, *40*, 189–193. [CrossRef]

124. Bayan, I.E.; Yulianda, F.; Setyobudiandi, I. Degradation analysis of mangrove ecological function as macrozoobenthos habitat and its management in the Angke Kapuk Coastal Area, Jakarta. *Biodiversitas (Bonorowo Wetl.)* **2016**, *6*, 1–11. [CrossRef]

Article

Ecological Impact of End-of-Life-Tire (ELT)-Derived Rubbers: Acute and Chronic Effects at Organism and Population Levels

Stefano Magni [1,*], Erica Tediosi [2,*], Daniela Maggioni [3], Riccardo Sbarberi [1], Francesca Noé [2], Fabio Rossetti [4], Daniele Fornai [5], Valentina Persici [6] and Maria Chiara Neri [2]

1   Department of Biosciences, University of Milan, Via Celoria 26, 20133 Milan, Italy; riccardo.sbarberi@unimi.it
2   ChemService Controlli e Ricerche s.r.l.—Lab Analysis Group, Via Fratelli Beltrami 15, 20026 Novate Milanese, Italy; francesca.noe@chemservice.it (F.N.); mariachiara.neri@chemservice.it (M.C.N.)
3   Department of Chemistry, University of Milan, Via Golgi 19, 20133 Milan, Italy; daniela.maggioni@unimi.it
4   Lab Analysis s.r.l., Via Europa 5, 27041 Casanova Lonati, Italy; fabio.rossetti@labanalysis.it
5   Ecopneus scpa, Via Messina 38, 20154 Milan, Italy; d.fornai@ecopneus.it
6   Waste and Chemicals s.r.l., Circonvallazione Gianicolense 216E, 00152 Rome, Italy; valentina.persici@wasteandchemicals.com
*   Correspondence: stefano.magni@unimi.it (S.M.); erica.tediosi@chemservice.it (E.T.); Tel.: +39-02-50314729 (S.M.); +39-02-3569961 (E.T.)

Citation: Magni, S.; Tediosi, E.; Maggioni, D.; Sbarberi, R.; Noé, F.; Rossetti, F.; Fornai, D.; Persici, V.; Neri, M.C. Ecological Impact of End-of-Life-Tire (ELT)-Derived Rubbers: Acute and Chronic Effects at Organism and Population Levels. *Toxics* **2022**, *10*, 201. https://doi.org/10.3390/toxics10050201

Academic Editor: Susanne M. Brander

Received: 21 March 2022
Accepted: 17 April 2022
Published: 19 April 2022

Publisher's Note: MDPI stays neutral with regard to jurisdictional claims in published maps and institutional affiliations.

Abstract: Considering the large amount of tires that reach the end of life every year, the aim of this study was the evaluation of both acute and chronic effects of end-of-life-tire (ELT)-derived rubber granules (ELT-dg) and powder (ELT-dp) on a freshwater trophic chain represented by the green alga *Pseudokirchneriella subcapitata*, the crustacean *Daphnia magna* and the teleost *Danio rerio* (zebrafish). Adverse effects were evaluated at the organism and population levels through the classical ecotoxicological tests. Acute tests on *D. magna* and *D. rerio* revealed a 50% effect concentration (EC_{50}) > 100.0 mg/L for both ELT-dg and ELT-dp. Chronic exposures had a lowest observed effect concentration (LOEC) of 100.0 mg/L for both ELT-dg and ELT-dp on *P. subcapitata* grow rate and yield. LOEC decreased in the other model organisms, with a value of 9.8 mg/L for *D. magna*, referring to the number of living offspring, exposed to ELT-dg suspension. Similarly, in *D. rerio*, the main results highlighted a LOEC of 10.0 mg/L regarding the survival and juvenile weight parameters for ELT-dg and a LOEC of 10.0 mg/L concerning the survival and abnormal behavior in specimens exposed to ELT-dp. Tested materials exhibited a threshold of toxicity of 9.8 mg/L, probably a *non*-environmental concentration, although further investigations are needed to clarify the potential ecological impact of these emerging contaminants.

Keywords: rubbers; tire particles; acute and chronic effects; freshwater species

1. Introduction

The advent of plastics is unclear but could be matched between the discovery of natural rubbers by Charles Marie de La Condamine (1736) and the subsequent introduction of vulcanization by Thomas Hancock and Charles Goodyear (1843–1844), which confers to rubbers resistance for a wide plethora of uses [1]. In this context, there is not a well-defined position by the scientific community regarding the inclusion of some types of elastomers in the classification of plastics. For example, rubbers are not plastics according to the definition of the International Organization for Standardization (ISO) [2]. However, some of these materials, including tire rubbers, contain from 40 to 60% synthetic polymers (i.e., styrene-butadiene and polybutadiene rubbers) and are characterized by physical proprieties such as solid state and insolubility in water; therefore, some agencies and environmental researchers classify them among plastics [3–6]. Tires, in addition to synthetic polymers, contain silica, oil, carbon black, sulfur compounds and zinc oxide [7]. Zinc (Zn) represents 1–2% of the total weight of tires, and the leaching of this element in water represents an important environmental concern [8]. In turn, tires are considered by some authors a considerable source of plastic debris due to, e.g., their mechanical abrasion during the activity of transport [9–13]. This aspect originates tire-wear particles (TWPs) and tire-road-wear particles (TRWPs) when TWPs are aggregated with other road debris/pollutants [6,14]. The contribution of tires to the total amount of released plastic debris in the environment ranges from 30 to 50% in Germany, Denmark and Norway [3,9,10,15–18]. In this context, a study by Knight et al. [19] reported a concentration of these contaminants ranging from 0.6 ± 0.33 to 65 ± 7.36 particles/5 mL of analyzed material from a natural area close to a road. However, relative to studies related to the monitoring of conventional (micro)plastics, few have reported the presence of tire particles in the environment [17,20]. Indeed, the main detected plastics, e.g., in aqueous matrices, are constituted by polymers, such as polystyrene (PS), polyethylene (PE), polypropylene (PP), polyester (PEST), polyacrylate (PAK) and polyamide [21–24]. In this context, it is important to consider that tire particles are difficult to detect with the identification methodologies of conventional (micro)plastics, such as the Fourier transform infrared spectroscopy (FT-IR), due to the lack of appropriate reference standards, as well as the presence of carbon black in tires, which absorbs in the infrared region [25]. Therefore, tire particle monitoring is hampered by analytical methods, currently represented by chromatography–mass spectrometry (GC/MS) and pyrolysis GC/MS [26]. The small amounts of TWPs and TRWPs identified in monitoring studies do not correlate with the massive use of tires around the world; in Europe alone every year, about 3.4 million tons of these products reach the end of life [27]. So-called end-of-life tires (ELTs) are normally recycled into ELT-derived rubber granules (ELT-dg) or ELT-derived rubber powder (ELT-dp) and transformed into other products, such as performance infills for artificial turfs and playground safety floorings [28–30]. However, regarding the potential toxicity of these substances, few studies (some of which are very old) have been conducted, and many research gaps currently affect the available knowledge, such as the lack of data on long-term toxicity; the differences in the results expression; and terminological inconsistency about TWP, TRWP and ELT [10,14]. For instance, a study on the crustacean *Hyallela azteca* revealed acute and chronic effects of TWPs after an exposure for 21 days to 500–2000 particles/mL as significant impacts on mortality, reproduction and growth [31], identifying 50% lethal concentration (LC_{50}) at 3426 ± 172 particles/mL. Another study revealed the absence of effects of ELTs at 10% sediment dry weight after 28 days of exposure in the freshwater crustaceans *Gammarus pulex* and *Asellus aquaticus* and in the Oligochaeta worms *Tubifex* spp. [32]. On the other hand, a larger number of studies have reported the toxicity of tire particle leachates [10,14]. A study on the rainbow trout *Oncorhynchus mykiss*, the crustacean *Daphnia magna* and the fathead minnow *Pimephales promelas* exposed to a leachate of worn, pristine and breakwater tires revealed no effects on *D. magna* and *P. promelas* and a 96 h LC_{50} for *O. mykiss* at 11.8–19.3% *v/v* for old tire leachate, which is more toxic than leachate from new tires [33]. On *O. mykiss*, an increase in ethoxyresorufin-O-deethylase (EROD) activity after exposure to water with whole tires was also observed [34]. Other studies highlighted a wide range of 50% effect concentration (EC_{50}) on *D. magna* after 48 h exposure to TWP leachates, ranging from 0.13 to 10 g/L [35,36], and 0.55 to 5 g/L for *Ceriodaphnia dubia* [37]. On the contrary, no significant effects were obtained in *H. azteca*, *P. promelas*, *C. dubia* and in the Diptera *Chironomus dilutus* exposed for 42 days to TRWP leachate (10 g/kg sediment) [38]. The effects of TRWPs on *Pseudokirchneriella subcapita*, *D. magna* and *P. promelas* were also evaluated using sediment elutriate and under standard test temperatures, reporting an EC_{50} greater than 10,000 mg/L [39]. Lastly, in another study, the toxicity of micro- (1–20 μm) and nano- (<1 μm) tire particles (from 0 to 3.0×10^9 particles/mL), as well as their leachate (from 0 to 100%), was assessed on *Danio rerio* and *D. magna*. This study, which combined the assessment of both particles and leachate toxicity, reported mortality and malformations on exposed specimens, especially those treated with nanoparticles [40].

Based on these complicated and heterogeneous pieces of evidence and considering the sparse information about the potential effects related to ELTs in particular, we evaluated the acute and chronic effects induced by aqueous suspensions of ELT-dg (size ranging from 0.8 to 2.5 mm) and ELT-dp (size < 0.8 mm) on freshwater organisms. The material used in the ecotoxicity tests were derived from the mechanical shredding of whole tires containing all compounds used in tire production (inner liners, treads, etc.), as well as the substances normally detected in TRWPs, such as bituminous residues and agglomerates of exhaust fumes. We tested different ELT-dg and ELT-dp concentrations, from 0.12 to 100.0 mg/L, using a battery of classical ecotoxicological tests on freshwater species, such as the unicellular green alga *P. subcapitata*, the crustacean *D. magna* and the teleost *D. rerio* (zebrafish), which constitute a simple aquatic trophic chain.

2. Materials and Methods

2.1. ELT-dg and ELT-dp Collection

Sampling of ELT-dg was conducted during July 2018 for one week in 20 different ELT treatment plants (19 in Italy and 1 in Switzerland) involved in the Ente Nazionale Italiano di Unificazione (UNI) project of GL14 group (ConfoReach-GVG group). Sampling was performed before the bagging step on particles ranging from 0.8 to 2.5 mm in size using a graduated jag to collect 400 g of selected material for each of the 5 samplings/day repeated for 5 days in a week (25 different samples). At the end of each sampling day, the collected materials were pooled and stored in glass bottles, obtaining 20 different samples (one for each selected plant) [41–45]. Samples were stored avoiding exposure to heat sources and UV rays. To reduce the amount of material, the ELT-dg samples were pooled and reduced to a primary sample of 2 kg by successive stages of splitting. Subsequently, 2 secondary samples of 1 kg each were obtained; one of them was used for ELT characterization, whereas the other one was used for ecotoxicity tests. To obtain the ELT-dp, a fraction of ELT-dg sample was subjected to gridding (cracker mill machine; mesh of 0.8 mm) to reach a powder size < 0.8 mm for ecotoxicity evaluation.

2.2. ELT Characterization and Preparation of Aqueous Suspensions

In the 20 different solid samples of ELT-dg, 152 chemicals were preliminarily quantified (Table S1, see references to the methods followed for the determination of each chemical). In addition, the ultrastructure of both ELT-dg and ELT-dp was investigated using scanning electron microscopy (SEM). Selected materials were placed on aluminum stubs with a carbon tape, gold-coated and observed with an FE-SEM SIGMA (Zeiss, Oberkochen, Germany) operating at 5 kV with a working distance of 20 mm [46].

According to the concentrations of chemicals detected in the ELT-dg (Table S1) and considering the theoretical ecotoxicity of these substances/elements as identified by the European Union CLP (classification, labelling and packaging) Regulation (e.g., hazard statements H400, H410, H411, H412 and H413), we selected 20 chemicals (Table S2) with a potential ecotoxicological implication. First, to evaluate the effective detection of these substances/elements in the exposure media from ELT-dg suspension, we calculated their theoretical concentrations in a water fraction of 100.0 mg/L ELT-dg. This concentration used for the ecotoxicity tests was chosen according to the Organization for Economic Co-operation and Development (OECD) guidelines [47–50]. According to the theoretical results (Table S2), the majority of the 20 selected chemicals were not detectable in water (concentrations < 1 µg/L), except for Zn, Cobalt (Co), Copper (Cu), 2-mercaptobenzothiazole, 4-t-octylphenol, aniline and N-cyclohexyl-cyclohexanamine. To confirm this aspect experimentally, we prepared a suspension of 100.0 mg/L ELT-dp from the 20 pooled samples of ELT-dg in ultrapure water, mixing for 48 h and filtering through 0.45 µm filters. The ELT-dp was chosen, assuming that its homogenization was better than that of ELT-dg due to the smaller size of particles. Subsequently, we quantified the selected 20 chemicals in the water fraction (Table S2, see references to the methods followed for the determination of each chemical). The analytical results showed that Zn was the most abundant solubilized component (53.4 µg/L; coefficient of variation or relative standard deviation, RSD = 2.0%;

Table S2). For this reason, Zn was chosen as the main solubilized and quantified component among the 20 selected chemicals with a possible ecotoxicological effect during the exposure.

The suspensions of ELT-dg and ELT-dp were prepared by mixing these materials in water and settling to separate the coarse components [51]. In particular, to identify the suitable mixing time, two suspensions of 100.0 mg/L of both ELT-dg and ELT-dp were mixed for 144 h in ultrapure water, and the concentration of Zn was assessed at 24, 48, 72 and 144 h. After 72 h of mixing, we obtained 80% Zn solubilization and decided to use this time to mix the suspensions for the preparation of test media. Regarding the settling time, two other suspensions of 100.0 mg/L of both ELT-dg and ELT-dp were mixed for 24 h and decanted for 144 h, and the total organic carbon (TOC; <3 mg/L, determination by cuvette test LCK 385, Hach), UV-VIS Spectrum (no significant absorption was observed in the UV-vis spectrum) and Zn concentrations were assessed. The quantification of Zn showed a stable concentration of this element after 48 h of settling, and this time was used for the suspension sedimentation. Lastly, considering that the majority of coarse ELT-dg and ELT-dp floated on the water surface and only a few amount of debris decanted, we prepared the tested media by serial dilutions of 100.0 mg/L, collecting the water from the center of suspensions using a peristaltic pump, avoiding the collection of coarse ELT-dg and ELT-dp from the surface or the bottom of used containers.

2.3. Characterization of ELT Suspensions

Because each model organism used in this study has a different exposure medium, we prepared the ELT suspensions separately for each exposure test. Zn concentration was measured during the entire exposures on one sample obtained on water mix from the different exposure tanks. For this reason, the standard deviation for these data was not calculated, and only the RSD, provided by instrumentation on 3 technical measurements, was reported. In detail, water samples were taken from the exposure tanks and acidified with 0.5% *v/v* nitric acid (HNO_3), stored at room temperature in dark conditions and analyzed within 7 days. The analyses of Zn were performed using a plasma optical emission spectrometer (ICP/OES; Agilent model 5110). Zn in a multi-element standard solution at 100.0 mg/L was used as reference material to prepare diluted standard solutions. In addition, to evaluate the performance of the method, two aliquots of 40 mL water medium were spiked with Zn concentrations from 1.5 to 80 µg/L. The limit of quantification of the analytical method depended on the basal level of Zn in aqueous media and was 1.5 µg/L in both fish and *D. magna* acute test media, 2.3 µg/L in the algal medium and 13.0 µg/L in the *D. magna* reproduction medium (see the Supplementary Materials for medium composition).

To investigate the presence and size of ELT particles in the test media, we performed both SEM and dynamic light scattering (DLS) analyses. In detail, SEM analysis was applied to 100.0 mg/L of ELT-dg and ELT-dp suspensions, sampling the aliquots from the water column and avoiding the collection of floating or decanted coarse particles. Samples were placed on aluminum stubs with a carbon tape, gold-coated and observed with an FE-SEM SIGMA (Zeiss, Oberkochen, Germany) operating at 5 kV with a working distance of 20 mm [46]. Once the presence of particles in the 100 mg/L suspensions was ascertained, we also investigated this aspect with DLS measurements, using a Malvern Zetasizer Nano ZS instrument (Malvern Instruments Ltd., Malvern, UK) equipped with a solid-state He-Ne laser with a wavelength of 633 nm, collecting the scattered light at 173°. The measurements on ELT suspensions were repeated 3 times using a disposable cuvette, leaving 30 s for temperature equilibration. The correlograms were fitted by a *non*-negative least square regression using multiexponential decay with Zetasizer Nano Series Software 7.02.

2.4. Toxicity Studies

The concentrations of ELT suspensions tested in the acute assays were chosen based on both OECD guidelines and on results of the range-finding tests (performed at 1.0, 10.0 and 100.0 mg/L of ELT-dg and ELT-dp), in which we did not observe significant effects. For this reason, as defined by OECD guidelines, we performed the limit tests only at 100.0 mg/L. For the subsequent chronic tests, the concentrations were determined in accordance with the results of acute tests. The ecotoxicity methods followed in this study are described in detail in the OECD guidelines [47–50,52], and for this reason we report only a brief description of them below.

2.4.1. Growth Inhibition Test on *P. subcapitata*

A chronic toxicity test was conducted on *P. subcapitata* with 72 h exposure under static conditions to 1.0, 3.1, 9.8, 31.3 and 100.0 mg/L of both ELT-dg and ELT-dp suspensions [47]. We evaluated the percentage of cell growth rate and yield inhibition at 24, 48 and 72 h. The organisms were cultured in the facility of ChemService Controlli e Ricerche s.r.l. in a climatic chamber at 24 ± 2 °C, 4440–8880 lx in the spectral range of 400–700 nm and maintained under continuous shaking to guarantee both the cell suspension and the correct concentration of carbon dioxide (CO_2). Only the algal cultures showing an exponential growth were used in the exposure, and the procedures were conducted under a laminar flow hood to avoid algal contamination. For the exposure, 100 mL capacity conical glass flasks with algal medium (see Supplementary Materials for composition) were used, with 3 replicates for each concentration and 6 replicates for the control. In each flask, 50 mL of test solution was poured and inoculated with the algae (10^4 cells/mL). Algal density was measured every 24 h by diluting an aliquot in 9 g/L sodium chloride (NaCl) solution and reading the algal cell amount with a cell counter (Beckman coulter Z2).

2.4.2. Acute and Chronic Tests on *D. magna*

The acute toxicity test on *D. magna* for both ELT-dg and ELT-dp suspensions was conducted in a 48 h exposure under static conditions to 100.0 mg/L [48]. In particular, we evaluated the immobilization, defined as organisms not able to swim within 15 s after gentle agitation. Specimens (daphnid age < 24 h) obtained by ChemService Controlli e Ricerche s.r.l. breeding were exposed in 40 mL beakers with pre-aerated (oxygenation > 5.62 mg/L) reconstituted water (see Supplementary Materials for composition), with 4 replicates *per* treatment (5 daphnids *per* beaker) at 20 °C, with 16 h of light and 8 h of darkness (1188–1401 lx). The immobilization was checked after 24 and 48 h. The chronic test on *D. magna* for both ELT-dg and ELT-dp was conducted in 21 days of exposure in semi-static condition (renewal 3 times *per* week) to 1.0, 3.1, 9.8, 31.3 and 100.0 mg/L [50]. In this context, after 21 days of exposure, we evaluated the mortality rate and reproduction parameters (total number of living offspring). Specimens (daphnid age < 24 h) were exposed in 50 mL beakers with pre-aerated (oxygenation > 5.00 mg/L) reconstituted water (see the Supplementary Materials for the composition) with 10 replicates *per* treatment (10 daphnids *per* treatment, 1 specimen *per* beaker) at 20 °C, with 16 h of light and 8 h of darkness (1236–1467 lx). During exposure, daphnids were feed with *P. subcapitata* after the water change.

2.4.3. Acute and Chronic Tests on *D. rerio*

The acute toxicity test on *D. rerio* for both ELT-dg and ELT-dp suspensions comprised a 96 h exposure to 100.0 mg/L under static conditions [49]. Fish were considered dead if no movements were visible and if no reaction was observable when touching the caudal peduncle; in addition, we checked for visible abnormalities after 2, 24, 48, 72 and 96 h. Juvenile fish were obtained by ChemService Controlli e Ricerche s.r.l. facility and fed up to 24 h before the test. The facilities of ChemService Controlli e Ricerche s.r.l. follow Italian laws, rules and regulations (Legislative Decree No. 116/92; authorization n. 30/2012-A of 25 January 2012). Fish sizes were checked before the exposure by measuring a representative group of fish of the same batch. For ELT-dg exposure, mean length was 2.40 ± 0.26 cm and mean weight was 0.25 ± 0.04 g, whereas for ELT-dp exposure, mean length was 1.80 ± 0.17 cm and mean weight was 0.30 ± 0.04 g. Organisms of 2 months were exposed in 60 L glass aquaria (7 specimens *per* treatment) filled with reconstituted water (see Supplementary Materials for composition), with one replicate *per* concentration at a temperature of 22 °C, with 12 h light (680–827 lx) and 12 h darkness daily photoperiod (2 daily 30 min transition periods; 754–827 lx), oxygenation > 60% and 6.60–7.77 pH range. The chronic toxicity test on *D. rerio*, for both ELT-dg and ELT-dp suspensions, comprised 30 days of exposure to 0.12, 0.37, 1.1, 3.3 and 10.0 mg/L under semi-static conditions, renewing the media 3 times *per* week [52]. The concentrations were chosen according to OECD 210 [52], and the highest concentration was adopted in accordance with both acute (LC_{50} > 100.0 mg/L) and fish embryo tests (FET; conducted as previous screening; no mortality was observed at 100.0 mg/L). During the test, we evaluated different end points, e.g., hatching, survival, abnormal behavior, weight and length of juvenile fish. For the embryo stage, fish were exposed in glass containers with 50 mL ELT-dg and ELT-dp suspensions, whereas for the juvenile stage, the volume was increased up to 1200 mL. Four replicates *per* treatment (20 eggs *per* well) were used. The exposure was conducted at a temperature between 25.1 and 26.0 °C, with 12 h light (602–853 lx) and 12 h darkness daily photoperiod, oxygenation > 60% and 7.75–7.88 pH range. Fish were fed *ad libitum* starting from 2 days after hatching (the amount was incremented during the test period; Gemma Micro 75, Skretting for larvae and Gemma Micro 150 for juveniles).

2.5. Statistical Analyses and Data Elaboration

To avoid confusion between the specific end points evaluated for each selected ecotoxicological test, we summarized and reported the main results as EC_{50}, no observed effect concentration (NOEC) and lowest observed effect concentration (LOEC). The variance and distribution of data were verified by Bartlett and Shapiro–Wilk tests. For the growth inhibition test on *P. subcapitata*, the EC_{50} was determined by linear interpolation method for growth rate and by *non*-linear regression analysis for yield. The NOEC and LOEC for growth rate and yield were determined by Bonferroni correction test. For acute and chronic tests on *D. magna*, the EC_{50} was determined with linear interpolation, and NOEC and LOEC were determined by the Steel many-one rank sum test. For acute and chronic tests on *D. rerio*, the EC_{50} was measured by linear interpolation, and NOEC and LOEC were calculated, where possible, by Fisher exact test and Dunnett's multiple comparison test. CETIS v.1.8.7.7 software was used to carry out these analyses.

In addition, as further data elaboration, we performed a comparison between treated and control, time *versus* time (where possible) of selected end points. For this purpose, we used one/two-way analysis of variance (ANOVA) or a *non*-parametric test if normality or homoscedasticity were not verified, followed by Bonferroni correction test. STATISTICA 7.1 Software was used in these analyses.

3. Results and Discussion

3.1. Zn and ELT Particle Detection in the Exposure Media

SEM analysis of ELT-dg and ELT-dp revealed a heterogeneous size of selected particles, with a complex ultrastructure represented by a wide plethora of rubber shapes (Figure 1). This aspect could be extremely important for understanding the releasing behavior of Zn by selected materials in the aqueous suspensions. In this context, the presence of Zn in ELTs is associated with its use as an activator in the vulcanization process [53], making this element an environmental marker of tire particles [10]. For this reason, the monitoring of Zn in the aqueous suspensions used in the present study could be pivotal for the interpretation of the ecotoxicity results. Regarding the chronic test on *P. subcapitata*, Zn concentrations at the beginning of the test were under the detection limit for the lowest concentrations of ELT-dg (1.0 and 3.1 mg/L) and of 4.1, 7.7 and 26.3 µg/L for ELT-dg concentrations of 9.8, 31.3 and 100.0 mg/L, respectively (Figure 2). At the end of the static exposure (72 h), the concentration of Zn related to the 9.8 mg/L of ELT-dg decreased under the detection limit and to 2.5 and 21.0 µg/L for the highest ELT-dg concentrations of 31.3 and 100.0 mg/L. For the ELT-dp at the beginning of *P. subcapitata* exposure, Zn concentrations were under the limits of detection/quantification in the groups from 1.0 to 9.8 mg/L of ELT-dp and 5.2 and 17.3 µg/L for the highest concentrations of 31.3 and 100.0 mg/L. Similarly, at the end of the exposure, these values remained under the detection/quantification limits for concentrations up to 9.8 mg/L, whereas we obtained values of 3.0 and 12.4 µg/L for 31.3 and 100.0 mg/L, respectively. Regarding the acute test on *D. magna*, the concentration of Zn in the aqueous suspension from 100.0 mg/L of ELT-dg at the beginning of the static test was 20.7 µg/L, decreasing to 12.3 µg/L at the end of exposure (48 h). On the contrary, for 100.0 mg/L of ELT-dp, we obtained the highest and most constant concentrations during the static exposure, with 45.1 µg/L at time 0 and 41.8 µg/L at the end of the test. Similarly, in the chronic test on *D. magna* under semi-static conditions, Zn concentration for the ELT-dg suspension was measured only in the highest tested concentration of 100.0 mg/L because preliminary analytical tests showed that at the lower tested concentrations, Zn was not quantifiable. During the 21 days of exposure, the concentration of Zn detected was between 14.6 µg/L (maximum measured concentration; MaMC) and 5.0 µg/L (minimum measured concentration; MiMC). Regarding the ELT-dp suspension, Zn concentration was measured, once again, only in the two highest tested concentrations of 31.3 and 100.0 mg/L. During the 21 days of exposure, the Zn concentration detected in 31.3 mg/L suspension was between 17.2 µg/L (MaMC) and 13.0 µg/L (MiMC), whereas the concentration of Zn detected in 100.0 mg/L was between 49.5 µg/L (MaMC) and 22.9 µg/L (MiMC). Due to the large amount of data, in Figure 2, we reported only the concentrations of Zn at the beginning and at the end of each exposure, but it is important to note that in the chronic tests, the media were renewed every 3 days, and every 3 days, the concentrations of Zn in fresh and spent solutions were measured. Therefore, for chronic tests, the concentrations of Zn at the beginning and at the end of the tests are not directly related. Moving to the acute test on *D. rerio*, the concentration of Zn in the suspension of 100.0 mg/L of ELT-dg was 8.4 µg/L at the beginning of static-exposure, with a slight reduction to 5.6 µg/L at the end of the test (96 h). Similarly, we obtained a decrease in Zn concentration in the suspension from ELT-dp, where the concentration was 28.1 µg/L at the beginning of the test and then reduced to 22.2 µg/L at the end of the exposure. Lastly, in the chronic test on *D. rerio* under semi-static conditions, for ELT-dg suspension of 10.0 mg/L, the concentration of Zn detected was between 16.3 µg/L (MaMC) and 1.5 µg/L (MiMC); for ELT-dg suspension of 3.3 mg/L, the concentration of Zn detected was between 9.4 µg/L (MaMC) and 0.5 µg/L (MiMC). For ELT-dp suspension of 10.0 mg/L, the concentration of Zn detected was between 11.0 µg/L (MaMC) and 0.5 µg/L (MiMC); for ELT-dp suspension of 3.3 mg/L, the concentration of Zn detected was between 6.2 µg/L (MaMC) and 0.5µg/L (MiMC). In this context, we also detected Zn in the control at the end of exposures (Figure 2). According to preliminary analytical stability tests, it can be stated that Zn is stable for 3 days; when the food supply started, Zn was influenced by the presence of dissolved food and/or uneaten food and feces.

In general, Zn began to be released in water at appreciable concentrations starting from 10.0 and 31.3 mg/L of ELT-dg and ELT-dp. A similar trend in the release of this element in water was observed at a concentration of 100.0 mg/L for both ELT-dg and ELT-dp, as represented by the histograms at t = 0 h (fresh solution; Figure 2). This aspect suggests that the smaller size of ELT-dp caused a higher release of Zn, probably due to the increase in the surface/volume ratio of particles. Indeed, despite it has been demonstrated that the leaching of some elements from crumb rubbers was size-independent [8], in the present study the release of Zn was higher by ELT-dp, as observed in the tests on *D. magna* and *D. rerio* (Figure 2). In this context, some evidence highlighted that the leaching of Zn from TWPs was reduced by an increase in salinity and pH and enhanced by fluorescent light compared to dark conditions [54]. However, because an unclear trend related to salinity or pH of water media was observed in our work, further investigations are needed to clarify the role of chemical/physical parameters in the release of Zn by both ELT-dg and ELT-dp.

In the context of Zn release, Capolupo et al. [55] reported that this element was the main chemical detected in car tire rubber leachate, together with benzothiazole and Co. The fluctuating values of Zn obtained at the beginning of each exposure were probably related to the heterogeneity of sizes and shapes of ELT particles (Figure 1). This aspect, which cannot be controlled during the preparation of aqueous suspensions (each weighing of ELT material contains very different particles), could cause a heterogeneous Zn release in water in a size/shape-dependent manner.

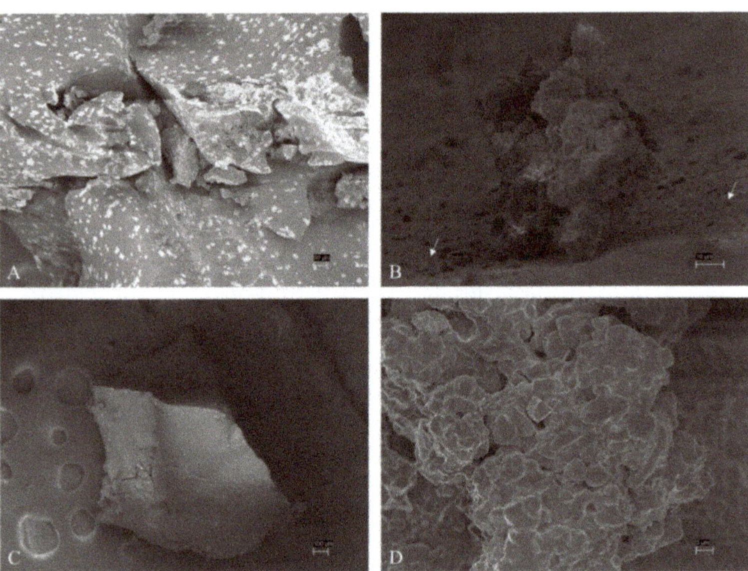

Figure 1. SEM analysis of both ELT-dg (**A,B**) and ELT-dp (**C,D**). There is wide range of sizes and shapes of selected materials, with nanoparticles (size < 1 μm, based on classification proposed by Hartmann et al. [6] on plastic size) on the surface of a single ELT-dg (**B**) debris (indicated by arrows).

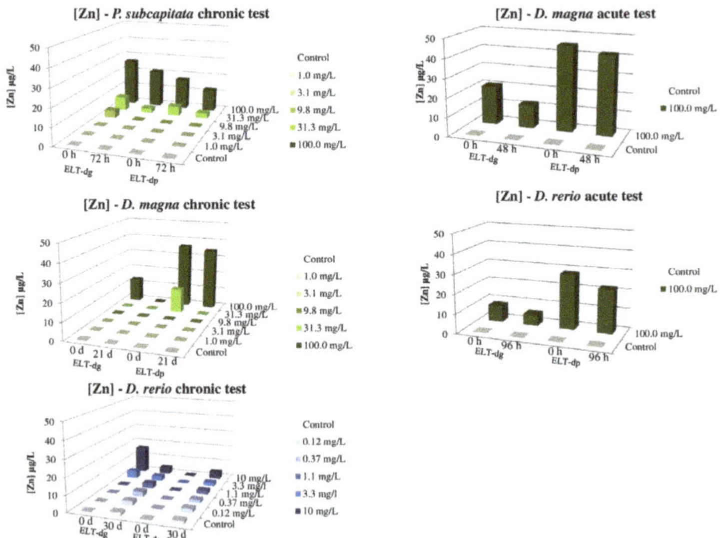

Figure 2. Zn *versus* ELT-dg and ELT-dp concentrations in the suspensions used in the different ecotoxicological tests at the beginning (t = 0) and at the end of each exposure. The concentration of Zn at t = 0 (fresh suspension) corresponds to Zn directly released by ELT-dg and ELT-dp in water (RSD range for groups of *P. subcapitata* = 0.22–1.43%; RSD range for groups of *D. magna* = 0.16–2.91%; RSD range for groups of *D. rerio* = 0.76–5.97%).

The presence of ELT particles was certified in some ELT suspensions through SEM and DLS analyses. At the highest concentration of 100.0 mg/L, we detected the presence of both ELT-dg and ELT-dp nanoparticles (size < 1 µm, based on classification proposed by Hartmann et al. [6] on plastic size; Figure 3) in the aqueous fraction. Consequently, we investigated this aspect with DLS in the other ELT suspensions. Concerning ELT-dg, we observed particles with a mean size of 810 ± 215 nm, 1290 ± 520 nm and 1508 ± 615 nm in the suspensions of 100.0, 31.3 and 10.0 mg/L, respectively. This evidence highlighted the presence of both nano- and microparticles in ELT-dg suspensions. For the other suspensions, no repeatable data were obtained, probably due to the serial dilutions, which decreased the number of particles under the detection limit of DLS instrumentation. Concerning ELT-dp, readings appeared more unstable than for ELT-dg, resulting in a change in the correlation function. We observed a value of 530 ± 109 nm in the 100 mg/L suspension. In the other dilution of 31.3 mg/L, correlograms were close to the detection limit, with two particle populations identified, one of 478 ± 106 nm and another one exceeding the range covered by the DLS instrument of about 8000 nm.

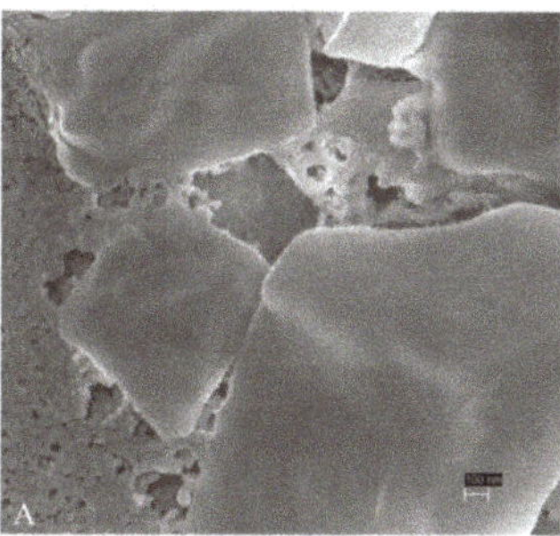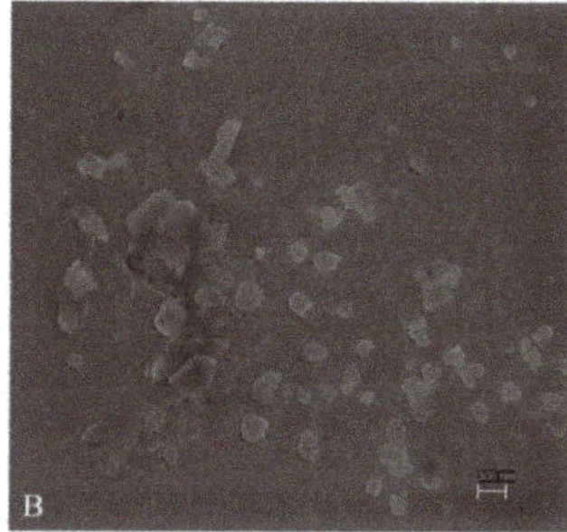

Figure 3. SEM analysis of both ELT-dg (**A**) and ELT-dp (**B**) aqueous suspensions of 100.0 mg/L. These images confirm the presence of some micro- and nanoparticles (size < 1 µm, based on classification proposed by Hartmann et al. [6] on plastic size) in the selected exposure media.

3.2. Acute and Chronic Effects of ELT-dg and ELT-dp Suspensions

Regarding the acute test on *D. magna*, we did not observe immobilization in the control group, and no daphnids were trapped on the water surface during the exposure. In addition, daphnids did not show signs of disease or stress in the controls during the test. We obtained a value of 24 and 48 h EC_{50} for ELT-dg and ELT-dp suspensions > 100.0 mg/L (Table 1). Based on Zn concentrations measured in the exposure media at the beginning and at the end of exposure, we performed a time-weighted arithmetic mean of Zn concentrations between time 0 and 48 h [50], obtaining an EC_{50} > 16.2 µg/L of Zn for ELT-dg and an EC_{50} > 43.5 µg/L of Zn for ELT-dp at both 24 and 48 h of exposure. In the same manner, no mortality or anomalies were observed in *D. rerio* in both control and treated groups. On *D. rerio*, we obtained an EC_{50} > 100.0 mg/L for both ELT-dg and ELT-dp suspensions from 24 to 96 h of exposure. Regarding the concentration of Zn, we obtained an EC_{50} > 6.9 µg/L of Zn for ELT-dg and an EC_{50} > 25.0 µg/L of Zn for ELT-dp suspension (see also the Zn concentrations in the water media reported in Figure 2).

Regarding the test on the green alga *P. subcapitata*, for both ELT-dg and ELT-dp, we obtained a LOEC of 100.0 mg/L from 0 to 72 h. In this context, as reported in Figure 4, we observed a significant effect of treatment on the cell density (we used the cell density to calculate the growth rate and yield) of *P. subcapitata* exposed to ELT-dg at the end of exposure to 100.0 mg/L ($p < 0.01$; Figure 4A) and in algae exposed to 100.0 mg/L of ELT-dp at 48 h ($p < 0.01$) and 72 h ($p < 0.01$; Figure 4B).

Table 1. Values of EC$_{50}$, NOEC and LOEC for the main tested end points obtained by the different OECD tests for both ELT-dg and ELT-dp suspensions. For EC$_{50}$ determination, the following conditions were evaluated: the 95% confidence interval did not contain zero and was not overly wide, the 95% confidence interval for the predicted mean did not contain the control mean, there was no significant lack of fit of regression model to the data. If the above conditions were not satisfied, the NOEC approach was used.

		Suspension from ELT-dg			Suspension from ELT-dp		
		EC$_{50}$	NOEC (mg/L)	LOEC (mg/L)	EC$_{50}$	NOEC (mg/L)	LOEC (mg/L)
P. subcapitata chronic test	Growth rate	93.7 (0–72 h)	31.3 (0–72 h)	100.0 (0–72 h)	>100.0 (0–72 h)	31.3 (0–72 h)	100.0 (0–72 h)
	Yeld	54.2 (0–72 h)	31.3 (0–72 h)	100.0 (0–72 h)	73.6 (0–72 h)	31.3 (0–72 h)	100.0 (0–72 h)
D. magna acute test	Immobilization	>100.0 (24 and 48 h)	100.0 (24 and 48 h)		>100.0 (24 and 48 h)	100.0 (24 and 48 h)	
D. rerio acute test	Mortality	>100.0 (24, 48, 72 and 96 h)	100.0 (24, 48, 72 and 96 h)		>100.0 (24, 48, 72 and 96 h)	100.0 (24, 48, 72 and 96 h)	
D. magna chronic test	Reproduction	>100.0 (21 d)	3.1 (21 d)	9.8 (21 d)	>100.0 (21 d)	100.0 (21 d)	>100.0 (21 d)
	Parental mortality		100.0 (21 d)	>100.0 (21 d)		100.0 (21 d)	>100.0 (21 d)
D. rerio chronic test	Hatching		10.0 (96 hpf)	>10.0 (96 hpf)		10.0 (96 hpf)	>10.0 (96 hpf)
	Juvenile survival		3.3 (30 d)	10.0 (30 d)		3.3 (30 d)	10.0 (30 d)
	Juvenile weight		3.3 (30 d)	10.0 (30 d)		10.0 (30 d)	>10.0 (30 d)
	Juvenile lenght		10.0 (30 d)	>10.0 (30 d)		10.0 (30 d)	>10.0 (30 d)
	Abnormal behaviour					3.3 (30 d)	10.0 (30 d)

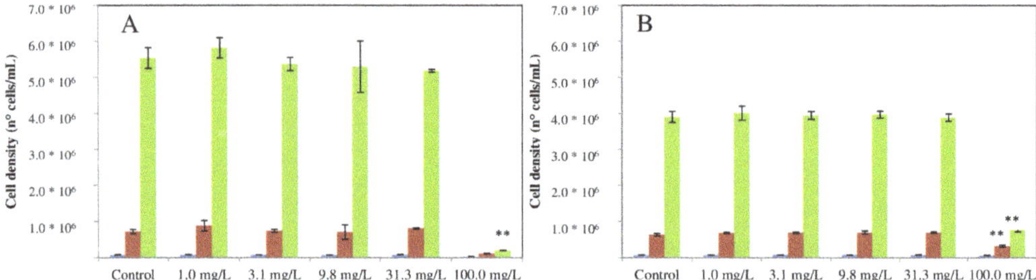

Figure 4. Significant effects indicated by asterisks (** $p < 0.01$), control *versus* treated, induced by both ELT-dg (**A**) and ELT-dp (**B**) suspensions in *P. subcapitata* (cell density) at 24 (blue bars), 48 (red bars) and 72 h (green bars).

At the higher trophic level, in the chronic test on *D. magna*, we observed a value of 21 days LOEC in the exposure to the suspension from ELT-dg, referring to the reproduction, of 9.8 mg/L. In this context, we obtained the following total number of living offspring produced *per* parental animal: 87.4 ± 7.6 in the control, 97.1 ± 9.6 at 1.0 mg/L, 86.4 ± 17.0 at 3.1 mg/L, 60.2 ± 12.0 at 9.8 mg/L, 58.8 ± 12.5 at 31.3 mg/L and 58.2 ± 26.9 at 100 mg/L. Coherently with the LOEC, we observed a significant effect of treatment on the number of living offspring, with a significant reduction compared to the control ($p < 0.01$) at the three highest concentrations (9.8, 31.3 and 100.0 mg/L; Figure 5). Conversely, for all other considered parameters (mean number of dead offspring, aborted eggs and body length of parent animals), no significant effects were observed (data not shown). Regarding the ELT-dp, we did not observe any significant effect compared to the control, whereas we obtained a LOEC > 34.5 µg/L of Zn. In this context, for *P. subcapitata*, *D. magna* and *D. rerio* chronic tests, toxicity results expressed as mean Zn concentrations (time-weighted arithmetic mean, measured at the beginning and at the end of the test) were not determined because it was not possible to quantify their concentration for some samples, as reported in Section 3.1.

In *D. rerio*, for ELT-dg and ELT-dp suspensions, the test started at 4–8 cells, corresponding to 1.25 hpf. In both tests, we observed hatching of 100% (LOEC > 10.0 mg/L; Table 1). For the survival parameter, we obtained a LOEC of 10.0 mg/L on the specimens exposed to ELT-dg and ELT-dp. In addition, for the juvenile fish weight parameter, the LOEC for specimens exposed to ELT-dg was 10.0 mg/L, as well as for the abnormal behavior, only in fish exposed to ELT-dp (Table 1). In this context, we observed a significant reduction in survival ($p < 0.05$) and fish weight ($p < 0.05$) in *D. rerio* exposed to 10.0 mg/L ELT-dp and ELT-dg suspensions, respectively (Figure 6A,B).

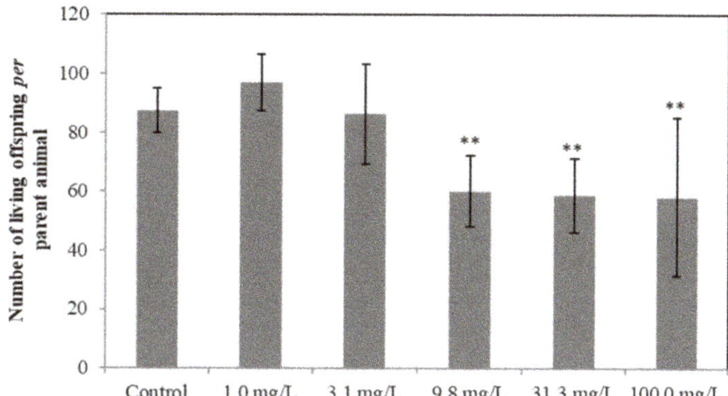

Figure 5. Significant effects indicated by asterisks (** $p < 0.01$), control *versus* treated, induced by ELT-dg suspension in *D. magna* (number of living offspring).

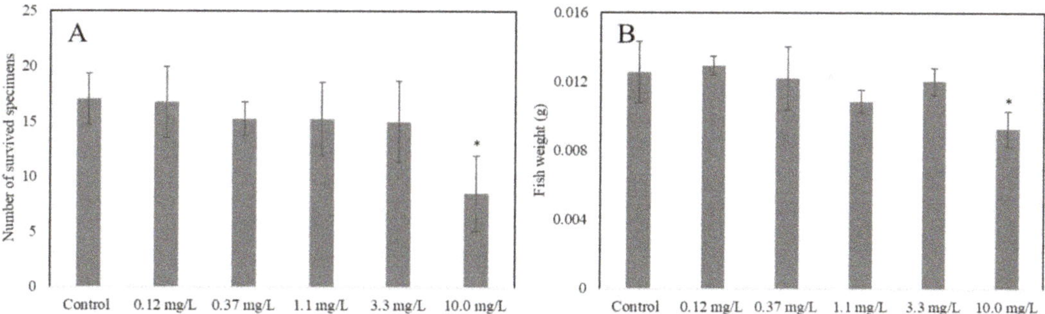

Figure 6. Significant effects indicated by asterisks (* $p < 0.05$), control *versus* treated, induced in *D. rerio* by both ELT-dp ((**A**); survival) and ELT-dg ((**B**); fish weight) suspensions.

In some studies, leachates from tires generally show low toxicity on aquatic species, with high values of EC_{50} between 0.1 g/L and 100 g/kg [56]. As observed in the present study, the toxicity of ELT suspensions might be associated with the release of metals, such as Zn [56], as well as ELT nano- and microparticles detected in the exposure media. In this context, in future studies, it will also be important to consider the role played by other chemicals released by ELTs, such as 4-(dimethylbutylamino)diphenylamine. In the aquatic environment, this nitrosamine used as an antioxidant in tires produces a very toxic quinone able to induce acute effects on aquatic species at concentrations raging from <0.3 to 19 μg/L, as recently observed in the salmonid *Oncorhynchus kisutch* (LC_{50} 0.8 ± 0.16 μg/L) [57]. However, coherently with our work, Marwood et al. [39] investigated the effects of TRWP sediment elutriate from Michelin, Pirelli and Bridgestone tires using a road simulator laboratory, reporting an EC_{50} higher than 10,000 mg/L for *P. subcapitata*, *D. magna* and *P. promelas*. In this work, Zn and aniline were identified as the main toxic chemicals [39]. The study of Halsband et al. [58] highlighted how crumb rubber granules from ELTs, used as performance infill in synthetic turf pitches as well as their leaching, could pose a potential threat for wildlife, reporting acute effects of leachates within 24 h at the highest tested concentrations (100 and 50 g/L), with 48 h LC_{50} of 35 g/L for *Calanus* sp. and <5 g/L for *Acartia* sp. Once again, benzothiazole and Zn were the main components of the leachates in this study. In this context, some evidence has suggested that when salinity and pH increase, the leaching of Zn is reduced [54] and leachates decrease their toxicity [59]. This aspect could suggest the key role of Zn in the toxicity modulation of leachates. At the same time, when the leachates are obtained at pH < 7, Zn increases together with toxicity [60]. Considering this evidence and that obtained in our study concerning the trends in the release of Zn in water by ELT-dg and ELT-dp, other investigations on the role of Zn in ELT toxicity are necessary. Indeed, without a mechanistic approach, e.g., by the application of biomarkers, such as metallothioneins, cellular stress and oxidative damage end points, it is difficult to determine whether Zn is the main driver of ELT toxicity. Therefore, the effects observed at the organism and population level in this study on reproduction, survival and growth of exposed specimens represent a starting point for further mechanistic studies on ELT impact.

4. Conclusions

Considering the obtained results, neither ELT-dg nor ELT-dp can be classified in the context of CLP regulation, either for short-term aquatic hazard (because we observed an $EC_{50} > 1$ mg/L in acute toxicity tests) or for long-term aquatic hazard (because we observed NOEC > 1 mg/L in chronic tests) for all trophic levels.

The obtained results suggested that ELT suspensions exhibit a threshold of toxicity of 9.8 mg/L for the tested end point at the organism and population level. In this context, future studies on ELTs should focus on the chronic toxicity of these contaminants, as well as characterizing their possible infiltration in the biota tissues. Currently, there is little information in the scientific literature about the ecotoxicological implication of ELTs. This also affects the comparative evaluations between different works and experimental approaches. For this reason, other investigations are urgently needed using, e.g., more sensitive methodologies, such as biomarkers or "omics" techniques, to evaluate ELT effects at the biochemical, molecular and cellular level, delineating their potential mechanism of action. Therefore, the characterization of the ecological impact of ELTs needs more explanations, and this study represents a preliminary work to fill the gap concerning the impact of these materials in freshwater ecosystems.

Supplementary Materials: The following supporting information can be downloaded at: https://www.mdpi.com/article/10.3390/toxics10050201/s1, Supplementary Methods; Table S1: Chemicals (mg/Kg) quantified in the 20 different ELT-dg samples (solid fraction). Zinc oxide, Titanium oxide and Magnesium oxide were not detected analytically in the samples and for this reason their presence in the ELT-dg was not certified.; Table S2: Twenty selected chemicals with the concentration detected in the ELT suspensions. In the table are reported the CAS number, the used methods for the chemical detection, the CLP classification as well as the theoretical and analytical concentration of considered substances/elements. In particular, in the column (A) it is reported the max concentration measured in the 100.0 mg/L of ELT-dg suspensions (mg of each component in 1 Kg of ELT-dg, corresponding to µg in 1 g of ELT-dg; Table S1), in the column (B) the µg component in 100.0 mg of sample (calculated form max concentration), in the column (C) the theoretical concentration calculated assuming the hypothesis that all the amount present in 100.0 mg of ELT-dg was dissolved/solubilized in 1 L of water and, lastly, in the column (D) the analytical concentration measured in the 100.0 mg/L of ELT-dp.

Author Contributions: Conceptualization, M.C.N.; methodology, S.M., E.T., F.N. and F.R.; validation, E.T., F.N. and F.R.; formal analysis, E.T., R.S., F.N. and F.R.; investigation, S.M., E.T., D.M., F.N. and F.R.; data curation, S.M. and E.T.; writing—original draft preparation, S.M.; writing—review and editing, E.T., D.M., D.F. and V.P.; supervision, S.M. and M.C.N.; project administration, M.C.N.; funding acquisition, D.F. All authors have read and agreed to the published version of the manuscript.

Funding: This research was funded by Ecopneus scpa.

Institutional Review Board Statement: Regarding the use of vertebrates (*D. rerio*) in this study, the facilities of ChemService Controlli e Ricerche s.r.l.—Lab Analysis Group follow Italian laws, rules and regulations: Legislative Decree No. 116/92, authorization n. 30/2012-A of 25 January 2012.

Informed Consent Statement: Not applicable.

Data Availability Statement: Not applicable.

Conflicts of Interest: Authors of ChemService and Lab Analysis Group (E.T., F.N., F.R. and M.C.N.) report a relationship with Ecopneus scpa (represented by D.F.), who financed the present work. V.P. of Waste and Chemicals srl reports a relationship with Ecopneus scpa (represented by D.F.) that includes consulting activity. Considering these aspects, we assure that no inappropriate data were presented in this work.

References

1. Crawford, C.B.; Quinn, B. *Microplastic Pollutants*, 1st ed.; Elsevier Limited: Amsterdam, The Netherlands, 2016; p. 330.
2. *ISO 472*; Plastics—Vocabulary, 4th ed. ISO: Geneva, Switzerland, 2013; p. 406.
3. Lassen, C.; Foss Hansen, S.; Magnusson, K.; Norén, F.; Bloch Hartmann, N.I.; Rehne Jensen, P.; Torkel, G.N.; Brinch, A. *Microplastics—Occurrence, Effects and Sources of Releases to the Environment in Denmark. Environmental Project No. 1793*; Danish Environmental Protection Agency: Copenhagen, Denmark, 2015; p. 205.
4. Eunomia. *Plastics in the Marine Environment*; Eunomia: Bristol, UK, 2016.
5. Verschoor, A.; De Poorter, L.; Dröge, R.; Kuenen, J.; de Valk, E. *Emission of Microplastics and Potential Mitigation Measures—Abrasive Cleaning Agents, Paints and Tyre Wear*; National Institute for Public Health and the Environment: Bilthoven, The Netherlands, 2016; p. 73.
6. Hartmann, N.B.; Huffer, T.; Thompson, R.C.; Hasselov, M.; Verschoor, A.; Daugaard, A.E.; Rist, S.; Karlsson, T.; Brennholt, N.; Cole, M.; et al. Are we speaking the same language? Recommendations for a definition and categorization framework for plastic debris. *Environ. Sci. Technol.* **2019**, *53*, 1039–1047. [CrossRef]
7. Balbay, S. Effects of recycled carbon-based materials on tyre. *J. Mater. Cycles Waste Manag.* **2020**, *22*, 1768–1779. [CrossRef]
8. Rhodes, E.P.; Ren, Z.; Mays, D.C. Zinc leaching from tire crumb rubber. *Environ. Sci. Technol.* **2012**, *46*, 12856–12863. [CrossRef] [PubMed]
9. Kole, P.J.; Löhr, A.J.; Van Belleghem, F.G.A.J.; Ragas, A.M.J. Wear and tear of tyres: A stealthy source of microplastics in the environment. *Int. J. Environ. Res. Public Health* **2017**, *14*, 1265. [CrossRef]
10. Wagner, S.; Hüffer, T.; Klöckner, P.; Wehrhahn, M.; Hofmann, T.; Reemtsma, T. Tire wear particles in the aquatic environment-a review on generation, analysis, occurrence, fate and effects. *Water Res.* **2018**, *139*, 83–100. [CrossRef] [PubMed]
11. Hüffer, T.; Wagner, S.; Reemtsma, T.; Hofmann, T. Sorption of organic substances to tire wear materials: Similarities and differences with other types of microplastic. *Trends Anal. Chem.* **2019**, *113*, 392–401. [CrossRef]
12. Ziajahromi, S.; Drapper, D.; Hornbuckle, A.; Rintoul, L.; Leusch, F.D. Microplastic pollution in a stormwater floating treatment wetland: Detection of tyre particles in sediment. *Sci. Total Environ.* **2020**, *713*, 136356. [CrossRef] [PubMed]
13. Koski, M.; Søndergaard, J.; Christensen, A.M.; Nielsen, T.G. Effect of environmentally relevant concentrations of potentially toxic microplastic on coastal copepods. *Aquat. Toxicol.* **2021**, *230*, 105713. [CrossRef]
14. Halle, L.L.; Palmqvist, A.; Kampmann, K.; Khan, F.R. Ecotoxicology of micronized tire rubber: Past, present and future considerations. *Sci. Total Environ.* **2020**, *706*, 135694. [CrossRef]
15. Sundt, P.; Schulze, P.-E.; Syversen, F. Sources of Microplastics-pollution to the Marine Environment. *Nor. Environ. Agency Miljødirektoaret* **2014**, *86*, 20.
16. Bertling, J.; Bertling, R.; Hamann, L. Kunststoffe in der Umwelt. Mikro- und Makroplastik. In *Ursachen, Mengen, Umweltschicksale, Wirkungen, Lösungsansätze, Empfehlungen. Kurzfassung der Konsortialstudie*; Fraunhofer-Institut für Umwelt Oberhausen: Oberhausen, Germany, 2018.
17. Leads, R.R.; Weinstein, J.E. Occurrence of tire wear particles and other microplastics within the tributaries of the Charleston Harbor Estuary, South Carolina, USA. *Mar. Pollut. Bull.* **2019**, *145*, 569–582. [CrossRef] [PubMed]
18. Unice, K.M.; Weeber, M.P.; Abramson, M.M.; Reid, R.C.D.; van Gils, J.A.G.; Markus, A.A.; Vethaak, A.D.; Panko, J.M. Characterizing export of land-based microplastics to the estuary—Part I: Application of integrated geospatial microplastic transport models to assess tire and road wear particles in the Seine watershed. *Sci. Total Environ.* **2019**, *646*, 1639–1649. [CrossRef]
19. Knight, L.J.; Parker-Jurd, F.N.; Al-Sid-Cheikh, M.; Thompson, R.C. Tyre wear particles: An abundant yet widely unreported microplastic? *Environ. Sci. Pollut. Res.* **2020**, *27*, 18345–18354. [CrossRef] [PubMed]
20. Vogelsang, C.; Lusher, A.; Dadkhah, M.E.; Sundvor, I.; Umar, M.; Ranneklev, S.B.; Eidsvoll, D.; Meland, S. *Microplastics in Road Dust-Characteristics, Pathways and Measures*; NIVA-Rapport: Oslo, Norway, 2019.
21. Magni, S.; Binelli, A.; Pittura, L.; Avio, C.G.; Della Torre, C.; Parenti, C.C.; Gorbi, S.; Regoli, F. The fate of microplastics in an Italian Wastewater Treatment Plant. *Sci. Total Environ.* **2019**, *652*, 602–610. [CrossRef] [PubMed]
22. Binelli, A.; Pietrelli, L.; Di Vito, S.; Coscia, L.; Sighicelli, M.; Della Torre, C.; Parenti, C.C.; Magni, S. Hazard evaluation of plastic mixtures from four Italian subalpine great lakes on the basis of laboratory exposures of zebra mussels. *Sci. Total Environ.* **2020**, *699*, 134366. [CrossRef] [PubMed]
23. Magni, S.; Nigro, L.; Della Torre, C.; Binelli, A. Characterization of plastics and their ecotoxicological effects in the Lambro River (N. Italy). *J. Hazard. Mater.* **2021**, *412*, 125204. [CrossRef]

24. Binelli, A.; Della Torre, C.; Nigro, L.; Riccardi, N.; Magni, S. A realistic approach for the assessment of plastic contamination and its ecotoxicological consequences: A case study in the metropolitan city of Milan (N. Italy). *Sci. Total. Environ.* **2022**, *806*, 150574. [CrossRef]

25. Liu, F.; Olesen, K.B.; Borregaard, A.R.; Vollertsen, J. Microplastics in urban and highway stormwater retention ponds. *Sci. Total Environ.* **2019**, *671*, 992–1000. [CrossRef]

26. Baensch-Baltruschat, B.; Kocher, B.; Stock, F.; Reifferscheid, G. Tyre and road wear particles (TRWP)-A review of generation, properties, emissions, human health risk, ecotoxicity, and fate in the environment. *Sci. Total Environ.* **2020**, *733*, 137823. [CrossRef]

27. EASME. Recycling Rubber to Reduce Noise. 2015. Available online: https://ec.europa.eu/easme/en/news/recycling-rubber-reduce-noise (accessed on 3 February 2017).

28. Depaolini, A.R.; Bianchi, G.; Fornai, D.; Cardelli, A.; Badalassi, M.; Cardelli, C.; Davoli, E. Physical and chemical characterization of representative samples of recycled rubber from end-of-life tires. *Chemosphere* **2017**, *184*, 1320–1326. [CrossRef]

29. Grammelis, P.; Margaritis, N.; Dallas, P.; Rakopoulos, D.; Mavrias, G. Review on management of end of life tires (ELTs) and alternative uses of textile fibers. *Energies* **2021**, *14*, 571. [CrossRef]

30. Global ELT Management. *A Global State of Knowledge on Regulation, Management Systems, Impacts of Recovery and Technologies*; Global ELT Management, 2019.

31. Khan, F.R.; Halle, L.L.; Palmqvist, A. Acute and long-term toxicity of micronized car tire wear particles to *Hyalella azteca*. *Aquat. Toxicol.* **2019**, *213*, 105216. [CrossRef] [PubMed]

32. Redondo-Hasselerharm, P.E.; de Ruijter, V.N.; Mintenig, S.M.; Verschoor, A.; Koelmans, A.A. Ingestion and chronic effects of car tire tread particles on freshwater benthic macroinvertebrates. *Environ. Sci. Technol.* **2018**, *52*, 13986–13994. [CrossRef]

33. Day, K.E.; Holtze, K.E.; Metcalfe-Smith, J.L.; Bishop, C.T.; Dutka, B.J. Toxicity of leachate from automobile tires to aquatic biota. *Chemosphere* **1993**, *27*, 665–675. [CrossRef]

34. Stephensen, E.; Adolfsson-Erici, M.; Celander, M.; Hulander, M.; Parkkonen, J.; Hegelund, T.; Sturve, J.; Hasselberg, L.; Bengtsson, M.; Förlin, L. Biomarker responses and chemical analyses in fish indicate leakage of polycyclic aromatic hydrocarbons and other compounds from car tire rubber. *Environ. Toxicol. Chem.* **2003**, *22*, 2926–2931. [CrossRef] [PubMed]

35. Wik, A.; Dave, G. Environmental labeling of car tires—Toxicity to *Daphnia magna* can be used as a screening method. *Chemosphere* **2005**, *58*, 645–651. [CrossRef] [PubMed]

36. Wik, A.; Dave, G. Acute toxicity of leachates of tire wearmaterial to *Daphnia magna*—Variability and toxic components. *Chemosphere* **2006**, *64*, 1777–1784. [CrossRef] [PubMed]

37. Wik, A.; Nilsson, E.; Källqvist, T.; Tobiesen, A.; Dave, G. Toxicity assessment of sequential leachates of tire powder using a battery of toxicity tests and toxicity identification evaluations. *Chemosphere* **2009**, *77*, 922–927. [CrossRef]

38. Panko, J.M.; Kreider, M.L.; McAtee, B.L.; Marwood, C. Chronic toxicity of tire and road wear particles to water- and sediment-dwelling organisms. *Ecotoxicology* **2013**, *22*, 13–21. [CrossRef]

39. Marwood, C.; McAtee, B.; Kreider, M.; Ogle, R.S.; Finley, B.; Sweet, L.; Panko, J. Acute aquatic toxicity of tire and road wear particles to alga, daphnid, and fish. *Ecotoxicology* **2011**, *20*, 2079–2089. [CrossRef]

40. Cunningham, B.; Harper, B.; Brander, S.; Harper, S. Toxicity of Micro and Nano Tire Particles and Leachate for Model Freshwater Organisms. *J. Hazard. Mater.* **2022**, *429*, 128319. [CrossRef] [PubMed]

41. *UNI ISO 2859–1*; Procedimenti di Campionamento Nell'ispezione per Attributi—Parte 1: Schemi di Campionamento Indicizzati Secondo il Limite di Qualità Accettabile (AQL) Nelle Ispezioni Lotto per Lotto. ISO: Geneva, Switzerland, 2007.

42. *UNI EN 14243–2*; Materiali da Recupero di Pneumatici Fuori Uso—Parte 2: Granuli e Polverini—Metodi per Determinare la Distribuzione Delle Dimensioni Delle Particelle e Delle Impurità, Compresi il Contenuto di Ferro Libero e di Tessile Libero. ISO: Geneva, Switzerland, 2019.

43. *UNI ISO 11648–1*; Aspetti Statistici del Campionamento da Materiali Sfusi—Parte 1: Principi Generali. ISO: Geneva, Switzerland, 2008.

44. *UNI EN 15442*; Combustibili Solidi Secondari—Metodi di Campionamento. ISO: Geneva, Switzerland, 2011.

45. *ISO 3165*; Sampling of Chemical Products for Industrial Use—Safety in Sampling. ISO: Geneva, Switzerland, 1976.

46. Magni, S.; Bonasoro, F.; Della Torre, C.; Parenti, C.C.; Maggioni, D.; Binelli, A. Plastics and biodegradable plastics: Ecotoxicity comparison between polyvinylchloride and Mater-Bi® micro-debris in a freshwater biological model. *Sci. Total Environ.* **2020**, *720*, 137602. [CrossRef] [PubMed]

47. *OECD 201*; Freshwater Alga and Cyanobacteria, Growth Inhibition Test. OECD Guidelines for the Testing of Chemicals, Section 2. OECD Publishing: Paris, France, 2011.

48. *OECD 202*; Daphnia sp. Acute Immobilization Test. OECD Guidelines for the Testing of Chemicals, Section 2. OECD Publishing: Paris, France, 2004.

49. *OECD 203*; Fish, Acute Toxicity Test. OECD Guidelines for the Testing of Chemicals, Section 2. OECD Publishing: Paris, France, 2019.

50. *OECD 211*; Daphnia magna Reproduction Test. OECD Guidelines for the Testing of Chemicals, Section 2. OECD Publishing: Paris, France, 2012.

51. *OECD 23*; Guidance Document on Aquatic Toxicity Testing of Difficult Substances and Mixtures. OECD Series on Testing and Assessment. OECD Publishing: Paris, France, 2019.

52. *OECD 210*; Fish, Early-Life Stage Toxicity Test. OECD Publishing: Paris, France, 1992.

53. Banasiak, L.; Chiaro, G.; Palermo, A.; Granello, G. Environmental implications of the recycling of End-of-Life Tires in seismic isolation foundation systems. In *Advances in Sustainable Construction and Resource Management*; Springer: Singapore, 2021; Volume 144, pp. 43–52.
54. Degaffe, F.S.; Turner, A. Leaching of zinc from tire wear particles under simulated estuarine conditions. *Chemosphere* **2011**, *85*, 738–743. [CrossRef] [PubMed]
55. Capolupo, M.; Sørensen, L.; Jayasena, K.D.R.; Booth, A.M.; Fabbri, E. Chemical composition and ecotoxicity of plastic and car tire rubber leachates to aquatic organisms. *Water Res.* **2020**, *169*, 115270. [CrossRef] [PubMed]
56. Luo, Z.; Zhou, X.; Su, Y.; Wang, H.; Yu, R.; Zhou, S.; Xu, E.G.; Xing, B. Environmental occurrence, fate, impact, and potential solution of tire microplastics: Similarities and differences with tire wear particles. *Sci. Total Environ.* **2021**, *795*, 148902. [CrossRef] [PubMed]
57. Tian, Z.; Zhao, H.; Peter, K.T.; Gonzalez, M.; Wetzel, J.; Wu, C.; Hu, X.; Prat, J.; Mudrock, E.; Hettinger, R.; et al. A ubiquitous tire rubber-derived chemical induces acute mortality in coho salmon. *Science* **2021**, *371*, 185–189. [CrossRef]
58. Halsband, C.; Sørensen, L.; Booth, A.M.; Herzke, D. Car tire crumb rubber: Does leaching produce a toxic chemical cocktail in coastal marine systems? *Front. Environ. Sci.* **2020**, *8*, 125. [CrossRef]
59. Hartwell, S.I.; Jordahl, D.M.; Dawson, C.E.O. The effect of salinity on tire leachate toxicity. *Water Air Soil Pollut.* **2000**, *121*, 119–131. [CrossRef]
60. Gualtieri, M.; Andrioletti, M.; Vismara, C.; Milani, M.; Camatini, M. Toxicity of tire debris leachates. *Environ. Int.* **2005**, *31*, 723–730. [CrossRef]

Article

Acute and Chronic Toxicity of Binary Mixtures of Bisphenol A and Heavy Metals

Jun Yang [1,2], Anqi Liao [1], Shulin Hu [1,2], Yiwen Zheng [1,2], Shuli Liang [1,2], Shuangyan Han [1,2] and Ying Lin [1,2,*]

1   School of Biology and Biological Engineering, South China University of Technology, Guangzhou 510006, China; bijunyang@mail.scut.edu.cn (J.Y.); 202021049679@mail.scut.edu.cn (A.L.); 201921046830@mail.scut.edu.cn (S.H.); 201921047078@mail.scut.edu.cn (Y.Z.); shuli@scut.edu.cn (S.L.); syhan@scut.edu.cn (S.H.)

2   Guangdong Provincial Key Laboratory of Fermentation and Enzyme Engineering, South China University of Technology, Guangzhou 510006, China

*   Correspondence: feylin@scut.edu.cn; Tel.: +86-020-39380698

Abstract: Bisphenol A (BPA) and heavy metals are widespread contaminants in the environment. However, the combined toxicities of these contaminants are still unknown. In this study, the bioluminescent bacteria *Vibrio qinghaiensis* Q67 was used to detect the single and combined toxicities of BPA and heavy metals, then the joint effects of these contaminants were evaluated. The results show that chronic toxicities of chromium (Cr), cadmium (Cd), lead (Pb), arsenic (As), mercury (Hg), nickel (Ni), and BPA were time–dependent; in fact, the acute toxicities of these contaminants were stronger than the chronic toxicities. Furthermore, the combined toxicities of BPA and heavy metals displayed BPA + Hg > BPA + Cr > BPA + As > BPA + Ni > BPA + Pb > BPA + Cd in the acute test and BPA + Hg > BPA + Cd > BPA + As > BPA + Cd in the chronic test, which suggested that the combined toxicity of BPA and Hg was stronger than that of other mixtures in acute as well as chronic tests. Additionally, both CA and IA models underestimated the toxicities of mixtures at low concentrations but overestimated them at high concentrations, which indicates that CA and IA models were not suitable to predict the toxicities of mixtures of BPA and heavy metals. Moreover, the joint effects of BPA and heavy metals mainly showed antagonism and additive in the context of acute exposure but synergism and additive in the context of chronic exposure. Indeed, the difference in the joint effects on acute and chronic exposure can be explained by the possibility that mixtures inhibited cell growth and luminescence in chronic cultivation. The chronic toxicity of the mixture should be considered if the mixture results in the inhibition of the growth of cells.

Keywords: mixture toxicity; bisphenol A; heavy metal; joint effect; *Vibrio qinghaiensis* Q67

Citation: Yang, J.; Liao, A.; Hu, S.; Zheng, Y.; Liang, S.; Han, S.; Lin, Y. Acute and Chronic Toxicity of Binary Mixtures of Bisphenol A and Heavy Metals. *Toxics* **2022**, *10*, 255. https://doi.org/10.3390/toxics10050255

Academic Editors: Stefano Magni, Valerio Matozzo and François Gagné

Received: 12 April 2022
Accepted: 15 May 2022
Published: 17 May 2022

Publisher's Note: MDPI stays neutral with regard to jurisdictional claims in published maps and institutional affiliations.

1. Introduction

BPA is an important industrial chemical that has been widely used as the monomer of polycarbonate plastics and epoxy resins. Materials containing BPA appear in our daily life, especially in food contact materials, such as plastic bottles, cups, plates, goblets, and storage containers [1,2]. Heavy metals, which are natural substances in the Earth's crust, are widely spread in environment and foods [3]. The increasing demand for BPA has resulted in its accumulation in the environment [4–6]. BPA has been detected in packaged foods, drinking water, dust, sewage sludge, urine, serum, etc. [2,7]. A mean value of 8.99 ng/g BPA was tested in 45 vegetable samples [8]. Previous studies have shown that the concentration of BPA reached 174.6 ug/mL in a river sample [9] and 0.98 mg/L BPA was measured in drinking water [10]. Significantly, the concentration of BPA ranged from 0.1 to 3.9 ng/mL in 3021 human milk samples in 50 studies [11]. Therefore, the pollution of BPA has become serious. Furthermore, BPA and heavy metals widely co–exist in the environment and in foods. The toxicities of their mixtures must therefore be investigated.

The single toxicities of BPA and heavy metals have been well studied. BPA is not only well–known as an endocrine disrupter, but it also exhibits genotoxic activity [12] and reproductive [13,14], nervous [15,16], developmental [17,18], and immune toxicity [19,20]. Heavy metals can accumulate in the human body because of their long half–life, which results in diseases in target organs, such as the brain, kidney, and liver. [3]. Considering the long experimental time and poor repeatability for animal tests, especially the 3R principles regarding the ethical use of animals [21–23], a rapid and low–cost bioassay based on bioluminescent bacteria has been developed. Bioluminescent bacteria can rapidly respond to contaminants within a short time, and the light emission decreases with the increase in the concentration of contaminants [24,25]. To date, the acute toxicity of BPA evaluated by bioluminescent bacteria is lacking; however, the toxicities of heavy metals to bioluminescent bacteria are well studied in the environmental field. Most natural bioluminescent bacteria are isolated from the ocean, including four genera, namely *Photobacterium*, *Photorhabdus*, *Shewanella*, and *Vibrio* [25,26]. *Photobacterium phosphoreum* T3, *Vibrio fischeri*, and *V. qinghaiensis* Q67 are widely used to detect the acute toxicity of contaminants [25,27]. The acute toxicities of heavy metals, such as chromium, cadmium, copper, mercury, lead, and zinc were tested using *V. fischeri*, *V. qinghaiensis* Q67, and *P. phosphoreum* [28–30]. Indeed, it is known that high concentrations of NaCl (2–3%, w/v) impact the toxicities of contaminants, especially heavy metals [31–33]. Therefore, Q67 is suitable to measure the toxicities of heavy metals because it is the only strain isolated from freshwater so far. Thus, a lower concentration of NaCl (0.85%, w/v) needs to be added to samples [28].

In recent decades, binary- and ternary-mixture toxicities regarding BPA and other chemicals have been studied. However, the toxicities of mixtures of BPA and other chemicals have mainly focused on BPA and other endocrine disrupters (Eds) [16,34,35]. Studies regarding the combined toxicities of BPA and its analogues are increasing because BPA was replaced by its analogues due to its adverse effect [2,36–38]. In addition, the toxicities of mixtures of BPA and TiO$_2$ [15,17,39], Pb [16], and Cd [40] have been investigated using different types of wildlife; however, no clear synergism, additive, or antagonism effects were obtained in these works. The mixture toxicities of BPA and phthalates were studied using different organisms [41–44]. The joint effect of BPA and dibutyl phthalate was investigated, and the results showed that the co–exposure of these contaminants resulted in increased cytotoxicity, oxidative stress, and genotoxicity [40]. The mixtures of BPA and diethylhexyl phthalate (DEHP) and dibutyl phthalate (DBP) when exposed to human amniotic fluid reduced INSL3/RXFP2 signaling, which revealed an antagonist effect [45]. The antagonism effect of BPA and DEHP was observed in juvenile rats [46]. In addition, the joint effects of BPA and its analogues BPS, BPAF, BPB, and BPF were assessed by a highly sensitive micro–biosensor, and additive and synergistic effects were observed [47]. It is worth noting that *V. fischeri* was used to assess the joint effects of bisphenols and plasticizers and pharmaceuticals, and synergistic and antagonistic effects were observed [37]. Nonetheless, studies regarding the toxicities of mixtures of BPA and heavy metals are lacking, and their joint effects are still unknown. Bioluminescent bacteria have rarely been used to measure toxicities of mixtures of BPA and other contaminants until now. The rapid and low–cost approach based on bioluminescent bacteria should be used to estimate the joint effects of BPA and heavy metals in the environmental field.

In this study, the toxicities of binary mixtures of BPA and heavy metals were evaluated using bioluminescent bacteria *V. qinghaiensis* Q67. Further, the conventional models, CA and IA, were used to predict the toxicities of the mixtures. Moreover, the joint effects of BPA and heavy metals were assessed using the toxicity–unite method. This work proposes the application of Q67 for the detection of the toxicities of mixtures of BPA and other contaminants. Q67, which was isolated from Qinghai Lake in China, was the only strain isolated from the freshwater, and it might be the appropriate model from the aquatic system. The toxicity evaluated by Q67 might indicate the real toxicity of samples, which would contribute to ecotoxicology and risk assessment.

2. Materials and Methods

2.1. Bacterial Strains and Contaminants

The bioluminescent bacteria used in this work was *V. qinghaiensis* Q67. The freeze–dried Q67, used for the detection of the acute toxicity alone and in the mixture, was prepared according to a previous study [48]. The contaminants used in this study were bisphenol A (BPA), $K_2Cr_2O_7$, $Cd(NO_3)_2$, $ZnSO_4 \cdot 7H_2O$, $Pb(NO_3)_2$, arsenic (As), mercury (Hg), and nickel (Ni). BPA (GC, purity > 99%) was purchased from Macklin (Shanghai, China). All the other chemicals were purchased from Aladdin (Shanghai, China). All the concentrations of contaminants used in this work are listed in Table S1. It is worth mentioning that an aqueous BPA solution was prepared and compared with BPA dissolved in methanol. BPA was dissolved in deionized water at 75 °C for 2 h; then, the quantities of BPA dissolved in water and methanol at the same concentrations were compared using the e2695–series high–performance liquid chromatography (HPLC) system (Waters, Milford, MA, USA) [49]. Briefly, the column was ZORBAX Eclipse XDB–C18 (Agilent Technology, Santa Clara, CA, USA), the liquid phase was 60% (*v/v*) acetonitrile, the injection volume of the sample was 20 μL, the temperature was kept at 30 °C, and the detection wavelength was 276 nm.

2.2. Culture Medium and Culture Conditions

The fresh cultured *V. qinghaiensis* Q67 was used for the measurement of chronic toxicity. The medium for the Q67 contained tryptone (5 g/L), yeast extract (5 g/L), $MgCl_2$ (3.2 g/L), KBr (0.2 g/L), CaSO4 (0.1 g/L), KCl (4 g/L), NaCl (4 g/L), and glycerol (3 mL/L), with the pH adjusted to 8.5 [50]. *V. qinghaiensis* Q67 was cultured at 20 °C for 12 h with 180 rpm shaking; then, the harvested fresh culture was used to test the chronic toxicity.

2.3. Single Toxicity

The detection of acute and chronic single toxicity was performed according to the method used in a previous study [51,52]. For the test of the acute toxicity, the freeze–dried Q67 was resuscitated in a 0.85% NaCl solution for 15 min; then, 10 μL of resuscitated Q67 was transferred into a 150 μL contaminant solution. The contaminant solution was diluted using 0.85% NaCl over a range of concentrations. In terms of the chronic toxicity, 90 μL contaminant solution and 90 μL of 2–fold culture medium were first mixed; then, 20 μL of the fresh culture of Q67 was transferred into the last mixture. Additionally, the mixture was oscillated uniformly and incubated at 23 °C for 6 or 12 h. Finally, the bioluminescence of the mixtures was recorded by a SYNERGY H1 microplate reader (BioTek, Winooski, VT, USA). The arrangement of control and samples in a 96–well plate are shown in Figure S1. The inhibition ratio of the luminescence was calculated according to Equation (1). The concentration–inhibition data were fitted by the logistic (Equation (2)) and the dose–response model (Equation (3)) [52,53].

$$I = \frac{L_C - L_S}{L_C} \times 100\% \tag{1}$$

where I is the inhibition ratio of luminescence. L_C and L_S indicate the relative luminescence unit of control and contaminants solution, respectively.

$$I = A_2 + \frac{A_1 - A_2}{1 + (C/C_0)^P} \tag{2}$$

$$I = A_2 + \frac{A_2 - A_1}{1 + 10^{(logC_0 - C)P}} \tag{3}$$

where A_1 and A_2 are the bottom and top values, respectively, of the inhibition ratio. C is the concentration of tested samples, and C_0 indicates the value of C at 50% of the inhibition ratio of luminescence. P is the parameter of slope for the concentration–inhibition curve.

2.4. Mixture Toxicity

The toxicities of mixtures of BPA and heavy metals were analyzed. The test of the toxicity of the binary mixture was carried out using the same method as that for single toxicity. Based on the EC_{50} values of individual contaminants, the mixture toxicities of BPA and Cr, Pb, Hg, and Ni at equitoxic ratios were detected, while BPA and Cd and As at non–equitoxic ratios were prepared at $1:10^{-1.5}$, $1:10^{-1}$, $1:10^{-0.5}$, $1:1$, $10^{0.5}:1$, $10^{1}:1$, and $10^{1.5}:1$; then, the mixture toxicities were tested. The concentrations of individual contaminants in stocked mixtures are listed in Table S2. Stocked mixtures were diluted for further tests. The toxic unit of the mixture (TU_{mix}) was used to evaluate the joint effects of mixtures, which were calculated using Equation (4). According to a previous study, $0.8 < TU < 1.2$ was considered simple additivity, while $TU < 0.8$ revealed synergism; additionally, $TU > 1.2$ indicated antagonism [54,55].

$$TU_{mix} = \frac{C_A}{EC_{50A}} + \frac{C_B}{EC_{50B}} \tag{4}$$

where C_A and C_B are the individual concentrations of A and B, respectively, in a mixture that inhibits 50% of luminescence. EC_{50A} and EC_{50B} are effective concentrations of individual contaminants at 50% inhibition according to single toxicity.

2.5. Prediction of Mixture Toxicity

The quantitative structure–activity relationship (QSAR), independent action (IA), concentration addition (CA), and integrative models of the CA and IA such as two–stage prediction (TSP) are widely used models to predict the toxicity of mixtures. According to the different toxic modes of action, two conventional models are used to predict the joint effect of mixtures. One is independent action (IA, different mode of action, MOA), and the other is concentration addition (CA, similar to MOA) [56]. In this study, BPA and heavy metals indicate different toxic action modes, and both CA and IA were used to predict the mixture of toxicities. The mathematical functions of CA and IA are expressed as Equations (5) and (6), respectively:

$$EC_{x,m} = \left(\frac{P_A}{EC_{x,A}} + \frac{P_B}{EC_{x,B}} \right)^{-1} \tag{5}$$

where $EC_{x,m}$, $EC_{x,A}$, and $EC_{x,B}$ are the concentrations of the mixtures; A and B are the decreases in the $x\%$ luminescence, respectively; and P_A and P_B correspond the concentration ratios of A and B in the mixture.

$$E(C_M) = 1 - (1 - E(C_A)) \times (1 - E(C_B)) \tag{6}$$

where $E(C_M)$ is the toxic effect of the mixture, and $E(C_A)$ and $E(C_B)$ are the toxic effects of A and B in the individual toxicity tests, respectively.

2.6. Statistical Analysis

The statistical analysis was performed using Origin 8.5 and IBM SPSS 26.0 software. The difference between the acute and the chronic toxicity was analyzed, and $p < 0.05$ was considered statistically significant. The concentration–inhibition data were fitted using Origin 8.5, and the qualities of developed models were evaluated by the number of points, degrees of freedom, reduced chi–sqr, residual sum of squares, and R–squared (R^2).

3. Results

3.1. Single Toxicity of BPA and Heavy Metals

Before the examination of the binary–mixture toxicities of BPA and heavy metals, the single toxicities of BPA, Cr, Cd, Pb, As, Hg, and Ni to Q67 were tested (Figure 1). Considering that the acute toxicity of the aqueous BPA solution toward bioluminescent bacteria was detected for the first time, the quantities of solutions of aqueous BPA and

BPA dissolved in methanol at the same concentrations were tested using HPLC and then compared. The results showed that the aqueous BPA and methanol dissolved BPA solutions exhibited similar retention times and peak areas (Figure S2). In terms of toxicity detection, most of these concentration–inhibition data could be fitted by logistic and dose–response models. However, the concentration–inhibition data of Hg were fitted better than others by the bi–dose–response model. These models suggested good quality, and the values of R^2 were over 0.99 for all fitting models. The reduced chi–sqr values ranged from 1.57 to 30.85. The EC_{50} value of each contaminant was calculated by the fitting function, which indicated the toxicity of each contaminant.

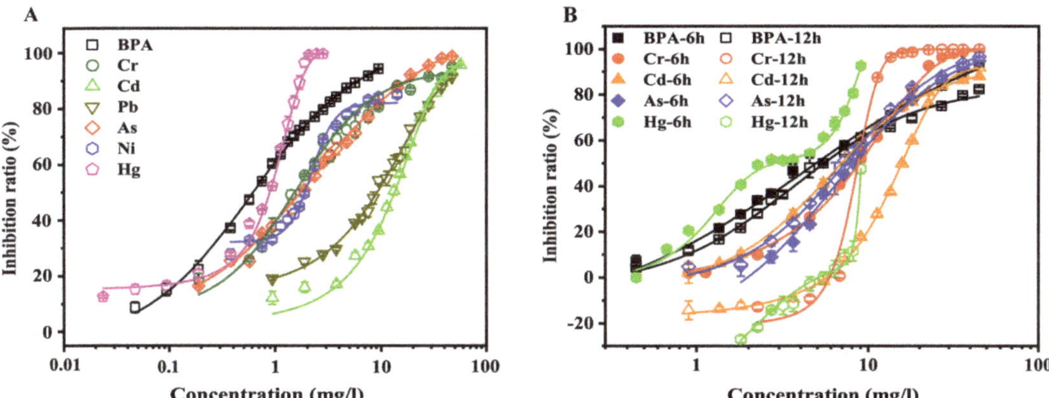

Figure 1. The acute (**A**) and chronic (**B**) single toxicity of BPA and Cr, Cd, Pb, As, Hg, and Ni. The concentration–inhibition data were fitted using the logistic and dose–response models. The error bars indicate the standard deviations from three independent experiments.

For the acute toxicity (Figure 1A), BPA and Cd displayed the strongest and weakest toxicity to Q67, respectively. The EC_{50} values of these contaminants ranged from 0.62 to 13.27 mg/L. The toxicities of these contaminants were in the following order: BPA > Hg > Cr > As > Ni > Pb > Cd. In the case of chronic toxicity, Pb and Ni revealed no obvious toxicity effects to Q67 in the range of the experimental concentration (data not shown). In addition, the chronic toxicity of these contaminants suggested a time–dependent nature (Figure 1B). The chronic toxicities of BPA, Cr, Cd, and Hg decreased with the increase in exposure time. All the concentration–inhibition data fit well with the logistic model for 6 h exposure but not for 12 h exposure; thus, the exposure time for chronic exposure was 6 h in other studies. The EC_{50} values of these contaminants (6 h exposure time) ranged from 2.81 to 8.79 mg/L. The chronic toxicities of these contaminants were as follows: Hg > BPA > Cd > As > Cr. The difference in the acute and chronic toxicities of these contaminants was significant ($p < 0.05$).

3.2. Toxicities of Binary–Mixtures of BPA and Heavy Metals

Based on the single toxicities of these contaminants, the toxicities of binary mixtures of BPA and heavy metals were analyzed at equitoxic and non–equitoxic ratios. Concentration–inhibition data were fitted using same mathematic models as the single toxicity test. For all models developed in this work, the values of R^2 (all over 0.99) and reduced chi–sqr (ranged from 0.70 to 15.86) indicated the good fits. In terms of acute toxicity of the mixtures, the EC_{50} of these binary mixtures revealed that their toxicities were as follows: BPA + Hg > BPA + Cr > BPA + As > BPA + Ni > BPA + Pb > BPA + Cd at equitoxic ratios (Figure 2A), which was the same as the single toxicities of heavy metals. In order to validate the toxicities of mixtures of BPA + Cd and BPA + As, the toxicities of mixtures of BPA + Cd and BPA + As at non–equitoxic were measured (Figures S3 and S4). The EC_{50}

values of BPA + Cd ranged from 0.91 to 17.42 mg/L. Indeed, the toxicities of mixtures of BPA + Cd decreased with the increase in the ratio of lg(BPA/Cd) (Figure 2B), and the same conclusion was obtained from BPA + As (Figure 2C). These results indicate that the toxicities of mixtures of BPA + Cd and BPA + As were affected by the ratio of Cd and As in the mixture, respectively.

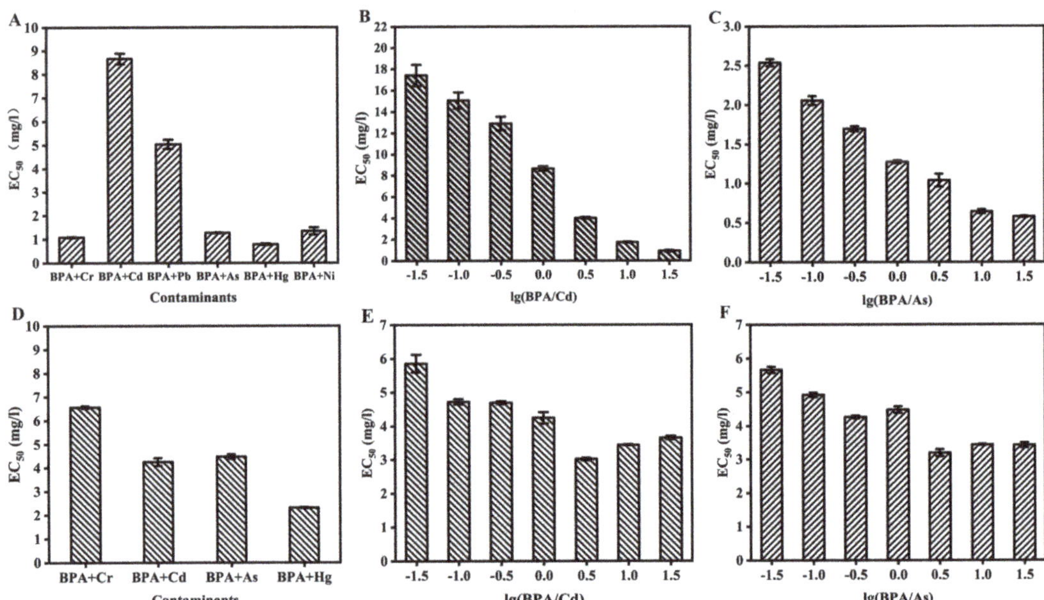

Figure 2. Toxicities of mixtures of BPA and heavy metals. (**A**) BPA and heavy metals at equitoxic ratios in acute toxicity tests; (**B,C**) BPA and Cd, BPA, and As at non–equitoxic ratios in the acute toxicity test, respectively; (**D**) BPA and heavy metals at equitoxic ratios in the chronic toxicity test; (**E,F**) BPA and Cd, BPA, and As at non–equitoxic ratios in the chronic toxicity test, respectively. The error bars indicate the standard deviations from three independent experiments.

In the case of chronic mixture toxicity, the toxicities of these mixtures were shown to be BPA + Hg > BPA + Cd > BPA + As > BPA + Cr at equitoxic ratios (Figures 2D and S5), which is the same as the single toxicities of Hg, Cd, As, and Cr. Furthermore, toxicities of mixtures of BPA + Cd and BPA + As at non–equitoxic ratios were measured (Figures S6 and S7). The EC_{50} values of BPA + Cd at non–equitoxic ratios ranged from 3.02 to 5.86 mg/L. When the ratio of lg(BPA/Cd) was under 0, toxicities of mixtures of BPA + Cd decreased with the ratio of lg(BPA/Cd) increasing, while it increased with the increase in the ratio of lg(BPA/Cd) when the ratio of lg(BPA/Cd) was over 0 (Figure 2E); the same conclusion could be obtained from BPA + As (Figure 2F). The differences between the acute and chronic toxicities of mixtures of BPA + Cr, BPA + Cd, BPA + As, and BPA + Hg were statistically significant at equitoxic ratios as well as non–equitoxic ratios ($p < 0.05$).

3.3. Toxicities of Mixtures of BPA and Heavy Metals Predicted by CA and IA Models

According to previous studies, the IA model was usually used to predict the mixture toxicity of a mixture that exhibited dissimilar MOA. BPA and heavy metals seemed to display dissimilar MOAs, which should be validated. Therefore, both CA and IA models were used to predict the toxicities of these mixtures. In this study, the acute toxicities of mixtures of BPA and heavy metals were measured at equitoxic ratios, and then the observed experimental toxicities were compared with the values predicted by the CA and IA models (Figure 3). The concentration–inhibition data were fitted using logistic and dose–response models, and the quality of all developed models was good ($R^2 > 0.99$). Due to the slow increase in the single toxicity of BPA with the increase in concentration, the mixture toxicities of BPA and heavy metals showed a similar relationship. However, the mixture of BPA + Hg revealed stronger toxicity than the other mixtures (Figure 3E). The mixture toxicity of BPA and Hg increased rapidly with the increase in concentration. In addition, the results showed that both CA and IA models underestimated the toxicity for all binary mixtures at low concentrations but overestimated the toxicity at high concentrations of the mixtures. Indeed, the value predicted by the IA model was lower than that predicted by the CA model, which was better close to the observed value. However, the IA model overestimated the toxicity of the mixture, and hence neither the CA nor the IA model was suitable to predict the toxicity of mixtures of BPA and heavy metals.

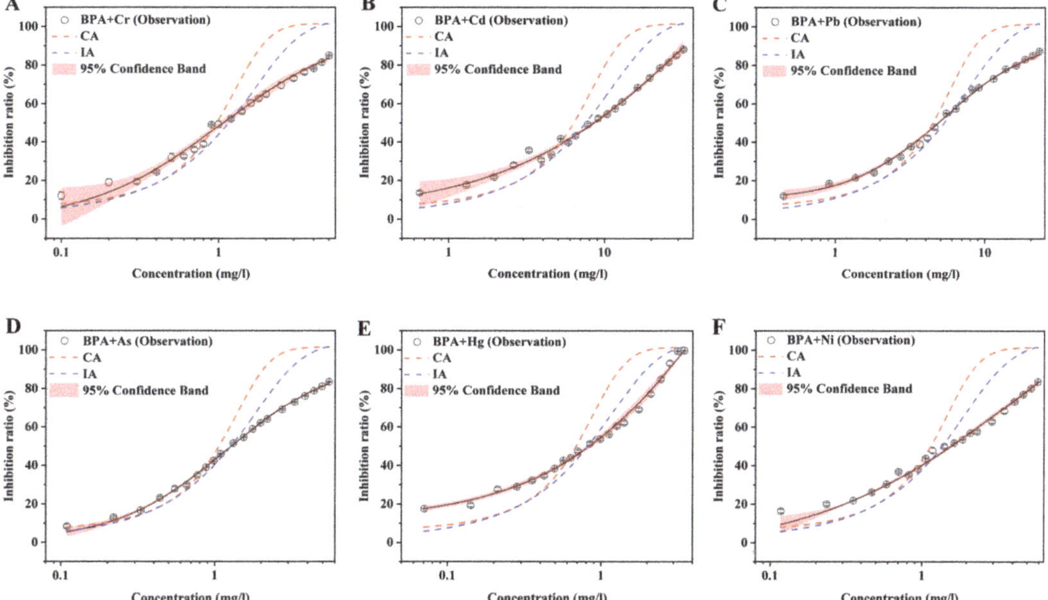

Figure 3. Acute toxicity of mixtures of BPA and Cr (**A**), Cd (**B**), Pb (**C**), As (**D**), Hg (**E**), and Ni (**F**) compared with the predicted value by the CA and IA models. The black circles are the experimental mean values of the toxicity of mixtures of BPA and heavy metals, and the black solid line is the fitting curve using the mathematical mode. The red shadow indicates the 95% confidence band. The red and blue dotted lines suggest the values predicted by the CA and IA models, respectively. The error bars indicate the standard deviations from three independent experiments.

3.4. Joint Effects of BPA and Heavy Metals

According to the toxicities of mixtures of BPA and heavy metals, the EC_{50} values of mixtures (EC_{50m}) were calculated using Equations (2) and (3). Based on the EC_{50} values of the mixtures, the TUs of mixtures were obtained using Equation (4). In the case of the acute toxicity of the mixtures, the joint effects of BPA and heavy metals exhibited simple additive and antagonisms, and the TU values of mixtures ranged from 1.06 to 1.33 (Figure 4A). BPA and Cd showed an antagonistic effect at the equitoxic ratio (the TU of the mixture was 1.33), whereas simple additive effects were observed among BPA and other heavy metals at the equitoxic ratio. In addition, according to the EC_{50} values of BPA + Cd and BPA + As at a non–equitoxic ratio, the joint effects of BPA + Cd and BPA + As exhibited antagonisms and simple additives (Figure 4B,C). The TU values of BPA + Cd at a non–equitoxic ratio ranged from 0.97 to 1.38, and the values decreased with the increase in the ratio of BPA in mixtures.

The same conclusion could be drawn for BPA + As, and the range of EC$_{50}$ values was from 0.94 to 1.59.

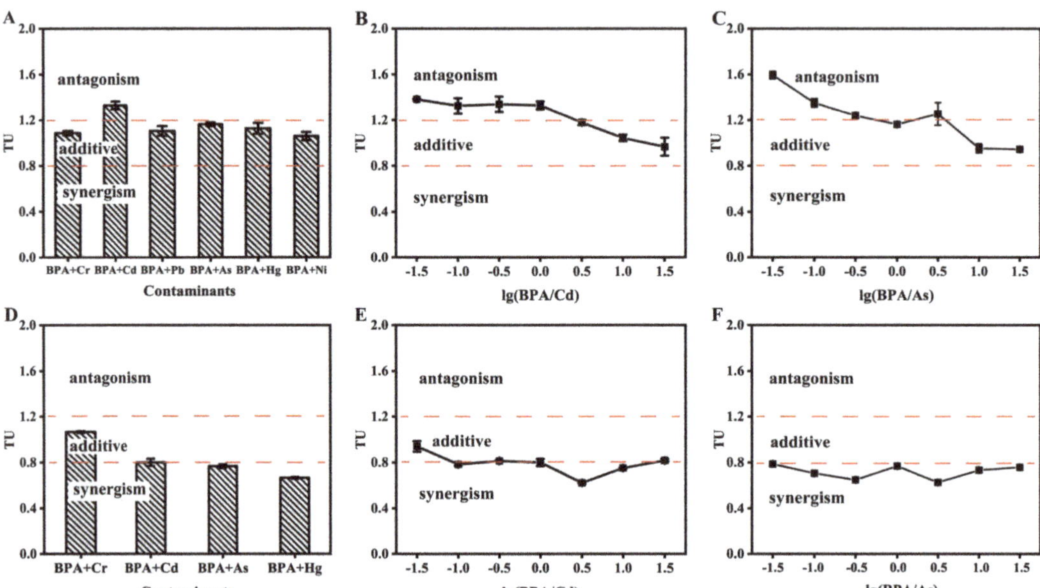

Figure 4. Joint effect of BPA and heavy metals. (**A–C**) BPA and heavy metals at equitoxic or non–equitoxic ratios in the acute toxicity test; (**D–F**) BPA and heavy metals at equitoxic or non–equitoxic ratios in a chronic toxicity test. The error bars indicate the standard deviations from three independent experiments.

In the case of the chronic toxicities of mixtures, the joint effects of BPA and heavy metals indicated a synergistic and additive effect (Figure 4D). The TU values of BPA + Cr, BPA + Cd, BPA + As, and BPA + Hg were 1.07, 0.80, 0.77, and 0.67, respectively, at an equitoxic ratio. The joint effect of BPA and Cr suggested an additive effect, while the others revealed a synergism. Meanwhile, the joint effects of BPA + Cd and BPA + As suggested synergistic and additive effects at non–equitoxic ratios. The TU values of BPA + Cd ranged from 0.62 to 0.94, while the TU values of BPA + As were shown to range from 0.63 to 0.79. These results suggest that the joint effects of BPA and heavy metals between acute and chronic exposure were different. The joint effects of BPA and heavy metals on acute exposure were shown to be antagonisms and additives, whereas additive effects and synergism were obtained with chronic exposure.

3.5. Mechanism of the Synergistic Effects of BPA and Heavy Metals on V. qinghaiensis Q67

In terms of the joint effects, BPA and heavy metals exhibited different effects between the acute and the chronic test. For example, the joint effects of BPA and Cd showed antagonistic and additive effects for the acute test at non–equitoxic ratios but synergistic and additive effects for the chronic test. To investigate whether the synergistic effect was due to the inhibition of growth and luminescence, the impact of the contaminants on cell growth and luminescence was tested. The concentrations of mixtures corresponded to the EC$_{50}$ values from the test of the mixtures' toxicities, and the concentrations of BPA and heavy metals corresponded to their individual concentrations in mixtures. The final concentrations of BPA + Cr, BPA + Cd, BPA + As, and BPA + Hg in mixtures were 6.57, 4.25, 4.48, and 2.31 mg/L, respectively. The results showed that BPA barely inhibited cell growth, but heavy metals significantly inhibited that. All the binary mixtures inhibited the growth of Q67 (Figure 5). Meanwhile, the luminescence inhibition ratio of mixtures was close to 50% after 6 h exposure but below 50% at earlier cultivation times (0.25, 2, and 4 h) except for BPA + Cr. The luminescence inhibitions of BPA and Cr were constantly close to 50% throughout the exposure time. In other words, the toxicities of mixtures of BPA and Cd, As, and Hg were shown to be time–dependent but BPA + Cr was not. It is worth mentioning that the joint effect of BPA and Cr displayed simple additive effects not only for 15 min exposure in the acute test but also for 15 min, and 2, 4, and 6 h exposure in the chronic test.

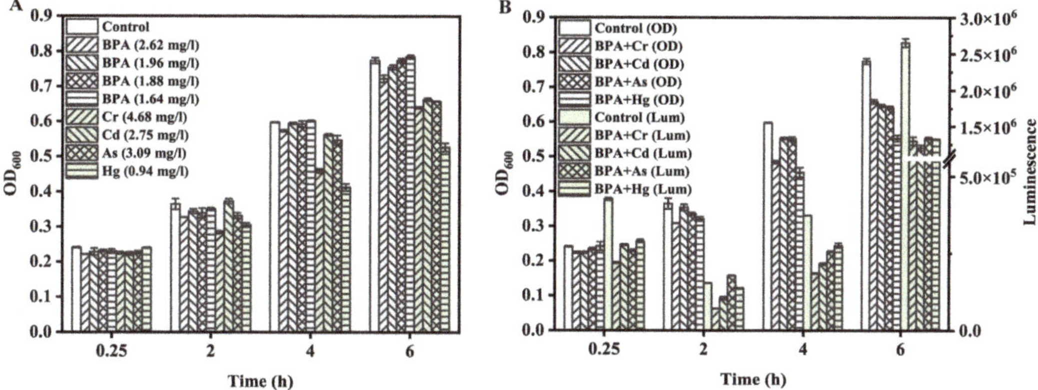

Figure 5. The impact of the mixtures of BPA and heavy metals (equitoxic ratio) on the growth and luminescence of *V. qinghaiensis* Q67. The concentrations of mixtures used in this work are the EC_{50} values of the mixtures in the test of the chronic toxicity of the mixtures, and the concentrations of BPA and heavy metals correspond to their individual concentrations in mixtures. The error bars indicate the standard deviations from three independent experiments. Impact of the individual contaminant on growth (**A**). The impact of mixtures on the growth and luminescence (**B**).

4. Discussion

BPA and heavy metals are widely spread in the environment. BPA is well known as an endocrine disrupter that appears in daily human life, even in human milk [11]. This study indicates that BPA exhibits a stronger acute toxicity than heavy metals. Studies regarding the toxicity of mixtures of BPA and other contaminants have mainly focused on BPA and its analogues, EDs. However, works on the toxicities of mixtures of BPA and heavy metals are lacking. The co–exposure of BPA and heavy metals in the environment should be further studied. In this study, the toxicities of binary mixtures of BPA and heavy metals were investigated using bioluminescent bacteria *V. qinghaiensis* Q67. The results indicated that BPA and heavy metals displayed an antagonistic and an additive effect in the context of acute exposure but synergistic and additive joint effects in the contexts of chronic exposure. Compared with other bioluminescent bacteria, such as *V. fischeri* and *P. phosphoreum* T3, Q67 is the only strain isolated from freshwater so far. Thus, the low concentration of NaCl added to samples decreased the impact of NaCl on the toxicities of heavy metals [31–33]. Q67 is therefore a suitable bioluminescent bacterial strain to detect the mixture toxicities of BPA and heavy metals. The results from bioluminescent bacteria might not be repeated in animal tests, but they could be considered a rapid screening approach used in the environmental field, especially in aqueous samples.

Studies regarding the mixture toxicities of BPA and heavy metals have been conducted in the past. The prenatal co–exposure of BPA and Pb has been associated with neurotoxicity, but the interaction of BPA and Pb should be further studied [16]. In addition, the toxicity of mixtures of BPA and Cd was investigated in HepG 2 cells, and the results showed that the co–exposure of BPA and Cd enhanced the cytotoxicity, oxidative stress, and genotoxicity, indicating an additive and synergistic effect [40]. The additive and synergistic effects of

BPA and Cd could be also obtained from a chronic test in this study. However, the acute joint effects of BPA and Cd were shown to be the antagonism and additive in this work. The chronic joint effect of BPA and Cd toward Q67 in this study was similar to a previous study. In fact, the detection of the toxicity of mixture contaminants to higher organisms and cells was often exhibited in the long term, which was similar to the chronic toxicity test of contaminants toward Q67. Otherwise, there are some similar studies that should be compared with this work. In one study, toxicities of mixtures of phthalate esters and Cd to *V. qinghaiensis* Q67 were investigated, and the results showed that mixtures of phthalate esters and Cd showed additive effects [57]. The EC_{50} value of Cd in that study (18.8 mg/L) was higher than in our study (13.274 mg/L). Further, the synergistic and antagonistic effects of the ternary mixtures of BPA, diethyl phthalate (DEP), and diglycidyl ether (BADGE) to *V. fischeri* were confirmed; however, BPA and other chemicals were dissolved in methanol in that study [37]. Indeed, methanol can affect the luminescence of bioluminescent bacteria. Therefore, the aqueous BPA solution was applicable for acute toxicity detection using bioluminescent bacteria.

The quantitative structure–activity relationship (QSAR) model, CA and IA models, and integrated CA and IA models such as TSP (two–stage prediction) have been used to predict the toxicity of mixtures of contaminants [56]. The CA model is usually used to predict the mixtures with the same MOA but IA with different MOA. The toxicities of mixtures of Cd, Pb, and Mn were predicted well by CA models [58]. However, the observed toxicities of these mixtures were often lower or higher than the predicted value. This was illustrated by the toxicities of mixtures of zinc oxide nanoparticles and chemicals with different MOAs, and neither CA nor IA models were suitable to predict the toxicities of mixtures of these contaminants [52]. Therefore, the prediction of the toxicity of mixtures in silicon was significant but should combined with the actual observed toxicity in experiments.

The difference in the joint effect of acute and chronic exposure was investigated (Figure 5), and the difference can be explained by mixtures inhibiting the growth of bioluminescent bacteria in 6 h long–term exposure. The luminescence in samples containing mixtures of BPA and heavy metals decreased by 50% relative to the control after 6 h of incubation; however, the biomass of these samples was lower than the control. In other words, the stronger inhibition of luminescence (over 50% ratio) should appear when the biomass of the samples equals the control. Consequently, mixtures of BPA and heavy metals inhibiting both cell growth and luminescence resulted in an antagonistic effect in acute exposure, which converted to a synergistic effect in the context of chronic exposure. The ecotoxicity of BPA to different classes of creatures, such as annelids, mollusks, crustaceans, insects, fish, and amphibians was investigated in past decades [20]. Additionally, such toxicity was also illustrated by BPA inhibiting algal and cyanobacterial growth [59,60]. Thus, it is reasonable that BPA inhibits the growth of bioluminescent bacteria. The effect of heavy metals on different creatures has been studied. Hg, Cd, Cu, Zn, Pb, and Cr affect the survival of ciliates [61]. Meanwhile, the influence of metals on the populations of microorganisms has been illustrated in the context of long–term exposure [62]. In this study, similar results were obtained in the chronic toxicity test (6 h exposure) (Figure 5A). Indeed, the mixture also inhibited the growth of Q67 (Figure 5B). Therefore, the joint effect of BPA and heavy metals revealed a synergism in the context of chronic detection. However, the joint effect of BPA and Cr exhibited an additive effect in both an acute and chronic test. In addition, the toxicities of mixtures of BPA and Cd, As, and Hg were shown to be time–dependent, but BPA and Cr were not, in the context of chronic exposure. BPA and Cr might present the same MOA, but this needs to be validated by further studies.

5. Conclusions

The acute and chronic toxicities of BPA and heavy metals, individually and in mixtures, were determined using bioluminescent bacteria *V. qinghaiensis* Q67 isolated from freshwater. The results showed that single chronic toxicities of BPA and heavy metals were time–dependent, and the toxicities of these contaminants were shown to be stronger in acute exposure than in chronic exposure, except for Cd. Furthermore, neither CA nor IA models were suitable for the prediction of the mixture toxicities of BPA and heavy metals. Moreover, the acute joint effects of BPA and heavy metals mainly displayed antagonistic and additive effects, whereas the chronic joint effects of these mixtures mainly exhibited synergistic and additive effects. Indeed, the mixtures inhibiting both cell growth and luminescence resulted in antagonistic effects in the context of acute exposure, shifting to a synergistic effect in the context of chronic exposure.

Supplementary Materials: The following supporting information can be downloaded at: https://www.mdpi.com/article/10.3390/toxics10050255/s1, Figure S1: Arrangement of control and test samples in 96–well microplate. Gray solid circles are the control, and the blue solid circles are test samples. Each concentration of sample locates in four wells in same column; Figure S2: Comparison of aqueous BPA solution and methanol dissolved BPA using HPLC. Black line, methanol dissolved BPA solution. Red line, aqueous BPA solution; Figure S3: Observed acute mixture toxicity of BPA and Cd at non–equitoxic ratio; Figure S4: Observed acute mixture toxicity of BPA and As at non–equitoxic ratio; Figure S5: Observed chronic mixture toxicity of BPA and heavy metals at equitoxic ratio; Figure S6: Observed chronic mixture toxicity of BPA and Cd at non–equitoxic ratio; Figure S7: Observed chronic mixture toxicity of BPA and As at non–equitoxic ratio; Table S1: Concentration of individual contaminant used in single toxicity test; Table S2: Concentrations of individual contaminants in stocked mixtures.

Author Contributions: Conceptualization, J.Y. and Y.L.; methodology, J.Y., S.L., and Y.L.; software, J.Y.; validation, J.Y. and S.H. (Shulin Hu).; formal analysis, J.Y.; investigation, J.Y., A.L., S.H. (Shulin Hu), and Y.Z.; resources, J.Y. and Y.L.; data curation, J.Y., S.L., and Y.L.; writing—original draft preparation, J.Y.; writing—review and editing, J.Y., Y.L., and S.L.; visualization, J.Y.; supervision, Y.L. and S.L.; project administration, Y.L. and S.H. (Shuangyan Han); funding acquisition, Y.L. All authors have read and agreed to the published version of the manuscript.

Funding: This research was funded by the Key Areas Research Development Projects of Guangdong Province (Department of Science and Technology of Guangdong Provience: 2019B020210001).

Institutional Review Board Statement: Not applicable.

Informed Consent Statement: Not applicable.

Data Availability Statement: The dataset is available upon the request of the corresponding authors.

Acknowledgments: We thank senior scientist Lingyun Mo from the Guilin University of Technology for providing the bioluminescent bacteria *V. qinghaiensis* Q67. The first author, Jun Yang, thanks Song Fan for his financial assistance for this work.

Conflicts of Interest: The authors declare that they have no competing interests.

References

1. Geens, T.; Aerts, D.; Berthot, C.; Bourguignon, J.P.; Goeyens, L.; Lecomte, P.; Maghuin-Rogister, G.; Pironnet, A.M.; Pussemier, L.; Scippo, M.L.; et al. A review of dietary and non-dietary exposure to bisphenol-A. *Food Chem. Toxicol.* **2012**, *50*, 3725–3740. [CrossRef] [PubMed]
2. Caballero-Casero, N.; Lunar, L.; Rubio, S. Analytical methods for the determination of mixtures of bisphenols and derivatives in human and environmental exposure sources and biological fluids. A review. *Anal. Chim. Acta* **2016**, *908*, 22–53. [CrossRef] [PubMed]
3. Mansour, S.A. *Monitoring and Health Risk Assessment of Heavy Metal Contamination in Food*; John Wiley & Sons: Hoboken, NJ, USA, 2014; pp. 235–255.
4. Yang, M.; Fan, Z.Q.; Xie, Y.J.; Fang, L.; Wang, X.R.; Yuan, Y.; Li, R.X. Transcriptome analysis of the effect of bisphenol A exposure on the growth, photosynthetic activity and risk of microcystin-LR release by Microcystis aeruginosa. *J. Hazard. Mater.* **2020**, *397*, 122746. [CrossRef]
5. Chen, P.; Yang, J.; Xiao, B.; Zhang, Y.; Liu, S.; Zhu, L. Mechanisms for the impacts of graphene oxide on the developmental toxicity and endocrine disruption induced by bisphenol A on zebrafish larvae. *J. Hazard. Mater.* **2021**, *408*, 124867. [CrossRef]
6. Huang, Y.Q.; Wong, C.K.C.; Zheng, J.S.; Bouwman, H.; Barra, R.; Wahlstrom, B.; Neretin, L.; Wong, M.H. Bisphenol A (BPA) in China: A review of sources, environmental levels, and potential human health impacts. *Environ. Int.* **2012**, *42*, 91–99. [CrossRef]

7. Banaderakhshan, R.; Kemp, P.; Breul, L.; Steinbichl, P.; Hartmann, C.; Furhacker, M. Bisphenol A and its alternatives in Austrian thermal paper receipts, and the migration from reusable plastic drinking bottles into water and artificial saliva using UHPLC-MS/MS. *Chemosphere* **2022**, *286*, 131842. [CrossRef]
8. Liao, C.Y.; Kannan, K. Concentrations and Profiles of Bisphenol A and Other Bisphenol Analogues in Foodstuffs from the United States and Their Implications for Human Exposure. *J. Agric. Food Chem.* **2013**, *61*, 4655–4662. [CrossRef]
9. Ronderos-Lara, J.G.; Saldarriaga-Norena, H.; Murillo-Tovar, M.A.; Vergara-Sanchez, J. Optimization and Application of a GC-MS Method for the Determination of Endocrine Disruptor Compounds in Natural Water. *Separations* **2018**, *5*, 33. [CrossRef]
10. Parto, M.; Aazami, J.; Shamsi, Z.; Zamani, A.; Savabieasfahani, M. Determination of bisphenol-A in plastic bottled water in markets of Zanjan, Iran. *Int. J. Environ. Sci. Technol.* **2021**, *19*, 3337–3344. [CrossRef]
11. Iribarne-Duran, L.M.; Peinado, F.M.; Freire, C.; Castillero-Rosales, I.; Artacho-Cordon, F.; Olea, N. Concentrations of bisphenols, parabens, and benzophenones in human breast milk: A systematic review and meta-analysis. *Sci. Total Environ.* **2022**, *806*, 150437. [CrossRef]
12. Balabanic, D.; Filipic, M.; Klemencic, A.K.; Zegura, B. Genotoxic activity of endocrine disrupting compounds commonly present in paper mill effluents. *Sci. Total Environ.* **2021**, *794*, 148489. [CrossRef] [PubMed]
13. Baralic, K.; Jorgovanovic, D.; Zivancevic, K.; Djordjevic, A.B.; Miljakovic, E.A.; Miljkovic, M.; Kotur-Stevuljevic, J.; Antonijevic, B.; Dukic-Cosic, D. Combining in vivo pathohistological and redox status analysis with in silico toxicogenomic study to explore the phthalates and bisphenol A mixture-induced testicular toxicity. *Chemosphere* **2021**, *267*, 129296. [CrossRef] [PubMed]
14. Dagher, J.B.; Hahn-Townsend, C.K.; Kaimal, A.; Al Mansi, M.; Henriquez, J.E.; Tran, D.G.; Laurent, C.R.; Bacak, C.J.; Buechter, H.E.; Cambric, C.; et al. Independent and combined effects of Bisphenol A and Diethylhexyl Phthalate on gestational outcomes and offspring development in Sprague-Dawley rats. *Chemosphere* **2021**, *263*, 128307. [CrossRef] [PubMed]
15. Fu, J.J.; Guo, Y.Y.; Yang, L.H.; Han, J.; Zhou, B.S. Nano-TiO_2 enhanced bioaccumulation and developmental neurotoxicity of bisphenol a in zebrafish larvae. *Environ. Res.* **2020**, *187*, 109682. [CrossRef] [PubMed]
16. Guo, J.Q.; Wu, C.H.; Zhang, J.M.; Qi, X.J.; Lv, S.L.; Jiang, S.; Zhou, T.; Lu, D.S.; Feng, C.; Chang, X.L.; et al. Prenatal exposure to mixture of heavy metals, pesticides and phenols and IQ in children at 7 years of age: The SMBCS study. *Environ. Int.* **2020**, *139*, 105692. [CrossRef]
17. Chen, L.G.; Hu, C.Y.; Guo, Y.Y.; Shi, Q.P.; Zhou, B.S. TiO_2 nanoparticles and BPA are combined to impair the development of offspring zebrafish after parental coexposure. *Chemosphere* **2019**, *217*, 732–741. [CrossRef]
18. Zhou, R.; Xia, M.; Zhang, L.; Cheng, W.; Yan, J.; Sun, Y.; Wang, Y.; Jiang, H. Individual and combined effects of BPA, BPS and BPAF on the cardiomyocyte differentiation of embryonic stem cells. *Ecotoxicol. Environ. Saf.* **2021**, *220*, 112366. [CrossRef]
19. Gowder, S.J. Nephrotoxicity of bisphenol A (BPA)—An updated review. *Curr. Mol. Pharmacol.* **2013**, *6*, 163–172. [CrossRef]
20. Oehlmann, J.; Schulte-Oehlmann, U.; Kloas, W.; Jagnytsch, O.; Lutz, I.; Kusk, K.O.; Wollenberger, L.; Santos, E.M.; Paull, G.C.; Van Look, K.J.W.; et al. A critical analysis of the biological impacts of plasticizers on wildlife. *Philos. Trans. R. Soc. B-Biol. Sci.* **2009**, *364*, 2047–2062. [CrossRef]
21. Riebeling, C.; Luch, A.; Tralau, T. Skin toxicology and 3Rs-Current challenges for public health protection. *Exp. Dermatol.* **2018**, *27*, 526–536. [CrossRef]
22. Kudryasheva, N.S.; Tarasova, A.S. Pollutant toxicity and detoxification by humic substances: Mechanisms and quantitative assessment via luminescent biomonitoring. *Environ. Sci. Pollut. Res.* **2015**, *22*, 155–167. [CrossRef] [PubMed]
23. Abbas, M.; Adil, M.; Ehtisham-ul-Haque, S.; Munir, B.; Yameen, M.; Ghaffar, A.; Shar, G.A.; Tahir, M.A.; Iqbal, M. *Vibrio fischeri* bioluminescence inhibition assay for ecotoxicity assessment: A review. *Sci. Total Environ.* **2018**, *626*, 1295–1309. [CrossRef]
24. Ma, X.Y.Y.; Wang, X.C.C.; Ngo, H.H.; Guo, W.S.; Wu, M.N.N.; Wang, N. Bioassay based luminescent bacteria: Interferences, improvements, and applications. *Sci. Total Environ.* **2014**, *468*, 1–11. [CrossRef] [PubMed]
25. Thouand, G.; Marks, R. Bioluminescence: Fundamentals and Applications in Biotechnology—Volume 3 Preface. In *Bioluminescence: Fundamentals and Applications in Biotechnology*; Springer: Berlin/Heidelberg, Germany, 2016; Volume 154. [CrossRef]
26. Caccamo, D.; Di Cello, F.; Fani, R.; Gugliandolo, C.; Maugeri, T.L. Polyphasic approach to the characterisation of marine luminous bacteria. *Res. Microbiol.* **1999**, *150*, 221–230. [CrossRef]
27. Thouand, G.; Marks, R. (Eds.) *Bioluminescence: Fundamentals and Applications in Biotechnology—Volume 2*, 2014th ed.; Springer: Berlin/Heidelberg, Germany, 2014; Volume 145.
28. Ma, M.; Tong, Z.; Wang, Z.; Zhu, W. Acute toxicity bioassay using the freshwater luminescent bacterium *Vibrio-qinghaiensis* sp. Nov.-Q67. *Bull. Environ. Contam. Toxicol.* **1999**, *62*, 247–253. [CrossRef]
29. Deheyn, D.D.; Bencheikh-Latmani, R.; Latz, M.I. Chemical speciation and toxicity of metals assessed by three bioluminescence-based assays using marine organisms. *Environ. Toxicol.* **2004**, *19*, 161–178. [CrossRef] [PubMed]
30. Hassan, S.H.A.; Oh, S.E. Improved detection of toxic chemicals by *Photobacterium phosphoreum* using modified Boss medium. *J. Photochem. Photobiol. B-Biol.* **2010**, *101*, 16–21. [CrossRef] [PubMed]
31. Rodea-Palomares, I.; Gonzalez-Garcia, C.; Leganes, F.; Fernandez-Pinas, F. Effect of pH, EDTA, and Anions on Heavy Metal Toxicity Toward a Bioluminescent *Cyanobacterial* Bioreporter. *Arch. Environ. Contam. Toxicol.* **2009**, *57*, 477–487. [CrossRef]
32. Ankley, G.T.; Hoke, R.A.; Giesy, J.P.; Winger, P.V. Evaluation of the toxicity of marine sediments and dredge spoils with the Microtox bioassay. *Chemosphere* **1989**, *18*, 2069–2075. [CrossRef]
33. Hinwood, A.L.; Hinwood, A.L.; McCormick, M.J.; McCormick, M.J. The effect of ionic solutes on EC50 values measured using the microtox test. *Toxic. Assess.* **1987**, *2*, 449–461. [CrossRef]

34. Zhou, R.; Cheng, W.; Feng, Y.; Wei, H.Y.; Liang, F.; Wang, Y. Interactions between three typical endocrine-disrupting chemicals (EDCs) in binary mixtures exposure on myocardial differentiation of mouse embryonic stem cell. *Chemosphere* **2017**, *178*, 378–383. [CrossRef]

35. De Oliveira, K.M.G.; Carvalho, E.H.D.; dos Santos, R.; Sivek, T.W.; Tha, E.L.; de Souza, I.R.; Coelho, L.D.D.; Pimenta, M.E.B.; de Oliveira, G.A.R.; de Oliveira, D.P.; et al. Single and mixture toxicity evaluation of three phenolic compounds to the terrestrial ecosystem. *J. Environ. Manag.* **2021**, *296*, 113226. [CrossRef] [PubMed]

36. Liu, S.; Xu, G.F.; Huang, M.Q.; Fu, L.; Jiang, X.; Yang, M. Bisphenol A and bisphenol AF co-exposure induces apoptosis in human granulosa cell line KGN through intracellular stress-dependent mechanisms. *Arab. J. Chem.* **2021**, *14*, 103399. [CrossRef]

37. Jatkowska, N.; Kudlak, B.; Lewandowska, P.; Liu, W.; Williams, M.J.; Schioth, H.B. Identification of synergistic and antagonistic actions of environmental pollutants: Bisphenols A, S and F in the presence of DEP, DBP, BADGE and BADGE center dot 2HCl in three component mixtures. *Sci. Total Environ.* **2021**, *767*, 144286. [CrossRef] [PubMed]

38. Li, D.; Chen, H.X.; Bi, R.; Xie, H.B.; Zhou, Y.; Luo, Y.J.; Xie, L.T. Individual and binary mixture effects of bisphenol A and lignin-derived bisphenol in Daphnia magna under chronic exposure. *Chemosphere* **2018**, *191*, 779–786. [CrossRef]

39. Parsai, T.; Kumar, A. Effect of seawater acidification and plasticizer (Bisphenol-A) on aggregation of nanoparticles. *Environ. Res.* **2021**, *201*, 111498. [CrossRef]

40. Li, X.H.; Yin, P.H.; Zhao, L. Effects of individual and combined toxicity of bisphenol A, dibutyl phthalate and cadmium on oxidative stress and genotoxicity in HepG 2 cells. *Food Chem. Toxicol.* **2017**, *105*, 73–81. [CrossRef]

41. Ribeiro, E.; Ladeira, C.; Viegas, S. EDCs Mixtures: A Stealthy Hazard for Human Health? *Toxics* **2017**, *5*, 5. [CrossRef]

42. M'Rabet, C.; Pringault, O.; Zmerli-Triki, H.; Ben Gharbia, H.; Couet, D.; Yahia, O.K.D. Impact of two plastic-derived chemicals, the Bisphenol A and the di-2-ethylhexyl phthalate, exposure on the marine toxic dinoflagellate Alexandrium pacificum. *Mar. Pollut. Bull.* **2018**, *126*, 241–249. [CrossRef]

43. Baralic, K.; Zivancevic, K.; Javorac, D.; Djordjevic, A.B.; Andelkovic, M.; Jorgovanovic, D.; Miljakovic, E.A.; Curcic, M.; Bulat, Z.; Antonijevic, B.; et al. Multi-strain probiotic ameliorated toxic effects of phthalates and bisphenol A mixture in Wistar rats. *Food Chem. Toxicol.* **2020**, *143*, 111540. [CrossRef]

44. Baralic, K.; Djordjevic, A.B.; Zivancevic, K.; Antonijevic, E.; Andelkovic, M.; Javorac, D.; Curcic, M.; Bulat, Z.; Antonijevic, B.; Dukic-Cosic, D. Toxic Effects of the Mixture of Phthalates and Bisphenol A-Subacute Oral Toxicity Study in Wistar Rats. *Int. J. Environ. Res. Public Health* **2020**, *17*, 746. [CrossRef] [PubMed]

45. Suteau, V.; Briet, C.; Lebeault, M.; Gourdin, L.; Henrion, D.; Rodien, P.; Munier, M. Human amniotic fluid-based exposure levels of phthalates and bisphenol A mixture reduce INSL3/RXFP2 signaling. *Environ. Int.* **2020**, *138*, 105585. [CrossRef] [PubMed]

46. Dokmeci, A.H.; Karaboga, I.; Guzel, S.; Erboga, Z.F.; Yilmaz, A. Toxicological assessment of low-dose bisphenol A, lead and endosulfan combination: Chronic toxicity study in male rats. *Environ. Sci. Pollut. Res.* **2021**, *29*, 10558–10574. [CrossRef]

47. Zhu, X.L.; Wu, G.L.; Xing, Y.; Wang, C.Z.; Yuan, X.; Li, B.K. Evaluation of single and combined toxicity of bisphenol A and its analogues using a highly-sensitive micro-biosensor. *J. Hazard. Mater.* **2020**, *381*, 120908, Erratum in *J. Hazard. Mater.* **2021**, *403*, 123924. [CrossRef] [PubMed]

48. Yang, J.; Hu, S.L.; Liao, A.Q.; Weng, Y.T.; Liang, S.L.; Lin, Y. Preparation of freeze-dried bioluminescent bacteria and their application in the detection of acute toxicity of bisphenol A and heavy metals. *Food Sci. Nutr.* **2022**, 1–13. [CrossRef]

49. Zhao, J.; Zeng, S.Q.; Xia, Y.; Xia, L.M. Expression of a thermotolerant laccase from Pycnoporus sanguineus in Trichoderma reesei and its application in the degradation of bisphenol A. *J. Biosci. Bioeng.* **2018**, *125*, 371–376. [CrossRef]

50. Zhu, X.W.; Liu, S.S.; Ge, H.L.; Liu, Y. Comparison between the short-term and the long-term toxicity of six triazine herbicides on photobacteria Q67. *Water Res.* **2009**, *43*, 1731–1739. [CrossRef]

51. Zou, X.M.; Lin, Z.F.; Deng, Z.Q.; Yin, D.Q.; Zhang, Y.L. The joint effects of sulfonamides and their potentiator on Photobacterium phosphoreum: Differences between the acute and chronic mixture toxicity mechanisms. *Chemosphere* **2012**, *86*, 30–35. [CrossRef]

52. Chen, F.; Wu, L.G.; Xiao, X.Y.; Rong, L.L.; Li, M.; Zou, X.M. Mixture toxicity of zinc oxide nanoparticle and chemicals with different mode of action upon *Vibrio fischeri*. *Environ. Sci. Eur.* **2020**, *32*, 41. [CrossRef]

53. Backhaus, T.; Froehner, K.; Altenburger, R.; Grimme, L.H. Toxicity testing with *Vibrio fischeri*: A comparison between the long term (24 h) and the short term (30 min) bioassay. *Chemosphere* **1997**, *35*, 2925–2938. [CrossRef]

54. Broderius, S.J.; Kahl, M.D.; Hoglund, M.D. Use of joint toxic response to define the primary mode of toxic action for diverse industrial organic chemicals. *Environ. Toxicol. Chem.* **1995**, *14*, 1591–1605. [CrossRef]

55. Xu, S.; Nirmalakhandan, N. Use of QSAR models in predicting joint effects in multi-component mixtures of organic chemicals. *Water Res.* **1998**, *32*, 2391–2399. [CrossRef]

56. Wang, D.; Wang, S.; Bai, L.M.; Nasir, M.S.; Li, S.S.; Yan, W. Mathematical Modeling Approaches for Assessing the Joint Toxicity of Chemical Mixtures Based on Luminescent Bacteria: A Systematic Review. *Front. Microbiol.* **2020**, *11*, 1651. [CrossRef] [PubMed]

57. Ding, K.K.; Lu, L.P.; Wang, J.Y.; Wang, J.P.; Zhou, M.Q.; Zheng, C.W.; Liu, J.S.; Zhang, C.L.; Zhuang, S.L. In vitro and in silico investigations of the binary-mixture toxicity of phthalate esters and cadmium (II) to *Vibrio qinghaiensis* sp.-Q67. *Sci. Total Environ.* **2017**, *580*, 1078–1084. [CrossRef] [PubMed]

58. Zhang, J.; Ding, T.T.; Dong, X.Q.; Bian, Z.Q. Time-dependent and Pb-dependent antagonism and synergism towards *Vibrio qinghaiensis* sp.-Q67 within heavy metal mixtures. *RSC Adv.* **2018**, *8*, 26089–26098. [CrossRef]

59. Czarny, K.; Krawczyk, B.; Szczukocki, D. Toxic effects of bisphenol A and its analogues on cyanobacteria Anabaena variabilis and Microcystis aeruginosa. *Chemosphere* **2021**, *263*, 128299. [CrossRef]

60. Elersek, T.; Notersberg, T.; Kovacic, A.; Heath, E.; Filipic, M. The effects of bisphenol A, F and their mixture on algal and cyanobacterial growth: From additivity to antagonism. *Environ. Sci. Pollut. Res.* **2021**, *28*, 3445–3454. [CrossRef]
61. Vilas-Boas, J.A.; Cardoso, S.J.; Senra, M.V.X.; Rico, A.; Dias, R.J.P. Ciliates as model organisms for the ecotoxicological risk assessment of heavy metals: A meta-analysis. *Ecotoxicol. Environ. Saf.* **2020**, *199*, 110669. [CrossRef]
62. McGrath, S.P.; Chaudri, A.M.; Giller, K.E. Long-term effects of metals in sewage sludge on soils, microorganisms and plants. *J. Ind. Microbiol.* **1995**, *14*, 94–104. [CrossRef]

Journal of
Marine Science and Engineering

Article

Effects of the Ionic Liquid [BMIM]Cl on the Baltic Microphytobenthic Communities

Zuzanna Sylwestrzak, Aleksandra Zgrundo * and Filip Pniewski

Faculty of Oceanography and Geography, University of Gdańsk, Al. Piłsudskiego 46, 81-378 Gdynia, Poland
* Correspondence: aleksandra.zgrundo@ug.edu.pl

Abstract: Ionic liquids (IL) are regarded as the solution to the modern world's need to create and use compounds that exhibit a range of desirable properties while having a low environmental impact. However, recent reports are shattering the image of ionic liquids as environmentally friendly substances, especially in relation to the aquatic environment, revealing their potentially toxic effects. To assess the potential environmental impact of ILs, we conducted an experiment involving 1-butyl-3-methylimidazolium chloride ([BMIM]Cl), a substance considered to be the least hazardous among the imidazolium chloride ILs, on Baltic microphytobenthic communities. Microphytobenthos collected from the environment was tested under controlled laboratory conditions, and both the cell counts and the chloroplast condition were used as endpoints. It was shown that [BMIM]Cl at concentrations of 10^{-3} and 10^{-2}, considered safe based on a cumulative impact assessment, has a negative effect on the condition of the microalgal cells and causes a reduction in population size. Although, under the influence of [BMIM]Cl, only a small proportion of the species was eliminated from the communities, only two species among those important to the communities showed resistance to this compound and eventually began to dominate the communities.

Keywords: ionic liquid; IL; [BMIM]Cl; microphytobenthos; microalgal communities; microphytobenthic communities; toxic effect; ecotoxicological test; environmental pollution

Citation: Sylwestrzak, Z.; Zgrundo, A.; Pniewski, F. Effects of the Ionic Liquid [BMIM]Cl on the Baltic Microphytobenthic Communities. *J. Mar. Sci. Eng.* **2022**, *10*, 1223. https://doi.org/10.3390/jmse10091223

Academic Editors: François Gagné, Stefano Magni and Valerio Matozzo

Received: 15 July 2022
Accepted: 28 August 2022
Published: 1 September 2022

Publisher's Note: MDPI stays neutral with regard to jurisdictional claims in published maps and institutional affiliations.

1. Introduction

The concept of sustainable development was introduced at the end of the 20th century, and it is based on the idea of social and economic development, which assumes that while meeting the needs of contemporary societies it will not limit the development opportunities for future generations. Sustainable development assumes a parallel development of the economy, society, and the environment. In line with these assumptions, more and more environmentally friendly substances, such as ionic liquids (ILs), have started to be used in industry. These are substances that are gaining increasing recognition and researchers in many scientific fields are interested in them as they are characterized by a set of specific properties, such as low vapor pressure, non-flammability, thermal and electrochemical stability, good conductivity, and catalytic properties [1,2]. They are used in various chemical processes, where they represent a new alternative to traditional organic solvents [3]. ILs are often called 'designer solvents'; the appropriate selection of anions and cations allows for the creation of a suitable chemical compound, depending on the future application [4,5]. However, with regard to ILs' specific features (e.g., high solubility, thermal stability, and/or poor biodegradability in water), they may potentially pollute the aquatic environment [6]. Science has known of ILs as solvents since 1914 [7]. However, the first stable ILs were described in 1995 [5]. Since then, there has been a rapid increase in interest in these substances, especially in terms of their effects on human health and the environment, as in, e.g., [6–15]. A major threat from ionic liquids is their low degradation rate. For example, a 28-day experiment showed a complete lack of biodegradation of [BMIM]Cl [7]. As a result, after years of research, their "green" status has been questioned [8–10]. Inadequate wastewater treatment, accidental spillage, or improper storage of waste contaminated with ILs can lead to the release of these substances into the environment where they subsequently cause negative effects in the ecosystem [1]. Numerous studies have proven the deleterious effects of ILs towards microalgae [16–21]. The toxicity of ILs depends on temperature and pH. Under conditions described as moderate, i.e., room temperature and pH close to neutral, these substances are stable. During industrial processes, the physicochemical conditions can change, and it has been shown that an acidic condition (pH about 3) and high temperature (of the range between 60 °C and 100 °C) accelerate the hydrolysis of the ILs, causing an increase in their toxicity. The IL used in this study was 1-butyl-3-methylimidazolium chloride ([BMIM]Cl), which is characterized by a relatively short alkyl chain [9]. The toxicity of ILs increases with the elongation of the alkyl chain [2]; hence, the IL used here is not considered to be a highly toxic substance [22]. However, its toxicity is comparable to that of chlorinated organic substances, such as dichloromethane and chloroform; thus, it is more dangerous to the environment than common solvents of organic origin (e.g., acetone, methanol, and ethanol). Due to its properties, [BMIM]Cl is used in cellulose processing, among other areas. Although the ionic liquid itself can be recovered to a very high degree during industrial applications (such as the one mentioned above), the very process of creating the imidazole cation involves the use of large amounts of natural organic materials, energy, and solvents, causing harmful emissions to both air and water as well [7].

The photosynthetic organisms that form the microphytobenthos are extremely valuable elements of aquatic ecosystems due to their role as primary producers, combined with oxygen production and CO_2 reduction. Understanding the joint response of photosynthetic organisms forming a microphytobenthic community is extremely important in order to reliably estimate the changes that may occur following the introduction of this increasingly common substance into the environment and to assess the associated risks.

Monitoring the response of organisms to potentially toxic substances introduced into the ecosystem is an important part of environmental quality control. To date, investigations into the toxicity of ionic liquids have provided information on the response of single algal strains under laboratory conditions, e.g., [21–23]. Previous ecotoxicological studies conducted on marine microphytobenthic communities in the Baltic Sea have tested substances such as irgarol 1051, Sea-Nine™211 (DCOIT), and TBT (trin-butyltin) [24,25]. However, most of the studies used only communities developed at salinities typical of marine waters, i.e., 32–36 PSU. Only studies conducted to determine the impact of glyphosate and copper ions on microphytobenthic communities were performed on organisms collected from environments with salinities around 8 [26,27]. Further ecotoxicological tests on other potentially toxic substances carried out on microphytobenthic communities typical of brackish waters, which are considered species-minimum waters for macrozoobenthos and macroalgae and aquatic higher plants, but which support an abundance and diversity of planktonic microorganisms [28], are an interesting contribution to the existing knowledge. In the case of ILs, the salinity aspect is extremely important because the toxicity of ILs increases in inverse proportion to the salinity [16–21]. At high salinity values, the toxicity of ILs decreases, probably due to the reduced permeability of the microalgal cell membranes which limits the migration of harmful cations [29]. Our study was designed to provide a general picture of the response of multispecies microalgal communities to an IL considered to be of relatively low toxicity, i.e., 1-butyl-3-methylimidazolium chloride—[BMIM]Cl, under brackish water conditions, and to complement the existing knowledge on the potential risks arising from the widespread use of this substance. For that purpose, observations were made at the population level, i.e., the change in species composition and the community dominance structure were determined, and at the cell level, i.e., the condition of the chloroplasts was analyzed.

2. Materials and Methods

2.1. Study Area and Field Works

The experiment investigating effects of the IL [BMIM]Cl on microphytobenthic communities is one of a series of tests based on the identical methodology described in detail in [26]. The experiments were conducted in parallel on communities with identical species composition to allow for the comparison of the results. In brief, the study material was collected from glass slides mounted on a dedicated culture panel (Figure 1b) exposed in the Gulf of Gdańsk waters at a distance of 300 m from the shore (54°26′49″ N, 8°34′24″ E) (Figure 1a) for two weeks. During this time, the temperature and salinity changed within limited ranges, i.e., 17–19 °C and 7.9–8.4 PSU, respectively. The 2-week incubation period allowed for the acquisition of a relatively rich and diverse microphytobenthic community but was still devoid of organisms such as fouling macroalgae and fauna that will eventually dominate the surface of any substrate in marine waters in the long term.

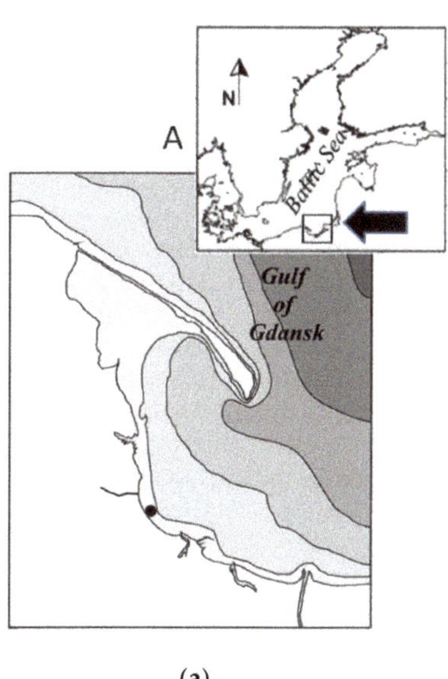
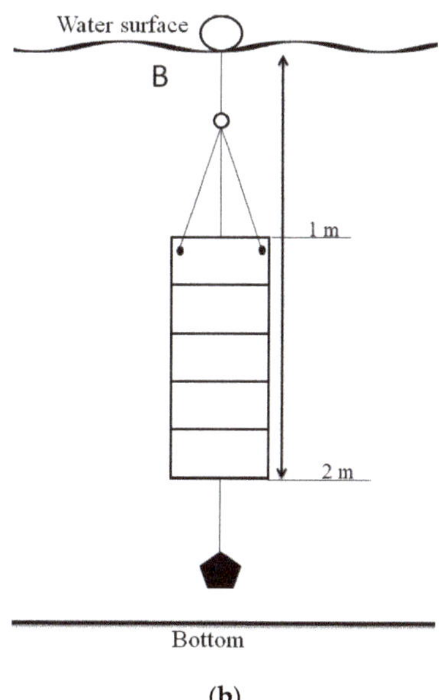

(a)

(b)

Figure 1. (**a**) Sampling site. The black dot indicates the location of the culture panel during exposure in the Gulf of Gdańsk (54°26′51″ N 18°34′33″ E). (**b**) The design of the culture panel used in this study.

2.2. Microalgal Material Preparation Procedure and Experimental Design

In the laboratory, the microphytobenthic communities were removed from the microscope slides by scraping them off with a scalpel. Subsequently, the microalgal cells were re-suspended in the seawater collected at the sampling site, which was first filtered through a glass filter (Whatman GF/C) and then autoclaved. The obtained microphytobenthos suspension was then sonicated, which allowed for the disruption and removal of cell aggregates. The sonication power was carefully chosen in order not to weaken or damage the cells [26].

The experiment was carried out in 250 mL flasks filled with 100 mL of microalgal suspension. Each microphytobenthos culture was insufflated with nitrogen for 30 s to remove heterotrophic microorganisms [30,31]. At the beginning of the experiment, the mean microalgal cell abundance was 38,800 cells/mL ± 700. Before the experiment, flasks with microalgal suspension were maintained in a thermostatic chamber for 72 h at constant light, temperature, and salinity conditions (i.e., 60 μmol photons·m^{-2}·s^{-1} with a photoperiod L:D 16:8 h, 18 °C ± 1 °C, and 8 PSU, respectively) to let the communities acclimate to the experimental conditions. The natural concentrations of the nutrient compounds in the sea water were: N-NH$_4$ 9.4 mg·m^{-3}, N-NO$_3$ 102 mg·m^{-3}, P-PO$_4$ 36 mg·m^{-3}, and Si-SiO$_4$ 600 mg·m^{-3}. As the nutrient concentrations in the natural Baltic water were sufficiently high to maintain the microphytobenthos community during the experiment, any kind of culture medium was not added. This was also dictated by the fact that the high nutrient content could facilitate the growth of random species rapidly responding to the increase in nutrient concentrations.

After the acclimation phase, the [BMIM]Cl toxicity tests were performed according to the following design: control—microphytobenthic assemblages kept in filtered sea water without the addition of the tested IL and test solutions—microphytobenthic assemblages treated with two [BMIM]Cl concentrations, i.e., 1.13×10^{-3} g·dm^{-3} and 1.75×10^{-2} g·dm^{-3}. The lower [BMIM]Cl concentration selected for the experiment was previously proven to have significant effects on the species composition of the Baltic microphytobenthic communities [32]. The higher concentration of [BMIM]Cl was inferred from previously published values indicated as having inhibitory effects on algae [18,33]. All experimental treatments were carried out in triplicates.

2.3. Microscopic Analysis

Qualitative and quantitative changes in assemblage composition and structure, i.e., changes in taxonomic composition and taxa abundance, were the primary parameters used to assess the changes in microphytobenthos. Microscopic analysis was conducted on microalgal material preserved in Lugol solution. The observations were conducted in all cultures on the third and seventh experiment day. Fifty fields of view were checked in Utermöhl chambers (2 mL) using a Nikon Eclipse TS100 inverted light microscope (magnifications of ×200 and ×400). All cells were counted and identified as laid out in the Utermöhl method [34] and Helcom [35] guidelines (cells or threads of 100 µm length are treated as units). Species were identified using appropriate keys and floras [36–43].

The analysis of the microalgal cell condition was also performed. For this purpose, the state of chloroplasts was observed and classified in one of three classes of cells: (1) live cells with normal chloroplasts, (2) live cells with abnormal chloroplasts, and (3) dead cells. Here, the results obtained for the two first cell classes are reported (Figure 2). The microalgal cell condition was evaluated in all cells counted in 50 fields of vision under a Nikon Eclipse 80i microscope fitted with a Nikon DSU2 camera at a magnification of ×400.

2.4. Statistical Analysis

Differences between means were verified with the Student's *t*-test using STATISTICA version 10 (StatSoft Polska Sp. z o.o., Kraków, Poland). Principal component analysis (PCA) was carried out with the Canoco 5 (Microcomputer Power, Ithaca, NY, USA) [44,45] and similarity percentage (SIMPER) with the PRIMER-e (PRIMER-e, Auckland, New Zealand).

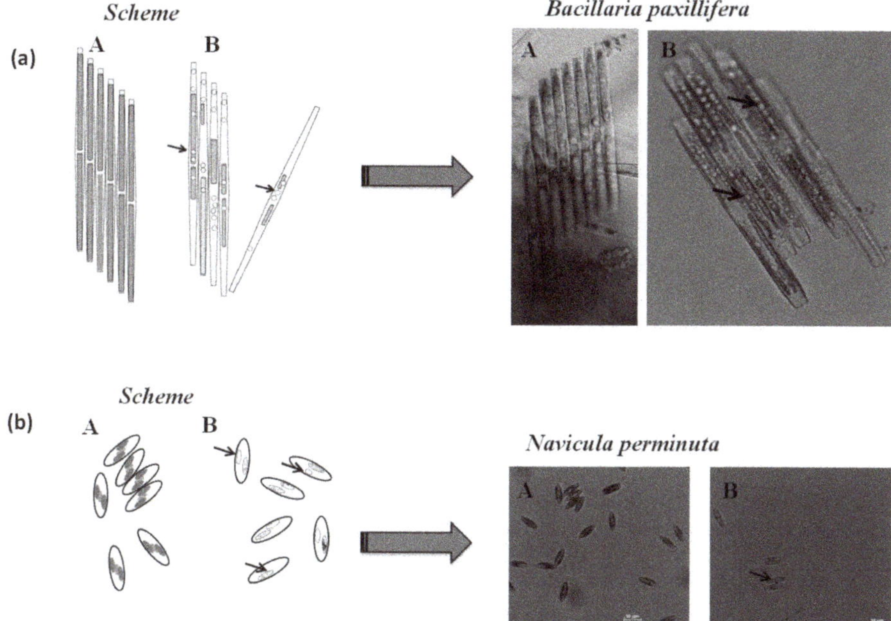

Figure 2. Examples of cells of *Bacillaria paxilifera* (**a**) and *Navicula perminuta* (**b**) with normal (A) and abnormal chloroplasts (B). Chloroplasts treated as abnormal have a deformed shape compared to the ones correctly formed. The shape of the chloroplast itself and thus its alternations are species- and/or genus-specific.

3. Results

3.1. Analysis of Taxonomic Composition and Structure

A total of 46 microalgae species were identified, including 35 diatoms, 6 cyanobacteria, and 2 green algae taxa, as well as representatives of dinoflagellates (*Peridinium* sp.) and haptophytes (*Prymnesium* sp.) (the list of all identified taxa is in the Appendix A (Table A1).

At the beginning of the experiment, the highest number of cells, 38,800 ± 700 cells/mL, was found (Figure 3). On the third day, a 47% decrease in the number of microalgae cells was observed in the concentration of 1.13×10^{-3} g·dm^{-3}, while in the concentration of 1.75×10^{-2} g·dm^{-3} [BMIM]Cl, a drop by only 21% was observed. On the seventh day of testing, a similar abundance of microalgae was observed in both concentrations, 23–26% less than in the control solution. All differences in abundance between the concentrations tested and the control solution were statistically significant ($p < 0.05$).

The microphytobenthic communities were heavily dominated by diatoms, constituting from 70% to 92% of all the observed photosynthetic microorganisms. In the control cultures, the abundance of cyanobacteria did not exceed 10% (the highest number was observed on the third day of the experiment). On the seventh day, at both [BMIM]Cl concentrations, i.e., 1.13×10^{-3} g·dm^{-3} and 1.75×10^{-2} g·dm^{-3}, they represented 23% and 28% of the total abundance, respectively. The abundance of *Prymnesium* sp. (Haptophyta) did not exceed 0.3% during the whole experiment, while the abundance of *Peridinium* sp. (Dinophyceae) in the control solution at the start of the tests was only 0.6%.

SIMPER similarity analysis showed a high similarity in community composition and structure at the control and both ionic liquid concentrations during the experiment. The average similarity was calculated as high as 71.92%. Based on the PCA analysis performed on the quantitative data, it was found that in addition to the concentration of the ionic liquid, the duration of the experiment also influenced the transformation of communities

(Figure 4). At the start of the experiment (point K_0, right part of the graph), the community was the richest (i.e., it was characterized by the highest number of species). On the third day of testing (upper left part of the graph), an increased proportion of cyanobacteria was observed (e.g., marked in graph as *cya_sp., mic_sp., spi_mai, wor_sp.*). The group of organisms dominating on the seventh day of testing (lower left part of the graph) included, among others, a diatom characterized by a relatively high resistance to ionic liquid—*Navicula perminuta* (marked as *nav_per*). However, based on the PCA analysis, it was not possible to delineate groups of organisms that were unequivocally sensitive or tolerant to the IL tested.

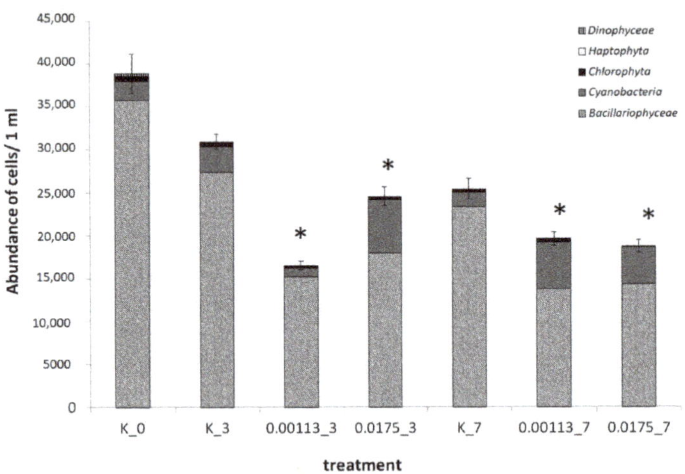

Figure 3. Abundance of microalgae. K indicates control cultures; the numbers 0.00113 and 0.0175 indicate cultures of lower (1.13×10^{-3} g·dm^{-3}) and higher (1.75×10^{-2} g·dm^{-3}) tested [BMIM]Cl concentrations, respectively. The number after the underline denotes the day of the experiment. Statistically significant differences were marked with the asterisk. In each case, the value from the experimental variant was compared with the value for the control sample on the same day ($p = 0.000002$ for 0.00113_3 vs. K_3, $p = 0.005617$ for 0.0175_3 vs. K_3, $p = 0.005178$ for 0.00113_7 vs. K_7, $p = 0.001357$ for 0.0175_7 vs. K_7).

Based on SIMPER analysis, the most important species in the communities were distinguished in terms of abundance (Appendix A, Table A2). The most abundant species was *Bacillaria paxilifera* (up to 36% on the third day in the control solution) (Figure 5). The second most abundant species was *Tabularia fasciculata* (up to 23% of all cells on the initial day of the experiment). The abundance of *Diatoma vulgaris* ranged from 9% to 17% throughout the experiment. For *Melosira nummuloides*, the highest number of cells was observed on the seventh day of testing in the control solution (20% of all cells). Interesting changes were observed in the case of *N. perminuta*; the proportion of this species in the initial community did not exceed 8%, but on the seventh day at both [BMIM]Cl concentrations, i.e., 1.13×10^{-3} g·dm^{-3} and 1.75×10^{-2} g·dm^{-3}, the share of this taxon increased to 15% and 16%, respectively. The highest share of *Cylindrotheca closterium* in the community was observed at the start of the experiment (13%), but on subsequent days the share did not exceed 8%. In the case of *Navicula gregaria*, a maximum share of 6% was observed at a concentration of 1.13×10^{-3} g·dm^{-3} on the seventh day of testing. While the share of the only representative of cyanobacteria, *Spirulina major*, did not exceed 1% during the whole experiment.

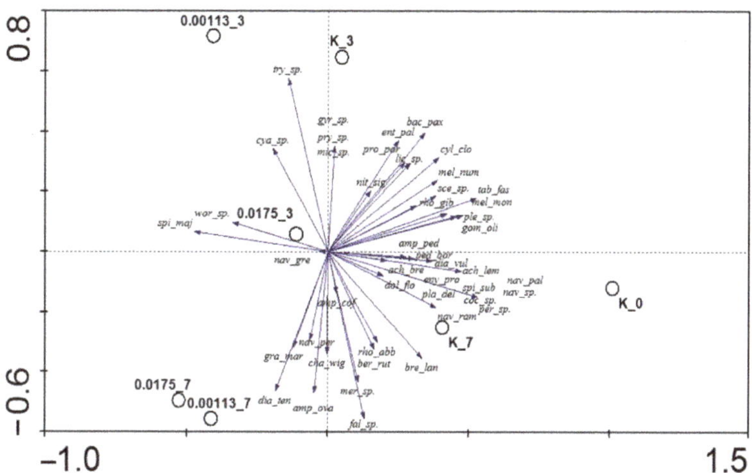

Figure 4. PCA (principal component analysis) of microphytobenthos species composition (based on the abundance data) from control and [BMIM]Cl treatments cultures. Percentage explained variation λ1 = 26%, λ2 = 22.4% λ3 = 17.9%. K indicates control cultures, 0.00113 and 0.0175 indicate cultures of lower (1.13×10^{-3} g·dm^{-3}) and higher (1.75×10^{-2} g·dm^{-3}) tested [BMIM]Cl concentrations, respectively. The number after the underline denotes the day of the experiment. Taxon codes are included in Table A1 in Appendix A.

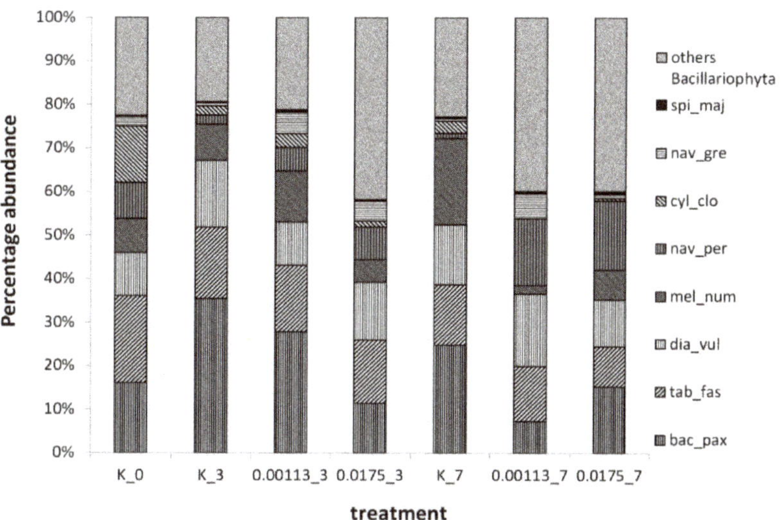

Figure 5. Percentage abundance of the 8 most important species distinguished in the tested communities based on SIMPER analysis. Labels used: spi_maj—*Spirulina major*, nav_gre—*Navicula gregaria*, cyl_clo—*Cylindrotheca closterium*, nav_per—*Navicula perminuta*, mel_num—*Melosira nummuloides*, dia_vul—*Diatoma vulgaris*, tab_fas—*Tabularia fasciculata*, bac_pax—*Bacillaria paxilifera*.

3.2. Abundance of Selected Taxa

In the presence of the IL, most of the taxa showed a statistically significant reduction in their abundance. Only for two taxa, i.e., *N. perminuta* and *Navicula ramosissima*, did the abundance in the [BMIM]Cl solutions increase compared to the control solution (Figure 6a,b). Hence, they were identified as tolerant species to this IL. The abundance of *N. perminuta* in the control solution on the seventh day decreased by 91% relative to the start of the experiment, while in both concentrations of the ionic liquid abundances of about 89% relative to the initial abundance were recorded. On the third day, in the lower [BMIM]Cl concentration cultures (1.13×10^{-3} g·dm^{-3}), the cell abundance was 142% of the number in the control solution, while at the concentration of 1.75×10^{-3} g·dm^{-3} it was as high as 289%. On day seven, as much as a tenfold rise in the cell abundance was observed for this taxon as compared to the control solution. A similar growth stimulation was noted on day three for *N. ramosissima*, but on day seven, the cell number was only 45% and 64% of that in the control depending on the [BMIM]Cl concentration.

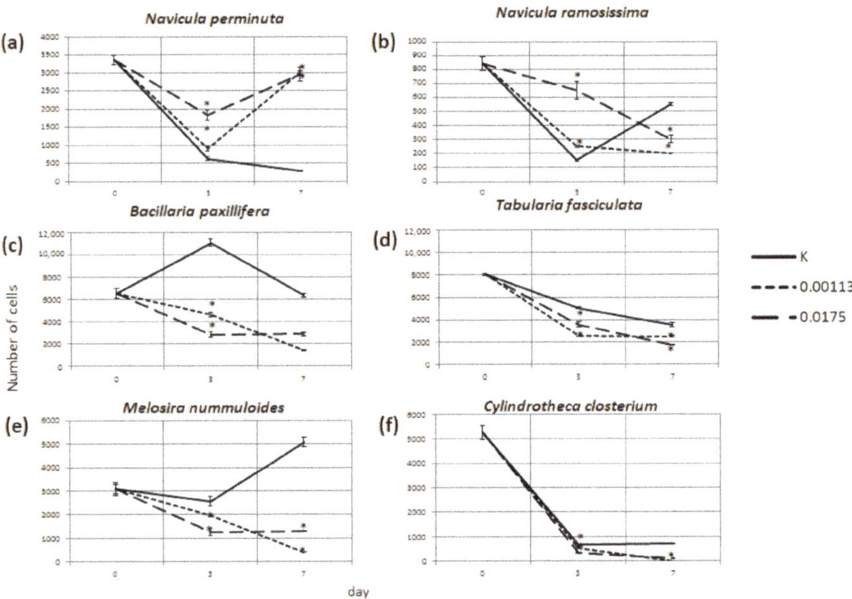

Figure 6. Number of cells of selected microalgae during experiment: K—control solution; 0.00113—the concentration of 1.13×10^{-3} g·dm^{-3} [BMIM]Cl; 0.0175—the concentration of 1.75×100^{-2} g·dm^{-3} [BMIM]Cl. Statistically significant differences between the control solution and [BMIM]Cl treatments are marked with the asterisk. (**a,b**)—tolerant species positively affected by the ionic liquid; (**c–f**)—sensitive species negatively affected by the ionic liquid.

Figure 6c–f presents changes in the abundance of selected taxa considered most important in the communities in relation to the abundance based on the SIMPER analysis. The reduction in cell numbers in the [BMIM]Cl solutions tested indicated statistically significant ($p < 0.05$) growth inhibition. Such a response was characteristic of almost all the organisms observed in the microphytobenthic communities. However, depending on the taxon and concentration used, either a gradual reduction in abundance over time (e.g., *B. paxillifera* and *M. nummuloides* in the solution of 1.13×10^{-3} g·dm^{-3} [BMIM]Cl) or a reduction in abundance followed by an initial increase (e.g., *B. paxillifera* and *M. nummuloides* in the solution of 1.75×10^{-2} g·dm^{-3} [BMIM]Cl or *T. fasciculata* in the solution of 1.13×10^{-3} g·dm^{-3} [BMIM]Cl) was observed. The most dramatic changes in the abundance of *C. closterium* were noted. In the solution of IL of 1.13×10^{-3} g·dm^{-3}, no living representatives of this species were observed on day seven, and in the 1.75×10^{-2} g·dm^{-3}, only 14% of the abundance in the control solution was noted.

3.3. Cell Condition in Selected Taxa

Analysis of the chloroplast state in the cells provided complementary information on the differences in the response of the various taxa to [BMIM]Cl. In species considered tolerant, i.e., *N. perminuta* and *N. ramosissima*, a small number of cells with abnormally shaped chloroplasts were observed (Figure 7a,b). Interestingly, in the case of *N. perminuta*, cells with abnormally shaped chloroplasts were mainly observed in the control solution (e.g., up to 38% of all cells on the third day) and were not present or were only in small proportions in the IL solutions (up to 20% in the solution of 1.13×100^{-3} g·dm^{-3} [BMIM]Cl). Similarly, in *N. ramosissima*, abnormally shaped chloroplasts were not observed in the cells at the beginning of the experiment. Deformed chloroplasts were present in the cells on the third day in the solution of 1.75×100^{-2} g·dm^{-3} [BMIM]Cl (15% of cells) and on the seventh day in the control solution (18% of cells) and in the solution of 1.13×100^{-3} g·dm^{-3} (50% of cells).

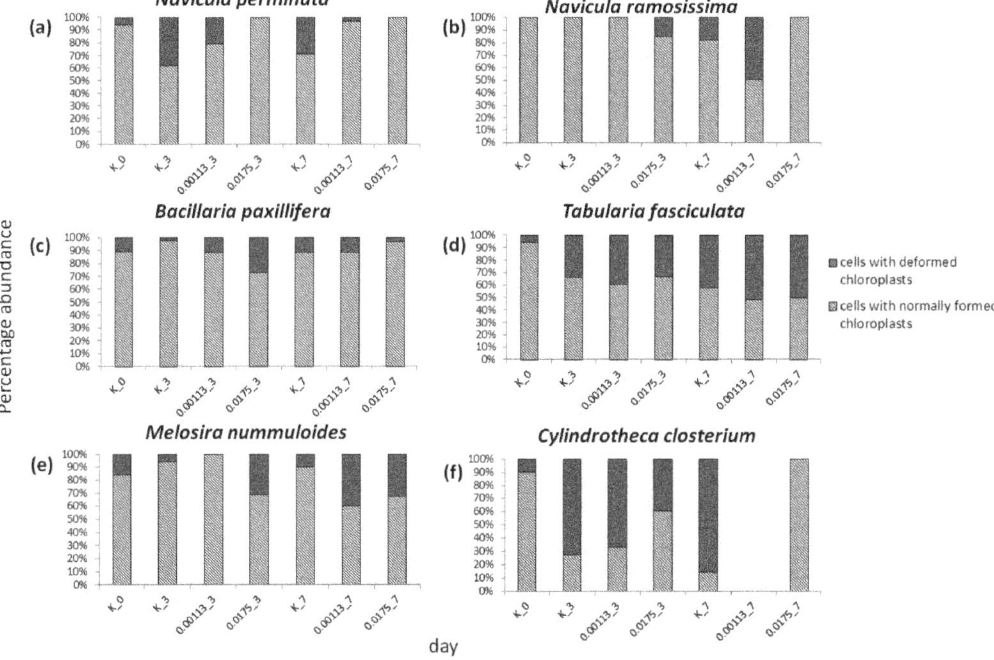

Figure 7. Condition of selected species shown as the percentage of cells with normal and abnormal chloroplasts. K indicates control cultures; 0.00113 and 0.0175 indicate cultures of lower (1.13×100^{-3} g·dm^{-3}) and higher (1.75×100^{-2} g·dm^{-3}) tested [BMIM]Cl concentrations, respectively. (**a,b**)—tolerant species positively affected by the ionic liquid; (**c–f**)—sensitive species negatively affected by the ionic liquid.

For the taxa considered sensitive, cells with deformed chloroplasts were observed irrespective of the solution and day of the experiment (Figure 7c–f). For example, in *B. paxillifera* less than 15% of cells were characterized by deformed chloroplasts. Only in the solution of 1.75×100^{-2} g·dm^{-3} [BMIM]Cl on the third day was chloroplast degradation observed in 27% of the living cells. In *T. fasciculata*, except for at the beginning of the experiment, cells with degraded chloroplasts accounted for about 30–40% of all the cells. However, the highest number of cells with degraded chloroplasts was observed on the last day of testing in both IL solutions (about 50% of all cells). Similarly, for *M. nummuloides* in the control solution and at the beginning of the experiment, cells with abnormally formed chloroplasts made up a small proportion of the population (0–16%). In contrast, in the solution of 1.13×100^{-3} g·dm^{-3} [BMIM]Cl, up to 40% of the cells with damaged chloroplasts were observed on the seventh day and in the solution of 1.75×100^{-2} g·dm^{-3}, 31% and 32% of cells on the third and seventh day, respectively. A significantly worse cell condition was observed in *C. closterium* as compared to the previously described species. In most of the solutions tested, cells with abnormally shaped chloroplasts accounted for about 40–85% of all the cells. Only at the solution of 1.75×100^{-2} g·dm^{-3} [BMIM]Cl did all of the cells have chloroplasts of normal shape, but the population size was low.

4. Discussion

ILs as solvents with salt structures have been known since 1914 [7]. However, the first stable ILs were described in 1995 [5]. The beginning of the 21st century brought the possibility of designing chemical compounds combining required biological properties with preferred physicochemical characteristics. Currently, these substances are being studied on a massive scale, as evidenced by the number of peer-reviewed publications e.g., [6,9,10,12–14]. Furthermore, the list of their potential applications as reaction media in many industrial fields is growing [9,12]. ILs have also found applications in medicine due to their antibacterial, antifungal, anticholinergic, and local anesthetic activities and agrochemistry as bactericides, fungicides, herbicides, plant growth stimulants, or wood preservatives [9,46]. Although ILs are popular in research and economics, this does not necessarily correspond to the amount of research related to the monitoring of these compounds in the environment and the subsequent risk assessment; the number of papers focusing on the presence of ILs or their compounds in the environment remains small [10,47–50]. In one of such studies, 1310 pollutants were identified in riverine waters in Germany, among which ca. 20 different compounds belonging to ILs were detected in concentrations of up to µg·dm^{-3} [48]. In addition, in the United States, based on analyses of sediments from lakes located within the state of Minnesota, it was shown that the concentration of the IL C4-PYR was 0.053 µg·dm^{-3} [51]. In this context, one of the ILs, i.e., 1-octyl-3-methyl imidazolium, is of particular significance as it was identified not only in environmental samples [52] but also in human blood [53].

The picture is completed by the fact that imadazolium-based ILs, such as [BMIM]Cl tested here, have a low rate of biodegradation and are resistant to photodegradation [54–56]. Previous studies have shown that ILs, after eventual emission into the environment, may behave similarly to some persistent organic pollutants [57]. An extremely important aspect is also the fact that the technology to effectively remove ILs from wastewater is still being developed [4]. Hence, it is to be expected that, due to the increasing popularity of ILs, they will be used widely and, consequently, will be uncontrollably introduced into the aquatic environment, remaining there for a long time due to the difficulties associated with water treatment and their poor biodegradability. Therefore, it is of paramount importance to investigate the ILs' toxicity under environmental conditions and not only under controlled laboratory conditions [58]. Our tests on the effects of the IL [BMIM]Cl, considered to be relatively harmless [32,33,57], on the whole communities of microphytobenthos collected from the environment, allowed us to determine the response of a wide spectrum of microorganisms and not just single strains as in standard ecotoxicological tests. Changes in the tested communities at the population and cellular level show in a more reliable way the direction of the changes to which the microphytobenthos, an important component of the marine ecosystem, is subjected.

In a study using cumulative impact assessment, it was found that the [BMIM]Cl turned out to be the least hazardous among the imidazolium chloride ionic liquids with the Safe Environmental Concentration (SEC) as high as 750×100^{-3} mmol/L, which corresponds to the concentration of 1.31×100^{-3} g·dm^{-3} [59]. In this study, however, it was shown that the concentration of 1.13×100^{-3} g·dm^{-3} already reduced the abundance of dominant species by 20% to 60% within 3 days and up to 75% within 7 days. Only two diatom species of higher abundances showed resistance and gained a quantitative advantage in the studied communities. The number of cells of *N. perminuta*, for example, increased, reaching almost 290% of the abundance in the control solution on the third and 1000% on the seventh day of the experiment. As a result, the total abundance of cells comprising the communities decreased by only 25%. Although the composition of the communities remained similar, the abundance structure changed due to the strong dominance of tolerant taxa—apart from diatoms, cyanobacteria also showed some resistance. Similar observations regarding the substitution of sensitive taxa by tolerant or indifferent taxa were found during experiments on the same Baltic microalgal communities, testing the effects of copper chloride [27] and glyphosate [26]. However, effects induced by the aforementioned substances were also manifested by a drastic shift in species composition, i.e., the increased contribution of cyanobacteria to the total community abundance.

Interesting changes were observed for the second taxon selected as resistant to the presence of the IL [BMIM]Cl—*N. ramosissima*. On the third day of the experiment, its cell numbers increased significantly, but on the seventh day, the cell abundance again decreased in the IL concentrations tested. Such a reaction may indicate depletion of the IL molecules, the presence of which has a stimulating effect on the test organisms. For example, in toxicological tests conducted by [20] on *Scenedesmus obliquus*, low concentrations of ILs have been shown to stimulate cells for biological activity (e.g., by changing the activity of catalase and superoxide dismutase). However, this may also be a response related to the competitive activity of other taxa, such as the predominant *N. perminuta*. Similarly, [21] selected from their study several species for which the IL [BMIM]Cl was practically harmless, i.e., the cyanobacterium *Anabaeana cylindrica* and the green alga *Chlorella pyrenoidosa*, and in the case of the green alga *Dunaliella salina*, they concluded that it was relatively harmless.

Typically, the reactions of taxa to toxicants tested in communities are milder than in laboratory tests conducted on monocultures [26,27,60]. In our studies, conducted on communities grown in nature, *B. paxillifera* cell counts decreased by 55% on the seventh day of the experiment in the solution of 1.75×100^{-2} g·dm^{-3} [BMIM]Cl. A very similar response was observed in toxicological tests conducted on a monoculture of *B. paxillifera* isolated from the Baltic Sea—the concentration of 1.75×100^{-2} g·dm^{-3} decreased the growth of the strain by 58% on the seventh day of testing [61]. A similarly large reduction was shown for the same concentration in *T. fasciculata* (50% reduction on the seventh day). Moreover, [19] observed an inhibition of 50% cell growth (EC50) in the planktonic diatom *Skeletonema marinoi* at the concentration of 0.1 mM [BMIM]Cl (1.745×100^{-2} g·dm^{-3}). In turn, for the green alga *Chlorella pyrenoidosa*, 50% inhibition of growth (IC50) was shown for the concentration of 21.4×100^{-2} g·dm^{-3} [21]. The phenomenon of the same species reacting the same way to a substance regardless of how it is cultured and tested (individually or in communities) may indicate that the communities as a whole, but also the individual components of the community, do not necessarily have mechanisms to protect them from the toxic effects of the IL under testing.

Exposure time was also shown to be a variable having an effect on the action of the IL because, as reported by [62], excessive accumulation of the IL in microorganisms increases its effect. In the case of *M. nummuloides*, which was part of the tested community in our study, a significant increase in cell number was observed in the control solution during the experiment, while on the third day at the [BMIM]Cl solution of 1.75×100^{-2} g·dm^{-3} 50% fewer cells were observed, while on the seventh day the abundance decreased to 8% of the abundance in the control solution. A similarly rapid abundance reduction response to [BMIM]Cl was observed for the green alga *Dunaliella salina* [21]. This implies that there is a group of sensitive species in natural marine phytobenthic communities that may be rapidly eliminated from the environment while exposed to ILs, e.g., the aforementioned *M. nummuloides* or *C. closterium*.

The cell condition index used in our study, which consists of an assessment of the chloroplast state, confirmed the observations based on cell abundance. In the species considered sensitive, the percentage of cells with deformed chloroplasts was much higher than in the resistant species, e.g., up to a half of the observed cells of *T. fasciculata* on the seventh day of testing had degraded chloroplasts at both solutions of [BMIM]Cl. The negative effect on the chloroplast condition was confirmed by toxicity studies on five ILs ([Cnmim]Cl, n = 6, 8, 10, 12, 16). Ultrastructural morphology performed during the study revealed IL negative effects on various cellular structures, e.g., chloroplast grana became loose and mitochondria and their intermembranes swelled [22]. Other studies also confirmed that chloroplast damage can be considered as an indicator of microalgal degradation [27,63].

Tests carried out on the Baltic Sea microphytobenthic communities from the Gulf of Gdańsk made it possible to assess the effect of the IL [BMIM]Cl on the microorganisms comprising this formation. The study showed that 1-butyl-3-methylimidazolium chloride is a relatively harmful substance for the entire community, which contradicts the assessment carried out by reports based on cumulative impact assessment [47,59,64]. Thus, we have confirmed that even ILs considered to be of relatively low hazard have a significant impact on the aquatic environment. Hence, we are convinced that this group of compounds requires special attention in the context of testing its effects on different ecosystem components, its bioaccumulation, and its fate in the environment, as suggested recently by a growing group of authors [9,11,64].

5. Conclusions

During this study, the toxic influence of [BMIM]Cl on the marine microphytobenthic communities was demonstrated. The majority of species comprising the tested community reacted negatively to the presence of [BMIM]Cl at concentrations between 10^{-3} and 10^{-2} g·dm^{-3}, with a reduction in cell abundance and a deterioration in cell condition. Only in the case of two dominant diatom species, *N. perminuta* and *N. ramosissima*, was a stimulation of growth observed. In conclusion, the IL [BMIM]Cl on a short time scale contributes to a reduction in the abundance of species representing diverse taxonomic groups, which translates into the decrease in the total abundance and biomass of the microphytobenthic communities. However, despite the elimination of individual taxa, it does not lead to the degradation of entire communities but to their transformation into communities strongly dominated by a few resistant taxa.

Author Contributions: Conceptualization, Z.S., A.Z. and F.P.; methodology, Z.S., A.Z. and F.P.; software, Z.S.; validation, Z.S., A.Z. and F.P.; formal analysis, Z.S., A.Z. and F.P.; investigation, Z.S.; resources, Z.S. and A.Z.; writing—original draft preparation, Z.S., A.Z. and F.P.; writing—review and editing, Z.S., A.Z. and F.P.; visualization, Z.S.; supervision, A.Z.; project administration, Z.S.; funding acquisition, Z.S. and A.Z. All authors have read and agreed to the published version of the manuscript.

Funding: This research was funded by a Research Project for Young Scientists from the Faculty of Oceanography and Geography, University of Gdansk (No. 538-G245-B209-16).

Institutional Review Board Statement: Not applicable.

Informed Consent Statement: Not applicable.

Data Availability Statement: All data are available upon request to the corresponding author.

Acknowledgments: The authors thank Adam Latała for his support in developing the idea for the experiments.

Conflicts of Interest: The authors declare no conflict of interest.

Appendix A

Table A1. List of taxa identified in the studied microphytobenthic communities with codes used in statistical analysis.

Group of Organisms	Taxon Code	Taxon Name	Author
Bacillariophyta	*ach_bre*	*Achnanthes adnata*	Bory
	ach_lem	*Achnanthes lemmermannii*	Hustedt
	amp_ova	*Amphora ovalis*	(Kützing) Kützing
	amp_ped	*Amphora pediculus*	(Kützing) Grunow
	bac_pax	*Bacillaria paxillifera*	(O.F. Müller) T. Marsson
	ber_rut	*Berkeleya rutilans*	(Trentepohl ex Roth) Grunow
	bre_lan	*Brebissonia lanceolata*	(C. Agardh) R.K. Mahoney & Reimer
	cha_wig	*Chaetoceros wighamii*	Brightwell
	coc_sp.	*Cocconeis* sp.	Ehrenberg
	cyl_clo	*Cylindrotheca closterium*	(Ehrenberg) Reimann & J.C. Lewin
	dia_ten	*Diatoma tenuis*	C. Agardh
	dia_vul	*Diatoma vulgaris*	Bory
	eny_pro	*Encyonema leibleinii*	(C. Agardh) W.J. Silva, R. Jahn, T.A.V. Ludwig, & M. Menezes
	ent_pal	*Entomoneis paludosa*	(W. Smith) Reimer
	fal_sp.	*Fallacia* sp.	Kützing
	gom_oli	*Gomphonella olivacea*	(Hornemann) Rabenhorst
	gram_mar	*Grammatophora marina*	(Lyngbye) Kützing
	gyr_sp.	*Gyrosigma acuminatum*	(Kützing) Rabenhorst
	amp_cof	*Halamphora coffeiformis*	(C. Agardh) Mereschkowsky
	lic_sp.	*Licmophora gracilis*	(Ehrenberg) Grunow
	mel_mon	*Melosira moniliformis*	C. Agardh
	mel_num	*Melosira nummuloides*	C. Agardh
	nav_gre	*Navicula gregaria*	Donkin
	nav_pal	*Navicula palpebralis*	Brébisson ex W. Smith
	nav_per	*Navicula perminuta*	Grunow
	nav_ram	*Navicula ramossisima*	(C. Agardh) Cleve
	nav_sp.	*Navicula* sp.	Bory
	nit_sig	*Nitzschia sigma*	(Kützing) W. Smith
	pla_del	*Planothidium delicatulum*	(Kützing) Round & Bukhtiyarova
	ple_sp.	*Pleurosigma* sp.	W. Smith
	pro_por	*Proschkinia poretzkajae*	(Koretkevich) D.G. Mann
	rho_abb	*Rhoicosphenia abbreviata*	(C. Agardh) Lange-Bertalot
	rho_gib	*Rhopalodia gibba*	(Ehrenberg) O. Müller
	tab_fas	*Tabularia fasciculata*	(C. Agardh) D.M. Williams & Round
	try_sp.	*Tryblionella*	W. Smith
Cyanobacteria	*dol_flo*	*Dolichospermum flosaquae*	(Brébisson ex Bornet & Flahault) P. Wacklin, L. Hoffmann & J. Komárek
	cya_sp.	*Cyanobacteria*	
	mer_sp.	*Merismopedia* sp.	(Turpin) Meneghini
	mic_sp.	*Microcystis* sp.	Lemmermann
	spi_maj	*Spirulina major*	Meyen
	spi_sub	*Spirulina subsalsa*	Oersted ex Gomont
	wor_sp.	*Woronichinia* sp.	A.A. Elenkin
Chlorophyta	*ped_bor*	*Pseudopediastrum boryanum*	(Turpin) E. Hegewald
	sce_sp.	*Scenedesmus* sp.	Meyen
Dinophyceae	*per_sp.*	*Peridinium* sp.	Ehrenberg
Haptophyta	*pry_sp.*	*Prymnesium* sp.	N. Carter

Table A2. Results of the similarity analysis calculated with SIMPER. Average similarity: 71.92%.

Species	Av. Abundance	Av. Similarity	Sim/SD	Contribution%	Cumulative%
Bacillaria paxillifera	6.06	6.13	8.84	8.52	8.52
Tabularia fasciculata	5.93	6.06	9.60	8.43	16.94
Diatoma vulgaris	5.73	5.89	9.08	8.19	25.13
Melosira nummuloides	5.09	5.02	7.19	6.98	32.12
Navicula perminuta	4.94	4.84	5.10	6.73	38.85
Merismopedia sp.	5.11	4.50	1.85	6.26	45.10
Halamphora coffeiformis	3.58	3.56	6.10	4.94	50.05
Navicula gregaria	3.73	3.52	4.99	4.90	54.94
Navicula ramossisima	3.54	3.40	6.27	4.72	59.67
Cylindrotheca closterium	3.99	3.17	1.78	4.40	64.07
Grammatophora marina	3.23	3.16	7.58	4.40	68.47
Spirulina major	2.38	2.55	7.16	3.55	72.01

References

1. Chen, B.; Xue, C.; Amoah, P.K.; Li, D.; Gao, K.; Deng, X. Impacts of Four Ionic Liquids Exposure on a Marine Diatom Phaeodactylum Tricornutum at Physiological and Biochemical Levels. *Sci. Total Environ.* **2019**, *665*, 492–501. [CrossRef] [PubMed]
2. Mai, N.L.; Ahn, K.; Koo, Y.-M. Methods for Recovery of Ionic Liquids—A Review. *Process Biochem.* **2014**, *49*, 872–881. [CrossRef]
3. Liu, D.; Liu, H.; Wang, S.; Chen, J.; Xia, Y. The Toxicity of Ionic Liquid 1-Decylpyridinium Bromide to the Algae Scenedesmus Obliquus: Growth Inhibition, Phototoxicity, and Oxidative Stress. *Sci. Total Environ.* **2018**, *622*, 1572–1580. [CrossRef]
4. Isosaari, P.; Srivastava, V.; Sillanpää, M. Ionic Liquid-Based Water Treatment Technologies for Organic Pollutants: Current Status and Future Prospects of Ionic Liquid Mediated Technologies. *Sci. Total Environ.* **2019**, *690*, 604–619. [CrossRef] [PubMed]
5. Szepiński, E. Synteza, Właściwości i Zastosowanie Cieczy Jonowych, Pochodnych Naturalnych Związków Organicznych. Ph.D. Thesis, Department of Organic Chemistry, Faculty of Chemistry, Gdansk University of Technology, Gdansk, Poland, 2019.
6. Cvjetko Bubalo, M.; Radošević, K.; Radojčić Redovniković, I.; Halambek, J.; Gaurina Srček, V. A Brief Overview of the Potential Environmental Hazards of Ionic Liquids. *Ecotoxicol. Environ. Saf.* **2014**, *99*, 1–12. [CrossRef] [PubMed]
7. Flieger, J.; Flieger, M. Ionic Liquids Toxicity—Benefits and Threats. *Int. J. Mol. Sci.* **2020**, *21*, 6267. [CrossRef]
8. Nikitenko, S.I.; Berthon, C.; Moisy, P. Instability of Actinide (IV) Hexachloro Complexes in Room-Temperature Ionic Liquid [BuMeIm] PF6 Due to Hydrolysis of the Hexafluorophosphate Anion. *Comptes Rendus Chim.* **2007**, *10*, 1122–1127. [CrossRef]
9. Freire, M.G.; Neves, C.M.S.S.; Marrucho, I.M.; Coutinho, J.A.P.; Fernandes, A.M. Hydrolysis of Tetrafluoroborate and Hexafluorophosphate Counter Ions in Imidazolium-Based Ionic Liquids. *J. Phys. Chem. A* **2010**, *114*, 3744–3749. [CrossRef]
10. Maculewicz, J.; Świacka, K.; Stepnowski, P.; Dołżonek, J.; Białk-Bielińska, A. Ionic Liquids as Potentially Hazardous Pollutants: Evidences of Their Presence in the Environment and Recent Analytical Developments. *J. Hazard. Mater.* **2022**, *437*, 129353. [CrossRef]
11. Kowalska, D.; Stolte, S.; Wyrzykowski, D.; Stepnowski, P.; Dołżonek, J. Interaction of Ionic Liquids with Human Serum Albumin in the View of Bioconcentration: A Preliminary Study. *Chem. Pap.* **2022**, *76*, 2405–2417. [CrossRef]
12. Frade, R.F.; Afonso, C.A. Impact of Ionic Liquids in Environment and Humans: An Overview. *Hum. Exp. Toxicol.* **2010**, *29*, 1038–1054. [CrossRef] [PubMed]
13. Costa, S.P.F.; Azevedo, A.M.O.; Pinto, P.; Saraiva, L. Environmental Impact of Ionic Liquids: An Overview of Recent (Eco) Toxicological and (Bio) Degradability Literature. *ChemSusChem* **2017**, *10*, 2321–2347. [CrossRef] [PubMed]
14. Singh, S.K.; Savoy, A.W. Ionic Liquids Synthesis and Applications: An Overview. *J. Mol. Liq.* **2020**, *297*, 112038. [CrossRef]
15. Armitage, J.M.; Erickson, R.J.; Luckenbach, T.; Ng, C.A.; Prosser, R.S.; Arnot, J.A.; Schirmer, K.; Nichols, J.W. Assessing the Bioaccumulation Potential of Ionizable Organic Compounds: Current Knowledge and Research Priorities. *Environ. Toxicol. Chem.* **2017**, *36*, 882–897. [CrossRef] [PubMed]
16. Latała, A.; Stepnowski, P.; Nędzi, M.; Mrozik, W. Marine Toxicity Assessment of Imidazolium Ionic Liquids: Acute Effects on the Baltic Algae Oocystis Submarina and Cyclotella Meneghiniana. *Aquat. Toxicol.* **2005**, *73*, 91–98. [CrossRef] [PubMed]
17. Latała, A.; Nędzi, M.; Stepnowski, P. Toxicity of Imidazolium Ionic Liquids towards Algae. Influence of Salinity Variations. *Green Chem.* **2010**, *12*, 60–64. [CrossRef]
18. Kulacki, K.J.; Lamberti, G.A. Toxicity of Imidazolium Ionic Liquids to Freshwater Algae. *Green Chem.* **2008**, *10*, 104–110. [CrossRef]
19. Samorì, C.; Sciutto, G.; Pezzolesi, L.; Galletti, P.; Guerrini, F.; Mazzeo, R.; Pistocchi, R.; Prati, S.; Tagliavini, E. Effects of Imidazolium Ionic Liquids on Growth, Photosynthetic Efficiency, and Cellular Components of the Diatoms Skeletonema Marinoi and Phaeodactylum Tricornutum. *Chem. Res. Toxicol.* **2011**, *24*, 392–401. [CrossRef]
20. Fan, H.; Liu, H.; Dong, Y.; Chen, C.; Wang, Z.; Guo, J.; Du, S. Growth Inhibition and Oxidative Stress Caused by Four Ionic Liquids in Scenedesmus Obliquus: Role of Cations and Anions. *Sci. Total Environ.* **2019**, *651*, 570–579. [CrossRef]

21. Zhu, Y.; Zhong, X.; Wang, Y.; Zhao, Q.; Huang, H. Growth Performance and Antioxidative Response of Chlorella Pyrenoidesa, Dunaliella Salina, and Anabaena Cylindrica to Four Kinds of Ionic Liquids. *Appl. Biochem. Biotechnol.* **2021**, *193*, 1945–1966. [CrossRef]

22. Liu, H.; Zhang, S.; Zhang, X.; Chen, C. Growth Inhibition and Effect on Photosystem by Three Imidazolium Chloride Ionic Liquids in Rice Seedlings. *J. Hazard. Mater.* **2015**, *286*, 440–448. [CrossRef] [PubMed]

23. Liu, H.; Zhang, X.; Chen, C.; Du, S.; Dong, Y. Effects of Imidazolium Chloride Ionic Liquids and Their Toxicity to Scenedesmus Obliquus. *Ecotoxicol. Environ. Saf.* **2015**, *122*, 83–90. [CrossRef] [PubMed]

24. Dahl, B.; Blanck, H. Pollution-Induced Community Tolerance (PICT) in Periphyton Communities Established under Tri-n-Butyltin (TBT) Stress in Marine Microcosms. *Aquat. Toxicol.* **1996**, *34*, 305–325. [CrossRef]

25. Porsbring, T.; Arrhenius, Å.; Backhaus, T.; Kuylenstierna, M.; Scholze, M.; Blanck, H. The SWIFT Periphyton Test for High-Capacity Assessments of Toxicant Effects on Microalgal Community Development. *J. Exp. Mar. Biol. Ecol.* **2007**, *349*, 299–312. [CrossRef]

26. Sylwestrzak, Z.; Zgrundo, A.; Pniewski, F. Ecotoxicological Studies on the Effect of Roundup® (Glyphosate Formulation) on Marine Benthic Microalgae. *Int. J. Environ. Res. Public. Health* **2021**, *18*, 884. [CrossRef]

27. Sylwestrzak, Z.; Zgrundo, A.; Pniewski, F. Copper Chloride (II) Effect on the Composition and Structure of Marine Microphyto-benthic Communities. *Environ. Monit. Assess.* **2022**, *194*, 1–15. [CrossRef]

28. Telesh, I.; Schubert, H.; Skarlato, S. Life in the Salinity Gradient: Discovering Mechanisms behind a New Biodiversity Pattern. *Estuar. Coast. Shelf Sci.* **2013**, *135*, 317–327. [CrossRef]

29. Pham, T.P.T.; Cho, C.-W.; Yun, Y.-S. Environmental Fate and Toxicity of Ionic Liquids: A Review. *Water Res.* **2010**, *44*, 352–372. [CrossRef]

30. Rosenberg, M.; Kulkarni, G.V.; Bosy, A.; McCulloch, C.A.G. Reproducibility and Sensitivity of Oral Malodor Measurements with a Portable Sulphide Monitor. *J. Dent. Res.* **1991**, *70*, 1436–1440. [CrossRef] [PubMed]

31. Connolly, N.M.; Crossland, M.R.; Pearson, R.G. Effect of Low Dissolved Oxygen on Survival, Emergence, and Drift of Tropical Stream Macroinvertebrates. *J. N. Am. Benthol. Soc.* **2004**, *23*, 251–270. [CrossRef]

32. Sylwestrzak, Z.; Zgrundo, A. Preliminary Results of Laboratory Tests Using Microphytobenthic Communities in the Gulf of Gdańsk for Toxicity Testing of Ionic Liquids. In *Metodologia Badań Wykorzystywana Przez Młodych Naukowców*; CREATIVETIME: Kraków, Poland, 2014; pp. 74–81. (In Polish)

33. Sylwestrzak, Z.; Zgrundo, A.; Latała, A. Effect of the Ionic Liquid [BMIM] Cl on the Baltic Diatom *Navicula ramosissma* (C. Agardh) Cleve in a Laboratory Experiment on Natural Microphytobenthic Communities of the Gulf of Gdansk. In *Zagadnienia Aktualnie Poruszane przez Młodych Naukowców 3*; CREATIVETIME: Kraków, Poland, 2015; pp. 227–231. (In Polish)

34. Utermöhl, H. Zur Vervollkommnung der Quantitativen Phytoplankton-MethodikMitteilungen Internationale Vereinigung Theoretische und Angewandte. *Limnologie* **1958**, *9*, 1–38.

35. HELCOM. *Manual for Marine Monitoring in the COMBINE Programme of HELCOM*; HELCOM: Helsinki, Finland, 2014.

36. Snoeijs, P.; Potapova, M. *Intercalibration and Distribution of Diatom Species in the Baltic Sea*; Opulus Press: Uppsala, Sweden, 1993; Volume 1.

37. Snoeijs, P.; Vilbaste, S. *Intercalibration and Distribution of Diatom Species in the Baltic Sea*; Opulus Press: Uppsala, Sweden, 1994; Volume 2.

38. Snoeijs, P.; Potapova, M. *Intercalibration and Distribution of Diatom Species in the Baltic Sea*; Opulus Press: Uppsala, Sweden, 1995; Volume 3.

39. Snoeijs, P.; Kasperovičiene, J. *Intercalibration and Distribution of Diatom Species in the Baltic Sea*; Opulus Press: Uppsala, Sweden, 1996; Volume 4.

40. Snoeijs, P.; Balashova, N. *Intercalibration and Distribution of Diatom Species in the Baltic Sea*; Opulus Press: Uppsala, Sweden, 1998; Volume 5.

41. Witkowski, A.; Lange-Bertalot, H.; Metzeltin, D. *Diatom Flora of Marine Coasts*; A.R.G. Gantner Verlag K.G.: Königstein, Germany, 2000; Volume I.

42. Pliński, M.; Hindák, F. *Zielenice-Chlorophyta (Green Algae):(With the English Key for the Identification to the Genus), Cz. 1, Zielenice Nienitkowate (Prasinophyceae & Chlorophyceae) = Pt. 1, Non-Filamentous Green Algae*; Wydawnictwo Uniwersytetu Gdańskiego: Gdańsk, Poland, 2010.

43. Pliński, M.; Komárek, J. *Sinice – Cyanobakterie (Cyanoprokaryota):(With the English Key for the Identification to the Genus) Vol. 1*; Wydawnictwo Uniwersytetu Gdańskiego: Gdańsk, Poland, 2007; p. 172.

44. Lepš, J.; Šmilauer, P. *Multivariate Analysis of Ecological Data Using CANOCO*; Cambridge University Press: Cambridge, UK, 2003.

45. Ter Braak, C.J.F.; Šmilauer, P. *Program CANOCO, Version 4.52. Biometris: Quantitative Methods in the Life and Earth Sciences*; Plant Research International, Wageningen University and Research Centre: Wageningen, The Netherlands, 2003.

46. Zajac, A.; Kukawka, R.; Pawlowska-Zygarowicz, A.; Stolarska, O.; Smiglak, M. Ionic Liquids as Bioactive Chemical Tools for Use in Agriculture and the Preservation of Agricultural Products. *Green Chem.* **2018**, *20*, 4764–4789. [CrossRef]

47. Kowalska, D.; Maculewicz, J.; Stepnowski, P.; Dołżonek, J. Ionic Liquids as Environmental Hazards–Crucial Data in View of Future PBT and PMT Assessment. *J. Hazard. Mater.* **2021**, *403*, 123896. [CrossRef] [PubMed]

48. Ahrens, L. Polyfluoroalkyl Compounds in the Aquatic Environment: A Review of Their Occurrence and Fate. *J. Environ. Monit.* **2011**, *13*, 20–31. [CrossRef]

49. Li, Y.; Si-Cong, L.; Shi, Y.; Yu-Ping, Q. Effects of Soil Characteristics on Sorption-Desorption of 1-Butyl-3-Methyl-Imidazolium-Based Ionic Liquids. *J. Agric. Resour. Environ.* **2017**, *34*, 30.
50. Neuwald, I.; Muschket, M.; Zahn, D.; Berger, U.; Seiwert, B.; Meier, T.; Kuckelkorn, J.; Strobel, C.; Knepper, T.P.; Reemtsma, T. Filling the Knowledge Gap: A Suspect Screening Study for 1310 Potentially Persistent and Mobile Chemicals with SFC-and HILIC-HRMS in Two German River Systems. *Water Res.* **2021**, *204*, 117645. [CrossRef]
51. Pati, S.G.; Arnold, W.A. Comprehensive Screening of Quaternary Ammonium Surfactants and Ionic Liquids in Wastewater Effluents and Lake Sediments. *Environ. Sci. Process. Impacts* **2020**, *22*, 430–441. [CrossRef]
52. Probert, P.M.; Leitch, A.C.; Dunn, M.P.; Meyer, S.K.; Palmer, J.M.; Abdelghany, T.M.; Lakey, A.F.; Cooke, M.P.; Talbot, H.; Wills, C. Identification of a Xenobiotic as a Potential Environmental Trigger in Primary Biliary Cholangitis. *J. Hepatol.* **2018**, *69*, 1123–1135. [CrossRef]
53. Leitch, A.C.; Ibrahim, I.; Abdelghany, T.M.; Charlton, A.; Roper, C.; Vidler, D.; Palmer, J.M.; Wilson, C.; Jones, D.E.; Blain, P.G. The Methylimidazolium Ionic Liquid M8OI Is Detectable in Human Sera and Is Subject to Biliary Excretion in Perfused Human Liver. *Toxicology* **2021**, *459*, 152854. [CrossRef]
54. Gathergood, N.; Scammells, P.J. Design and Preparation of Room-Temperature Ionic Liquids Containing Biodegradable Side Chains. *Aust. J. Chem.* **2002**, *55*, 557–560. [CrossRef]
55. Stepnowski, P.; Zaleska, A. Comparison of Different Advanced Oxidation Processes for the Degradation of Room Temperature Ionic Liquids. *J. Photochem. Photobiol. Chem.* **2005**, *170*, 45–50. [CrossRef]
56. Coleman, D.; Gathergood, N. Biodegradation Studies of Ionic Liquids. *Chem. Soc. Rev.* **2010**, *39*, 600. [CrossRef] [PubMed]
57. Hagiwara, R.; Lee, J.S. Ionic Liquids for Electrochemical Devices. *Electrochemistry* **2007**, *75*, 23–34. [CrossRef]
58. Amde, M.; Liu, J.-F.; Pang, L. Environmental Application, Fate, Effects, and Concerns of Ionic Liquids: A Review. *Environ. Sci. Technol.* **2015**, *49*, 12611–12627. [CrossRef] [PubMed]
59. Khan, M.I.; Mubashir, M.; Zaini, D.; Mahnashi, M.H.; Alyami, B.A.; Alqarni, A.O.; Show, P.L. Cumulative Impact Assessment of Hazardous Ionic Liquids towards Aquatic Species Using Risk Assessment Methods. *J. Hazard. Mater.* **2021**, *415*, 125364. [CrossRef] [PubMed]
60. Li, B.; Zhang, X.; Deng, J.; Cheng, Y.; Chen, Z.; Qin, B.; Tefsen, B.; Wells, M. A New Perspective of Copper-Iron Effects on Bloom-Forming Algae in a Highly Impacted Environment. *Water Res.* **2021**, *195*, 116889. [CrossRef]
61. Śliwińska-Wilczewska, S.; Sylwestrzak, Z.; Maculewicz, J.; Zgrundo, A.; Pniewski, F.; Latała, A. The Effects of Allelochemicals and Selected Anthropogenic Substances on the Diatom Bacillaria Paxillifera. *Eduk. Biol. Sr.* **2016**, *1*, 21–27.
62. Zhang, C.; Zhu, L.; Wang, J.; Wang, J.; Zhou, T.; Xu, Y.; Cheng, C. The Acute Toxic Effects of Imidazolium-Based Ionic Liquids with Different Alkyl-Chain Lengths and Anions on Zebrafish (Danio Rerio). *Ecotoxicol. Environ. Saf.* **2017**, *140*, 235–240. [CrossRef]
63. Kim, Y.; Ponomarev, A.V. Low-Dose Electron Beam Treatment of Red Tide Blooms Microalgae. *Radiat. Phys. Chem.* **2021**, *179*, 109201. [CrossRef]
64. Jungnickel, C.; Mrozik, W.; Markiewicz, M.; Luczak, J. Fate of Ionic Liquids in Soils and Sediments. *Curr. Org. Chem.* **2011**, *15*, 1928–1945. [CrossRef]

Review

Toxic Metals in a Paddy Field System: A Review

Yuanliang Duan, Qiang Li, Lu Zhang, Zhipeng Huang, Zhongmeng Zhao, Han Zhao, Jun Du and Jian Zhou *

Fisheries Institute, Sichuan Academy of Agricultural Sciences, Chengdu 611731, China; duan9104@126.com (Y.D.); liq7920@126.com (Q.L.); zhanglu425@163.com (L.Z.); h3392078@163.com (Z.H.); 18227552594@163.com (Z.Z.); zhaohan232323@163.com (H.Z.); dujun9100@126.com (J.D.)
* Correspondence: zhoujian980@126.com; Tel./Fax: +86-028-87955015

Abstract: The threat of toxic metals to food security and human health has become a high-priority issue in recent decades. As the world's main food crop source, the safe cultivation of rice has been the focus of much research, particularly the restoration of toxic metals in paddy fields. Therefore, in this paper, we focus on the effects of toxic metals on rice, as well as the removal or repair methods of toxic metals in paddy fields. We also provide a detailed discussion of the sources and monitoring methods of toxic metals pollution, the current toxic metal removal, and remediation methods in paddy fields. Finally, several important research issues related to toxic metals in paddy field systems are proposed for future work. The review has an important guiding role for the future of heavy metal remediation in paddy fields, safe production of rice, green ecological fish culture, and human food security and health.

Keywords: toxic metals; health risk; paddy field; bioremediation; rice–fish co-culture system

Citation: Duan, Y.; Li, Q.; Zhang, L.; Huang, Z.; Zhao, Z.; Zhao, H.; Du, J.; Zhou, J. Toxic Metals in a Paddy Field System: A Review. *Toxics* **2022**, *10*, 249. https://doi.org/10.3390/toxics10050249

Academic Editors: Stefano Magni, François Gagné and Valerio Matozzo

Received: 11 April 2022
Accepted: 13 May 2022
Published: 16 May 2022

Publisher's Note: MDPI stays neutral with regard to jurisdictional claims in published maps and institutional affiliations.

1. Introduction

Metals are typically classified as non-essential and essential metals. Non-essential metals (mercury (Hg), cadmium (Cd), lead (Pb), etc.) have no proven biological purpose, and their toxicity is a function of their concentration. In contrast, essential metals (iron (Fe), zinc (Zn), copper (Cu), etc.) have known biological effects, and toxicity occurs when the metal is metabolically deficient or at a high concentration [1]. In general, any metal or metalloid that is not used in the basic metabolism or is not biodegradable is considered a heavy metal [2]. Heavy metals are defined as those elements with a density greater than $5\ \text{g/cm}^3$ and an atomic number greater than 20, and include 53 metals [2] (gold (Au), silver (Ag), Cu, etc.). Despite being a metalloid element, arsenic (As) is listed as a heavy metal due to its multiple properties, similar to heavy metals. Heavy metals can be divided into toxic (e.g., Cd), precious (e.g., Au), and radionuclide (e.g., Uranium, U) metals [3]. The term "heavy metal" is considered meaningless at worst and imprecise at best [4,5]; this paper replaces it with "toxic metal" in later descriptions. When the intake of toxic metals exceeds a certain amount, it will do serious harm to humans' kidneys, nervous system, and fertility, and even cause cancer and death [6–8]. In China, Hg, Cd, Pb, chromium (Cr), and As are commonly known as the "five poisons" [9]. Numerous toxic metals (TM) (Cd, Pb, Hg, As, etc.) have been identified as non-essential elements for bodily functions [10] and have been included in the top 20 list of hazardous substances [11]. Therefore, the toxicity, and maximum allowable value of TMs in food and other related issues has been the focus of much attention.

Rice is an extremely important food crop, not only in China but also across the world, with a global cultivation area reaching 145 million hectares. The principle rice-producing areas are distributed in East, South, and Southeast Asia, followed by Mediterranean coastal countries, the United States and Brazil [12]. At present, more than half of the world's population relies on rice for their calory intake [13], and thus a decline in rice yield will have a significant global impact. By 2050, the growth in population and acceleration of urbanization will require each rice-producing hectare to feed at least 43 people, compared to the current value of 27 people per hectare [14]. Kaur et al. [15] predicted that rice production would need to increase by at least 40% by 2050 to ensure food security. Rice quality is particularly important due to the key role rice plays in the global diet. High-quality rice produced by artificial breeding has been observed to have therapeutic and preventive effects on several human diseases [16]. In addition to nutrients, the phytochemicals present in rice can play a biological role, with antioxidant, anticancer, anti-diabetic, and anti-inflammatory effects [17]. However, poor-quality rice can have a negative impact on human health, causing dizziness, thoracic stuffiness, nausea, vomiting, abdominal discomfort, etc. [18]. Therefore, ensuring the high quality and yield of rice is of great significance.

In 2020, a total of 720–811 million people worldwide were facing hunger, and nearly one-third of the global population (2.37 billion) were not able to get enough food [19]. The yield and quality of rice have been the subject of much interest. TM pollution not only reduces the quality and yield of rice, but also endangers human health and even causes death. The ecological cultivation of aquatic animals and plants has recently attracted the attention of scholars and governments, with its vigorous promotion and active application in practice by farmers. This has consequently improved the quality and yield of agricultural products as well as their taste and safety. An in-depth understanding of the relationships between TMs and rice, and their corresponding removal methods are important prerequisites for effectively improving the food safety of rice. To achieve this goal, in the current paper, we first introduce the effects of TMs on rice. Then, the sources, monitoring methods and measuring instruments of TM pollution, as well as the measures employed to remove TMs and reduce their bioavailability in paddy fields, are discussed in detail. Finally, the research trends and challenges of TM removal or remediation in paddy fields are introduced.

2. Effects of TMs on Rice

The relationships between TMs and the growth, development, metabolism, and nutrient composition of rice have been the subject of extensive research (Figure 1). As trace elements, the presence of TMs can have both positive and negative effects on organisms. Low TM concentrations can promote the metabolic activities of organisms and vice versa for high concentrations [20]. TM stress has been observed to have a significant effect on agronomic rice traits, including panicle number per plant, filled grain per panicle, 1000-grain weight, and grain yield per plant [21–25]. In addition, TM stress exerts a great impact on the molecular and gene level of rice, inducing changes of a higher complexity compared to those of the agronomic traits, which are mainly related to the physiological metabolism, variations in enzymes, and the regulation of gene expression [26,27]. Such examples are internal factors that affect the growth and development of rice. Thus, increasing our understanding of the TM mechanisms affecting rice is essential for the prevention and control of TMs.

2.1. Effects of TMs on Apparent Indexes and Body Composition of Rice

Highly available Zn concentrations in the soil not only result in rice chlorosis and inhibit plant growth but also reduce Fe concentrations in rice plant shoots to the level considered deficient [28]. Moreover, Cd can prevent the transformation of starch, affect seed germination by inhibiting amylase activity [29], and reduce the chlorophyll content of rice, thus affecting photosynthesis [23,30]. Hg can inhibit germination and seedling growth by damaging the embryo, while Pb destroys endosperm starch solubilization by inhibiting α-amylase activity, and obstructs seed germination and seedling growth [31]. Ni can affect the H-ATpase activity and lipid composition of rice cell membranes [32] and reduce the content of ions (Na, K, and Ca), photosynthetic pigments (chlorophyll and carotenoids), total protein, and organic nitrogen in seedlings [33]. The down-regulation of key metabolic enzymes with excessive Cu, such as α-amylase or enolase, not only affects the starvation absorption of water by seeds, but also leads to the failure of the reserve mobilization process [26]. Lanthanum (La) shoot contents exceeding the toxicity value result in a decline in plant growth and chlorophyll a/b, while peroxidase activity, cell membrane permeability, and proline content in the leaves increase [34–37]. Tables S1 and S2 report more effects of TMs on the apparent indexes and body composition of rice [38–52].

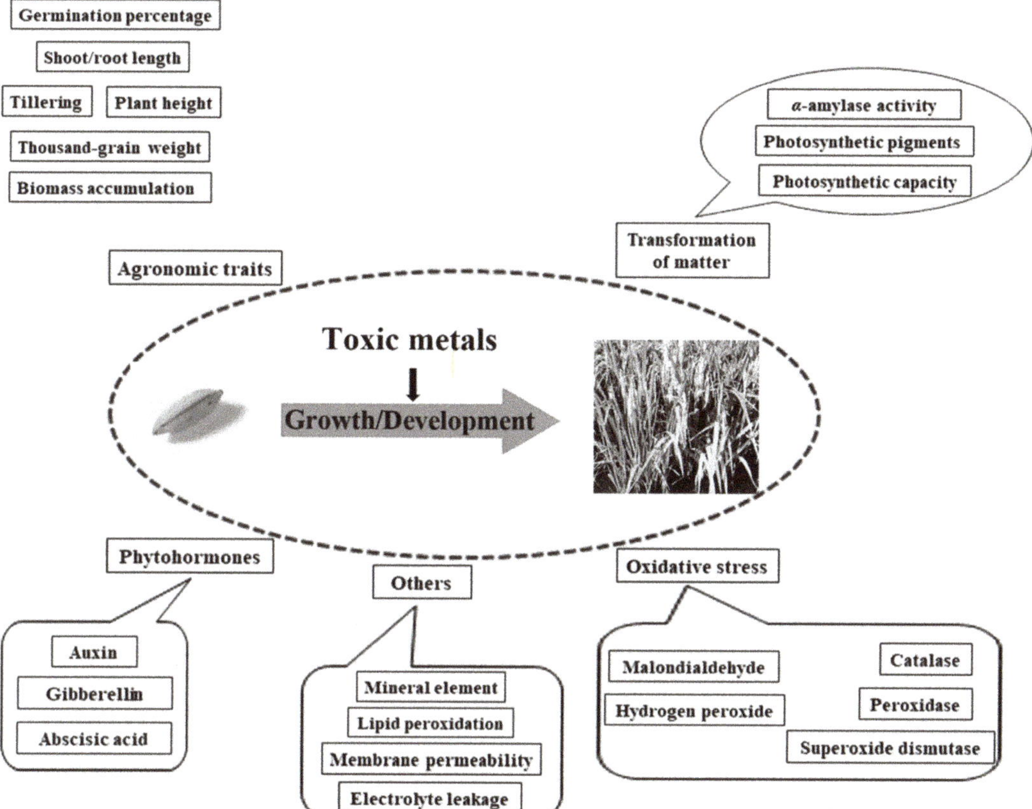

Figure 1. The relationships between toxic metals and the growth, development, metabolism, and nutrient composition of rice.

2.2. Effects of TMs on the Gene Expression of Rice

Genetic diversity is an important intrinsic factor determining the TM content in rice grains. The TM uptake of different cultivars is varied; for example, Cd-tolerant rice has a certain tolerance to Cd stress [53]. Metal exposure can result in the up- or downregulation of some genes/miRNAs in rice [54] and can consequently alter the corresponding physiological, biochemical, and metabolic properties. Furthermore, TMs can induce the expression of genes involved in numerous biological processes in rice [26,55–57], such as signal transduction, ion transport and binding, stress responses, metabolism, etc. However, the expression of genes can also affect the impact of TMs on rice (Table 1). The interaction between genes and TMs is highly complex, and more research is required to clarify the underlying mechanisms.

Table 1. Toxic metal-related genes.

Gene/Protein	Function	References
OsHMA1, OsHMA2, OsHMA3, OsHMA4, OsHMA5	involves the transfer of toxic metals and reduces toxic metals concentrations in rice grains	[55,57–59]
OsGSTL2	provides tolerance for toxic metals and other abiotic stresses	[60]
OsLEA4	contributes to toxic metal, drought and salt tolerance in rice plants	[61]
CAL1	reduces cytosolic Cd concentration by promoting the secretion of Cd into the extracellular space and chelating Cd in the cytosol	[56]
MTP11	Mn tolerance via intracellular Mn compartmentalization	[62]
OsATX1	promotes the redistribution of Cu from old leaves to developing tissues and seeds, as well as the root-to-shoot Cu translocation in rice	[63]
OsMTs	promotes rice multiple stress tolerance	[64]
WaarsM	induces As volatilization and methylation, thus reducing As content in grains	[65]
OsMRLK	promotes rice development and multiple stress tolerance	[66]
OsLCT1, OsHMA2, OsZIP3	alleviates the oxidative stress of Cd and Zn, decrease the translocation and accumulation of Cd to grains	[57]
OsVMT	enhances the content of Zn and Fe in polished rice	[67]
OsNRAMP1, OsNRAMP5	contributes to the uptake of Cd and Mn in rice	[68,69]

2.3. Effects of TMs with Types and Forms on Rice

The impacts of TMs on rice vary according to the types and forms of TMs, while the same TM can also have different effects on rice. When the concentration of TMs reaches a certain level, it can completely inhibit the germination and growth of rice. For example, high levels of Hg and Pb can inhibit rice germination and seedling growth [21,22] and different types of As can have similar effects [49,70]. The roots of rice are more sensitive to TMs compared to shoots [70]. The regularity effect of TMs on an enzyme is a function of the TM type. For example, Pb, Hg, Cd, and Zn can decrease the superoxide dismutase activity of rice [23,43,45,46], while the opposite has been reported for Cd and Pb [44,47]. Peroxidase is more sensitive to Pb and Cd stress than superoxide dismutase [71]. See Tables S1 and S2 for more information. Therefore, the relationship between TMs and the physiological and biochemical indexes of rice is highly complex and requires verification.

3. TM Sources in Paddy Fields

TMs originate from both natural and man-made sources. Natural sources include parent rock weathering [72], volcanic activity [73], and atmospheric deposition [74], etc. Anthropogenic sources refer to pollutants discharged into the environment via human activities, such as industrial emissions [75], pesticides, and chemical fertilizer residues [76], domestic wastewater [77], etc. Anthropogenic activity is the principal source of TMs in paddy fields. In particular, the concentration of TMs in industrialized areas is generally higher than that in non-industrialized areas [78]. The diffusion of TMs from the source to the surrounding areas typically depends on pollution sources, the geochemical status [79], and the surface runoff [80]. Moreover, TMs in the atmospheric phase are transferred to paddy fields via atmospheric deposition. In particular, wet atmospheric deposition of TMs is positively correlated with rainfall depth, while dry deposition is positively correlated with the number of dry days [81]. Sediments are a secondary pollution source of TMs, greatly threatening the safety of aquatic ecosystems. The accumulation or release of TMs in sediments to water bodies depends on the physical, chemical, and biological conditions of the sediment-water interface, such as changes in salinity and pH, biological disturbances, tidal currents, and floods [82].

4. TM Monitoring Methods

In view of the importance of food security and the crucial role of rice products in the human diet, TM monitoring and corresponding risk assessments are particularly important. Numerous methods have been developed to investigate the effects of TM pollution in rice paddy fields. Currently, the most popular methods are those using models based on advanced science and technology, for example, the back propagation neural-network [83], time-spectrum feature space [84], and World Food Study (WOFOST) models [85]. Brus et al. [86] employed a multiple linear model to successfully predict the content of TM s in rice grains. The generalized dynamic fuzzy neural network model combines spectral indices with environmental parameters to predict TM concentrations (Cu and Cd), outperforming adaptive-network-based fuzzy interference systems, back-propagation (BP) neural network models, and regression models [87]. Furthermore, the field-scale TM assessment model improves the monitoring of TM stress from its predecessor, the generalized dynamic fuzzy neural network model [88]. Compared with traditional methods (toxicity characteristic leaching, diethylenetriaminepentaacetic acid extraction, and HCl extraction), field capacity-derived soil solution extraction can successfully predict the total Cd content of rice from the tillering to mature stages [89]. The development of science and technology has permitted the gradual application of remote sensing technology to the TM monitoring of crops, for example, the collection of biochemical and hyperspectral data. The coupling of these two data types can be adopted for TM monitoring, indicating the relationship between the TM content in the soil and the cell structure and chlorophyll content in rice canopies or leaves [90,91]. Optical remote sensing typically monitors the internal structure, color, and additional characteristics of crop cells, while microwave techniques focus on the geometric characteristics and morphology of cells. Combining these two technologies can build a robust monitoring model for TM stress in rice, and can also be used to investigate a variety of environmental stresses [92]. The WOFOST model is the most suitable model for the application of remote sensing technology in this field. In particular, integrating remote sensing technology and statistical methods with the world food research model can greatly improve the accurate monitoring of crop TM stress.

5. TM Measuring Instruments

For TMs, most exist in nature at natural concentrations, which are relatively low and difficult to detect. However, TMs will be enriched into the human body via food, soil, water, air, and other means, resulting in the destruction of normal human physiological metabolism and a serious impact on human health. Therefore, in order to clarify the TM pollution existing in the current living and ecological environment, it is necessary to do a good job of TM detection with the help of various analytical instruments.

At present, the instruments for the determination of TMs mainly include atomic absorption spectrometer, inductively coupled plasma spectrometer, atomic fluorescence spectrometer, inductively coupled plasma mass spectrometer, voltammetric analyzer, etc. Because of its sensitivity, accuracy, and simplicity, atomic absorption spectrometers have been widely used in the analysis of TMs in agriculture, food, and environmental monitoring [93–95], etc. An inductively coupled plasma spectrometer can detect multiple elements (metal elements and non-metallic elements) in the sample [96,97], and its sensitivity is relatively high. Compared with atomic absorption spectrometry, an inductively coupled plasma spectrometer is suitable for the determination of more than three samples at the same time. Atomic fluorescence spectrometers have the advantages of low price, low detection cost, low detection limit and high sensitivity [98–100]. It is easier to popularize than other detection instruments. Inductively coupled plasma mass spectrometry has good detection limit, scanning ability, and relatively high sensitivity; moreover, it can determine multiple elements at the same time and can determine and identify isotopes [101,102]. However, compared with other analytical instruments, the detection cost of inductively coupled plasma mass spectrometers is relatively high, and the detection limit of some elements is limited. A voltammetric analyzer can have a better detection effect for the monitoring of trace metals [103,104]. It can have a certain sensitivity and precision for the determination of TMs in some substances by means of dissolution analysis, so the accuracy of the final monitoring results is relatively high. It is relatively simple to use and is an important analytical instrument in trace analysis.

6. Remediation of TMs from Paddy Fields

Reducing TM content in rice has been the focus of much research, typically via enhancing the stability of pollutants in the soil, reducing soil surface pollutant concentrations, modifying the ability of rice to absorb and transport TMs, and so on. Studies on reducing plant available TMs in the environment generally concentrate on soil remediation techniques: turnover and dilution, in situ stabilization via chemical improvers, and bioremediation [105].

6.1. Remediation of TMs with Soil Amendments

6.1.1. Metal Soil Amendments

At present, the majority of relevant research is directed toward soil amendments due to their rapid and efficient effects. In particular, soil amendments are a class of compounds containing Ca, Fe, etc. TM adsorption by soil amendments occurs via physical adsorption, surface complexation, and ion exchange [106], which convert TMs to non-biousable forms [107]. The bioavailability of TMs in soil is the result of the interaction of organic matter, ions, redox conditions, and soil pH [108]. Soil pH exerts a great influence on TM content in rice; for example, the optimal pH for the adsorption of Ni (II) and Cu (II) is 6 and 5, respectively [109]. pH levels in paddy fields can be increased by applying liming and red mud [106,110]. As well as impacting pH levels, red mud also improves the microbial composition in paddy fields and increases the activity of urease, acid phosphatase, and catalase in the soil [110]. Fe exhibits high bioavailability and does not exert adverse effects on rice quality and yield [111]. Adding Fe to paddy fields can reduce the absorption of TMs by rice and increase the elemental contents of Fe, Cu, Mn in rice grains, and Zn in rice plants [112]. Moreover, the application of Fe-containing materials can effectively reduce the concentration of As in soil solutions and rice grains, with zero-valent Fe demonstrated to be particularly powerful. Makino et al. [113] attributed this to the formation of arsenic sulfide. Moreover, Yu et al. [114] determined a significant positive correlation between As and Fe, suggesting that Fe-containing amendments may have an indirect influence on the fractionation of soil As and biological effects to ease the As for rice. The application of metals or additional complexes can enhance the amount of iron plaque, which is composed of crystallized and amorphous iron oxides, hydroxides, etc. [115], on the root surface [116]. This technique can also enhance the interception of TMs by rice roots.

6.1.2. Non-Metallic Soil Amendments

Non-metallic soil amendments are mainly compounds containing silicon (Si), organic matter, etc. Si soil amendments have been the focus of much research due to their ability to actively induce the molecular expression of Cd tolerance in rice leaves. The addition of Si to paddy fields can reduce the As and Cd content in rice, alleviate abiotic stress, and increase rice yields, while also significantly improving the uptake of N, P, and K by rice [117]. Immobilized metals are typically bound in soil organic matter components and exist as an organic binding state [118], thus facilitating research on organic soil amendments. Common biochar contains straw, hull, etc. [119,120]. Biochar can effectively fix TMs and reduce their bioavailability and mobility in soil [118]. Moreover, biochar can directly or indirectly affect indigenous microorganisms by changing the physical and chemical properties and TM content of sediments [121]. However, biochar has also been reported to induce oxidative stress in rice [122], and thus any potential negative effects of biochar must be considered in agricultural and environmental applications. Furthermore, the application of non-metallic elements (e.g., Se and S) can alleviate the toxic effects of TMs on rice [116,123].

6.1.3. Nanoscale Soil Amendments

Nanoscale soil amendments have recently been the focus of much research, achieving promising results. For example, Nano-Si has a positive impact on the yield and growth of rice in polymetallic contaminated soil and can reduce TM content in grains [124]. Moreover, CuO nanoparticles have been reported to accelerate the arrival of the rice heading stage, shorten the plant life cycle, and reduce As accumulation in grains [125]. Biochar nanoparticles have a high adsorption affinity for Cd, thus reducing the toxicity of Cd in rice. This is particularly true for biochar nanoparticles prepared under high temperature conditions, manifested as increased biomass, root activity, and chlorophyll content in rice plants [122]. However, the toxicity and outcome of co-existing metals with nanoparticles remain unclear. The negative impact exerted by CuO nanoparticles on plant growth is more significant than that of bulk particles [126]. High ZnO nanoparticle concentrations are able to enhance the content of bioavailable Cd in rhizosphere soil. The addition of ZnO nanoparticles at high concentrations to soil containing low levels of Cd can significantly promote Cd accumulation in rice [127].

6.1.4. Composite Soil Amendments

Multiple TMs are typically present in paddy fields; thus, applying composite soil amendments rather than a single component is required. The Cd bioavailability in the rhizosphere of rice can decrease by 92–100% from the tillering stage to maturity via the application of Ca-Si-rich composite minerals. In addition, Si deposition on the rice root cross-section has been observed to significantly increase following the application of a Ca-Si-rich composite mineral treatment, which consequently enhances the storage of Cd in roots and reduces the translocation of Cd from the root to the shoot [128]. Sulfur and iron-modified biochar amendments can significantly increase the amount of iron plaque on the root surface, facilitating the transition of Cd to binding states (such as Fe-Mn oxide) and reducing Cd concentration in contaminated soil and rice grains [116]. The combined application of biochar and lime reduces Pb availability in soil and Pb accumulation in brown rice at a greater rate compared to the corresponding single applications [129]. Furthermore, integrating ferric oxide and calcium sulfate into a single amendment can effectively reduce the bioavailability of Pb and Cd in soil and the content of Cd, As, and Pb in rice grains [130]. Moreover, Honma et al. [131] demonstrated the ability of prolonged flooding with short-range-order iron hydroxide and rainfed management combined with converter furnace slag to reduce both the Cd and As uptake of rice.

6.2. Bioremediation of TMs

Bioremediation is a low-cost technique that has a limited impact on the environment, and includes phytoremediation, microbial remediation, animal remediation, etc. Crop rotation and intercropping are common TM remediation methods that can effectively ensure the safety and yield of rice and restore metal-contaminated soil [107,132,133], with examples including wheat-rice rotation, oilseed rape-rice rotation, rice-water spinach intercropping, etc. The composition and sources of environmental microbiota play a key role in the health and productivity management of sustainable agriculture. The application of microorganisms to paddy fields contaminated with TMs is a newly developed remediation method. Scholars have identified a reduction in TMs with resistant bacteria in rice grains and have highlighted the potential of bioremediation for contaminated soil. For example, Cd transporters (OsHMA2 and OsNramp5) in rice roots can experience down-regulation following inoculation with *Stenotrophomonas maltophilia*. This may be an internal factor affecting Cd content in rice [134]. Lin et al. [135] determined *Stenotrophomonas acidaminiphila*, *Pseudomonas aeruginosa*, and *Delftia tsuruhatensis* to be Cd tolerant, effectively reducing the enrichment of Cd in rice grains. In particular, *P. aeruginosa* is considered to be a multi-metal-resistant bacterium. Animal remediation technology refers to the absorption, transfer, or degradation of TMs through the food chain of soil animals, and research in this field is relatively limited. The backbone of animal remediation is microbial remediation [136]. For example, earthworms have the ability to alter the structure and permeability of soil, and form the basis of the most commonly used remediation method for TM-contaminated soil.

6.3. Field Management

Water management approaches are easy to operate and are commonly adopted. For example, continuous culture flooding is an effective method for reducing TM content in rice grains [137], yet it has an increased risk of As accumulation [113]. In addition, the aerobic conditions created by the release of water [138] in aerobic treatments can increase Cd concentrations [139]. Flooding in paddy fields may cause sulfide mineral precipitation, significantly reducing trace metal solubility [140], as well as the affinity for metals in the rhizosphere and iron plaque on the root surface [137]. Although the drying-wetting cycles of soil promote the release of metals into the water, Honma et al. [141] determined that intermittent irrigation (3-day flooding and 5-day no-flooding) can simultaneously reduce the accumulation of As and Cd in grains. Current research on the combined benefits of intermittent and aerobic irrigation demonstrates the ability of intermittent irrigation to reduce the Cd content in grains and increase rice yields [142].

6.4. Planting Methods and Varieties

Furthermore, the cultivation, season, and variety of rice may affect the relationship between rice and TMs during the production process. Deng et al. [143] demonstrated greater Cd and Pb contents in brown rice, straw, and roots via the direct seeding method compared with manual transplanting and seedling throwing. Farooq and Zhu [144] and Yi et al. [145] determined that early and late planting of rice impacted the Cd content in white rice. Differences in rice varieties are attributed to gene differences. Under low and moderate soil Cd pollution, japonica rice cultivars are more suitable than indica rice [146]. Dry season varieties are more tolerant to arsenite or arsenate than rainy season varieties [147]. Studies have also reduced the toxicity of TMs to rice by inserting or removing certain genes via genetic engineering and cross breeding [57,61,65,68]. For example, low Cd accumulation may reduce Cd absorption by inhibiting the bioavailability of Cd in the rhizosphere or by decreasing Cd transport [148]. Thus, the genotype, environment, and their interaction are considered the most significant factors affecting the TM content in rice grains.

7. Recent Trends and Challenges

The importance of rice to human beings and TMs may cause great harm to human beings through bioaccumulation, which will lead to research on TMs in rice fields becoming a hot issue. The relevant research on the treatment of TM pollution in paddy fields around the world mainly includes: (1) changing the existing state of TMs in soil, making them fixed or stable, and then reducing the activity of TMs; (2) changing the planting system to reduce the absorption of TMs by rice; (3) extracting TMs from rice fields, so as to reduce the concentration of TMs in rice fields. Since 2011, the number of papers published in the field of TM remediation in paddy fields has increased rapidly. These studies mainly focused on five TMs: Cd, Pb, As, Cu, and Zn, which indirectly showed that rice fields around the world were mainly polluted by these heavy metals. In addition, there was also a small amount of research on rare earth metals. Among these research hotspots, the remediation and treatment of Cd pollution in paddy fields was the main, and the passivation remediation fixation/stabilization technology was the most. However, the best effect of reducing Cd was the planting of low-accumulation Cd varieties, which could reach 74.50% [149].

It is very difficult to completely remove TMs from rice fields. At present, scholars mainly focus on passivating TMs, reducing their activity, and reducing their absorption by rice. In addition, at present, most of the treatment technologies of TM pollution in rice fields are still in the stage of laboratory or field experiment and demonstration, and there are few real engineering applications, so they have great market potential in the future. In order to realize the simultaneous production and repair activities, we can develop environment-friendly, long-lasting, and high-efficiency passivators for moderate and mild TM pollution; for severe pollution, the combination of soil elution technology and passivation remediation technology or phytoremediation technology can be used; for microbial remediation, especially the screening and cultivation of microbial strains, further research is needed. Due to the differences in soil properties, there are few universal technologies that can be popularized in a large area. Therefore, it is particularly important to develop efficient and lasting comprehensive prevention and control technology for toxic metal pollution in paddy fields according to local conditions.

8. Conclusions

TMs may affect human growth and development, physiological metabolism, etc., and may cause diseases and even death. TMs enter the food chain via organisms located at the bottom of the food chain, and their concentration and toxicity are subsequently amplified as they move further up the food chain. Consuming a certain amount of food contaminated by TMs can threaten an individual's health. Thus, humans (who are at the top of the food chain) face great health risks, as they risk TM exposure principally through food intake. Rice is more important than fish in terms of the risk of metal exposure in the human diet, and arsenic requires particular attention. Grain crops (e.g., rice) that grow on soil/water polluted by TMs not only experience a reduction in yield and quality but also enrich a large amount of TMs. To reduce the threat of TMs to human health, measures must be taken from the source. In particular, uncontaminated soil and water bodies can guarantee the production of healthy food, which is key to human health. Therefore, the research and exploration of the technical methods of heavy metal removal or remediation in rice fields is of great significance to human food safety and health.

Supplementary Materials: The following supporting information can be downloaded at: https://www.mdpi.com/article/10.3390/toxics10050249/s1, Table S1: Effects of toxic metals on the apparent indexes and body composition of rice (Promote/Increase). Table S2: Effects of toxic metals on the apparent indexes and body composition of rice (Inhibit/Reduce).

Author Contributions: Conceptualization, Y.D., L.Z., Z.Z., J.D. and J.Z.; methodology, Y.D., Q.L., Z.H. and Z.Z.; software, Y.D. and H.Z.; validation, Y.D., Q.L. and J.Z.; formal analysis, J.Z.; investigation, Q.L., L.Z. and H.Z.; resources, Y.D. and J.Z.; data curation, Y.D. and Q.L.; writing—original draft preparation, Y.D.; writing—review and editing, J.D. and J.Z.; visualization, Y.D., Z.H. and J.D.; supervision, J.Z.; project administration, J.Z.; funding acquisition, Y.D. and J.Z. All authors have read and agreed to the published version of the manuscript.

Funding: This research was funded by [Y.D.] grant number [2021XKJS101] and [J.Z.] grant number [2021ZHCG0078, 21NZZH0049, 201203081]. The APC was funded by [J.Z.].

Institutional Review Board Statement: Not applicable.

Informed Consent Statement: Not applicable.

Data Availability Statement: All available data is included in the paper.

Acknowledgments: This work was supported by the Promotion Project of Modern Agricultural Discipline Construction of Sichuan Academy of Agricultural Sciences (2021XKJS101), the Sichuan Science and Technology Programs (2021ZHCG0078, 21NZZH0049), and the Ministry of Agriculture Public Welfare Research Project (201203081). The authors thank Glauco Favor for revising the paper.

Conflicts of Interest: The authors declare that they have no conflict of interest.

References

1. Sfakianakis, D.; Renieri, E.; Kentouri, M.; Tsatsakis, A. Effect of heavy metals on fish larvae deformities: A review. *Environ. Res.* **2015**, *137*, 246–255. [CrossRef] [PubMed]
2. Guevara-García, Á.A.; Juárez, K.; Herrera-Estrella, L.R. Heavy metal adaptation. In *eLS Encyclopedia of Life Sciences*; John Wiley & Sons, Ltd.: Chichester, UK, 2017; pp. 1–9.
3. Wang, J.; Chen, C. Biosorbents for heavy metals removal and their future. *Biotechnol. Adv.* **2009**, *27*, 195–226. [CrossRef] [PubMed]
4. Duffus, J.H. "Heavy metals" a meaningless term? (IUPAC Technical Report). *Pure Appl. Chem.* **2002**, *74*, 793–807, Erratum in *Pure Appl. Chem.* **2002**, *75*, 1357. [CrossRef]
5. Pourret, O.; Bollinger, J.C. 'Heavy metals'-what to do now: To use or not to use? (Letter to the Editor). *Sci. Total Environ.* **2018**, *610*, 419–420. [CrossRef] [PubMed]
6. Dickman, M.; Leung, K.; Koo, L. Mercury in Human Hair and Fish: Is there a Hong Kong Male Subfertility Connection? *Mar. Pollut. Bull.* **1999**, *39*, 352–356. [CrossRef]
7. Bandara, J.M.R.S.; Wijewardena, H.V.P.; Liyanege, J.; Upul, M.A.; Bandara, J.M.U.A. Chronic renal failure in Sri Lanka caused by elevated dietary cadmium: Trojan horse of the green revolution. *Toxicol. Lett.* **2010**, *198*, 33–39. [CrossRef]
8. Yu, X.-D.; Yan, C.-H.; Shen, X.-M.; Tian, Y.; Cao, L.-L.; Zhao, L.; Liu, J.-X. Prenatal exposure to multiple toxic heavy metals and neonatal neurobehavioral development in Shanghai, China. *Neurotoxicol. Teratol.* **2011**, *33*, 437–443. [CrossRef]
9. Jin, X. How many lands have been filled with poison—Investigation on soil heavy metal pollution in China. *Environ. Educ.* **2015**, *6*, 4–9. (In Chinese)
10. Rai, P.K.; Lee, S.S.; Zhang, M.; Tsang, Y.F.; Kim, K.-H. Heavy metals in food crops: Health risks, fate, mechanisms, and management. *Environ. Int.* **2019**, *125*, 365–385. [CrossRef]
11. ATSDR. *Toxicological Profile for Barium*; U.S. Department of Health and Human Services, Public Health Service: Atlanta, GA, USA, 2007.
12. Liu, S. A brief analysis of the rice production Status in the world and China. *Xin Nongye* **2020**, *20*, 14–16. (In Chinese)
13. CGIAR. The Global Staple. 2021. Available online: http://ricepedia.org/rice-as-food/the-global-staple-rice-consumers. (accessed on 3 March 2021).
14. Hibberd, J.M.; Sheehy, J.E.; Langdale, J.A. Using C4 photosynthesis to increase the yield of rice—rationale and feasibility. *Curr. Opin. Plant Biol.* **2008**, *11*, 228–231. [CrossRef] [PubMed]
15. Kaur, R.; Bhunia, R.K.; Rajam, M.V. MicroRNAs as potential targets for improving rice yield via plant architecture modulation: Recent studies and future perspectives. *J. Biosci.* **2020**, *45*, 116. [CrossRef] [PubMed]
16. Jung, J.; Choi, H.-Y.; Huy, N.-X.; Park, H.; Kim, H.H.; Yang, M.-S.; Kang, S.-H.; Kim, D.-I.; Kim, N.-S. Production of recombinant human acid β-glucosidase with high mannose-type N-glycans in rice gnt1 mutant for potential treatment of Gaucher disease. *Protein Expr. Purif.* **2019**, *158*, 81–88. [CrossRef]
17. Verma, D.K.; Srivastav, P.P. Bioactive compounds of rice (*Oryza sativa* L.): Review on paradigm and its potential benefit in human health. *Trends Food Sci. Technol.* **2020**, *97*, 355–365. [CrossRef]
18. Wang, Z.G.; Feng, J.N.; Tong, Z. Human toxicosis caused by moldy rice contaminated with fusarium and T-2 toxin. *Biomed. Environ. Sci.* **1993**, *6*, 65. [PubMed]
19. FAO. The State of Food Security and Nutrition in the World. 2021. Available online: https://www.fao.org/publications/sofi/en. (accessed on 5 April 2022).
20. Mahmood, T.; Islam, K.R.; Muhammad, S. Toxic effects of heavy metals on early growth and tolerance of cereal crops. *Pak. J. Bot.* **2007**, *39*, 451.
21. Mishra, A.; Choudhuri, M.A. Amelioration of lead and mercury effects on germination and rice seedling growth by antioxidants. *Biol. Plant.* **1998**, *41*, 469–473. [CrossRef]
22. Mishra, A.; Choudhuri, M.A. Monitoring of Phytotoxicity of Lead and Mercury from Germination and Early Seedling Growth Indices in Two Rice Cultivars. *Water Air Soil Pollut.* **1999**, *114*, 339–346. [CrossRef]
23. Hassan, M.J.; Shao, G.; Zhang, G. Influence of Cadmium Toxicity on Growth and Antioxidant Enzyme Activity in Rice Cultivars with Different Grain Cadmium Accumulation. *J. Plant Nutr.* **2005**, *28*, 1259–1270. [CrossRef]
24. He, J.; Ren, Y. Effect of Cadmium on Seed Germination, Seedling Growth and Amylolytic Activity of Rice. *Acta Agri-Cult. Boreali-Sin.* **2008**, *23*, 131–134.
25. Ashraf, U.; Hussain, S.; Anjum, S.A.; Abbas, F.; Tanveer, M.; Noor, M.A.; Tang, X. Alterations in growth, oxidative damage, and metal uptake of five aromatic rice cultivars under lead toxicity. *Plant Physiol. Biochem.* **2017**, *115*, 461–471. [CrossRef]
26. Ahsan, N.; Lee, D.-G.; Lee, S.-H.; Kang, K.Y.; Lee, J.J.; Kim, P.J.; Yoon, H.-S.; Kim, J.-S.; Lee, B.-H. Excess copper induced physiological and proteomic changes in germinating rice seeds. *Chemosphere* **2007**, *67*, 1182–1193. [CrossRef] [PubMed]
27. Ashraf, U.; Kanu, A.S.; Deng, Q.; Mo, Z.; Pan, S.; Tian, H.; Tang, X. Lead (Pb) Toxicity; Physio-Biochemical Mechanisms, Grain Yield, Quality, and Pb Distribution Proportions in Scented Rice. *Front. Plant Sci.* **2017**, *8*, 259. [CrossRef] [PubMed]
28. Silva, M.L.D.S.; Vitti, G.C.; Trevizam, A.R. Heavy metal toxity in rice and soybean plants cultivated in contaminated soil. *Rev. Ceres* **2014**, *61*, 248–254. [CrossRef]
29. He, J.-Y.; Ren, Y.-F.; Zhu, C.; Jiang, D. Effects of Cadmium Stress on Seed Germination, Seedling Growth and Seed Amylase Activities in Rice (*Oryza sativa*). *Rice Sci.* **2008**, *15*, 319–325. [CrossRef]

30. Rascio, N.; Vecchia, F.D.; La Rocca, N.; Barbato, R.; Pagliano, C.; Raviolo, M.; Gonnelli, C.; Gabbrielli, R. Metal accumulation and damage in rice (cv. Vialone nano) seedlings exposed to cadmium. *Environ. Exp. Bot.* **2008**, *62*, 267–278. [CrossRef]

31. Gautam, M.; Sengar, R.S.; Chaudhary, R.; Sengar, K.; Garg, S. Possible cause of inhibition of seed germination in two rice cultivars by heavy metals Pb^{2+} and Hg^{2+}. *Toxicol. Environ. Chem.* **2010**, *92*, 1111–1119. [CrossRef]

32. Ros, R.; Cook, D.T.; Martinez-Cortina, C.; Picazo, I. Nickel and Cadmium-related Changes in Growth, Plasma Membrane Lipid Composition, ATPase Hydrolytic Activity and Proton-pumping of Rice (*Oryza sativa* L. cv. Bahia) Shoots. *J. Exp. Bot.* **1992**, *43*, 1475–1481. [CrossRef]

33. Jamil, M.; Zeb, S.; Anees, M.; Roohi, A.; Ahmed, I.; Rehman, S.U.; Rha, E.S. Role of *Bacillus licheniformis* in Phytoremediation of Nickel Contaminated Soil Cultivated with Rice. *Int. J. Phytoremediat.* **2014**, *16*, 554–571. [CrossRef]

34. Zeng, Q.; Zhu, J.G.; Cheng, H.L.; Xie, Z.B.; Chu, H.Y. Phytotoxicity of lanthanum in rice in haplic acrisols and cambisols. *Ecotoxicol. Environ. Safe* **2006**, *64*, 226–233. [CrossRef]

35. Liu, D.; Lin, Y.; Wang, X. Effects of lanthanum on growth, element uptake, and oxidative stress in rice seedlings. *J. Plant Nutr. Soil. Sc.* **2012**, *175*, 907–911. [CrossRef]

36. Liu, D.; Wang, X.; Zhang, X.; Gao, Z. Effects of lanthanum on growth and accumulation in roots of rice seedlings. *Plant Soil Environ.* **2013**, *59*, 196–200. [CrossRef]

37. Liu, D.; Zheng, S.; Wang, X. Lanthanum regulates the reactive oxygen species in the roots of rice seedlings. *Sci. Rep.* **2016**, *6*, 31860. [CrossRef] [PubMed]

38. He, M.; Yang, J. Effects of different forms of antimony on rice during the period of germination and growth and antimony concentration in rice tissue. *Sci. Total Environ.* **1999**, *243–244*, 149–155. [CrossRef]

39. Abedin, J.; Cresser, M.S.; Meharg, A.A.; Feldmann, J.; Cotter-Howells, J. Arsenic Accumulation and Metabolism in Rice (*Oryza sativa* L.). *Environ. Sci. Technol.* **2002**, *36*, 962–968. [CrossRef]

40. Nautiyal, N.; Chatterjee, C.; Sharma, C. Copper stress affects grain filling in rice. *Commun. Soil Sci. Plant Anal.* **1999**, *30*, 1625–1632. [CrossRef]

41. Marin, A.R.; Pezeshki, S.R.; Masschelen, P.H.; Choi, H.S. Effect of dimethylarsenic acid (DMAA) on growth, tissue arsenic, and photosynthesis of rice plants. *J. Plant Nutr.* **1993**, *16*, 865–880. [CrossRef]

42. Wang, L.; Liu, B.; Wang, Y.; Qin, Y.; Zhou, Y.; Qian, H. Influence and interaction of iron and lead on seed germination in upland rice. *Plant Soil* **2020**, *455*, 187–202. [CrossRef]

43. Mishra, A.; Choudhuri, M. Effects of Salicylic Acid on Heavy Metal-Induced Membrane Deterioration Mediated by Lipoxygenase in Rice. *Biol. Plant.* **1999**, *42*, 409–415. [CrossRef]

44. Ashraf, U.; Tang, X. Yield and quality responses, plant metabolism and metal distribution pattern in aromatic rice under lead (Pb) toxicity. *Chemosphere* **2017**, *176*, 141–155. [CrossRef]

45. Huang, F.; Wen, X.-H.; Cai, Y.-X.; Cai, K.-Z. Silicon-Mediated Enhancement of Heavy Metal Tolerance in Rice at Different Growth Stages. *Int. J. Environ. Res. Public Health* **2018**, *15*, 2193. [CrossRef] [PubMed]

46. Kanu, A.S.; Ashraf, U.; Mo, Z.; Sabir, S.-U.; Baggie, I.; Charley, C.S.; Tang, X. Calcium amendment improved the performance of fragrant rice and reduced metal uptake under cadmium toxicity. *Environ. Sci. Pollut. Res.* **2019**, *26*, 24748–24757. [CrossRef] [PubMed]

47. Huang, D.-F.; Xi, L.-L.; Yang, L.-N.; Wang, Z.-Q.; Yang, J.-C. Comparison of Agronomic and Physiological Traits of Rice Genotypes Differing in Cadmium-Tolerance. *Acta Agron. Sin.* **2008**, *34*, 809–817. [CrossRef]

48. Fan, X.; Wen, X.; Huang, F.; Cai, Y.; Cai, K. Effects of silicon on morphology, ultrastructure and exudates of rice root under heavy metal stress. *Acta Physiol. Plant.* **2016**, *38*, 197. [CrossRef]

49. Begum, M.; Mondal, S. Relative Toxicity of Arsenite and Arsenate on Early Seedling Growth and Photosynthetic Pigments of Rice. *Curr. J. Appl. Sci. Technol.* **2019**, *33*, 1–5. [CrossRef]

50. Farooq, H.; Asghar, H.N.; Khan, M.Y.; Saleem-Mand, Z.A. Auxin-mediated growth of rice in cadmium-contaminated soil. *Turk. J. Agric. For.* **2015**, *39*, 272–276. [CrossRef]

51. Rehman, M.Z.U.; Rizwan, M.; Rauf, A.; Ayub, M.A.; Ali, S.; Qayyum, M.F.; Waris, A.A.; Naeem, A.; Sanaullah, M. Split application of silicon in cadmium (Cd) spiked alkaline soil plays a vital role in decreasing Cd accumulation in rice (*Oryza sativa* L.) grains. *Chemosphere* **2019**, *226*, 454–462. [CrossRef]

52. Moya, J.L.; Ros, R.; Picazo, I. Influence of cadmium and nickel on growth, net photosynthesis and carbohydrate distribution in rice plants. *Photosynth. Res.* **1993**, *36*, 75–80. [CrossRef]

53. Cao, F.; Cai, Y.; Liu, L.; Zhang, M.; He, X.; Zhang, G.; Wu, F. Differences in photosynthesis, yield and grain cadmium accumulation as affected by exogenous cadmium and glutathione in the two rice genotypes. *Plant Growth Regul.* **2015**, *75*, 715–723. [CrossRef]

54. Ding, Y.; Wang, Y.; Jiang, Z.; Wang, F.; Jiang, Q.; Sun, J.; Chen, Z.; Zhu, C. MicroRNA268 Overexpression Affects Rice Seedling Growth under Cadmium Stress. *J. Agric. Food Chem.* **2017**, *65*, 5860–5867. [CrossRef]

55. Takahashi, R.; Bashir, K.; Ishimaru, Y.; Nishizawa, N.K.; Nakanishi, H. The role of heavy-metal ATPases, HMAs, in zinc and cadmium transport in rice. *Plant Signal. Behave* **2012**, *7*, 1605–1607. [CrossRef] [PubMed]

56. Luo, J.-S.; Huang, J.; Zeng, D.-L.; Peng, J.-S.; Zhang, G.-B.; Ma, H.-L.; Guan, Y.; Yi, H.-Y.; Fu, Y.-L.; Han, B.; et al. A defensin-like protein drives cadmium efflux and allocation in rice. *Nat. Commun.* **2018**, *9*, 645. [CrossRef] [PubMed]

57. Tian, S.; Liang, S.; Qiao, K.; Wang, F.; Zhang, Y.; Chai, T. Co-expression of multiple heavy metal transporters changes the translocation, accumulation, and potential oxidative stress of Cd and Zn in rice (*Oryza sativa*). *J. Hazard. Mater.* **2019**, *380*, 120853. [CrossRef] [PubMed]

58. Deng, F.; Yamaji, N.; Xia, J.; Ma, J.F. A Member of the Heavy Metal P-Type ATPase OsHMA5 Is Involved in Xylem Loading of Copper in Rice. *Plant Physiol.* **2013**, *163*, 1353–1362. [CrossRef]

59. Huang, X.; Deng, F.; Yamaji, N.; Pinson, S.R.; Fujii-Kashino, M.; Danku, J.; Douglas, A.; Guerinot, M.L.; Salt, D.E.; Ma, J.F. A heavy metal P-type ATPase OsHMA4 prevents copper accumulation in rice grain. *Nat. Commun.* **2016**, *7*, 12138. [CrossRef]

60. Kumar, S.; Asif, M.H.; Chakrabarty, D.; Tripathi, R.D.; Dubey, R.S.; Trivedi, P.K. Expression of a rice Lambda class of glutathione S-transferase, OsGSTL2, in Arabidopsis provides tolerance to heavy metal and other abiotic stresses. *J. Hazard. Mater.* **2013**, *248–249*, 228–237. [CrossRef]

61. Hu, T.; Zhu, S.; Tan, L.; Qi, W.; He, S.; Wang, G. Overexpression of OsLEA4 enhances drought, high salt and heavy metal stress tolerance in transgenic rice (*Oryza sativa* L.). *Environ. Exp. Bot.* **2016**, *123*, 68–77. [CrossRef]

62. Tsunemitsu, Y.; Genga, M.; Okada, T.; Yamaji, N.; Ma, J.F.; Miyazaki, A.; Kato, S.-I.; Iwasaki, K.; Ueno, D. A member of cation diffusion facilitator family, MTP11, is required for manganese tolerance and high fertility in rice. *Planta* **2018**, *248*, 231–241. [CrossRef]

63. Zhang, Y.; Chen, K.; Zhao, F.-J.; Sun, C.; Jin, C.; Shi, Y.; Sun, Y.; Li, Y.; Yang, M.; Jing, X.; et al. OsATX1 Interacts with Heavy Metal P1B-Type ATPases and Affects Copper Transport and Distribution. *Plant Physiol.* **2018**, *178*, 329–344. [CrossRef]

64. Kim, Y.-O.; Kang, H. Comparative expression analysis of genes encoding metallothioneins in response to heavy metals and abiotic stresses in rice (*Oryza sativa*) and *Arabidopsis thaliana*. *Biosci. Biotechnol. Biochem.* **2018**, *82*, 1656–1665. [CrossRef]

65. Verma, S.; Verma, P.; Meher, A.K.; Bansiwal, A.K.; Tripathi, R.D.; Chakrabarty, D. A novel fungal arsenic methyltransferase, WaarsM reduces grain arsenic accumulation in transgenic rice (*Oryza sativa* L.). *J. Hazard. Mater.* **2018**, *344*, 626–634. [CrossRef] [PubMed]

66. Jing, X.-Q.; Shalmani, A.; Zhou, M.-R.; Shi, P.-T.; Muhammad, I.; Shi, Y.; Sharif, R.; Li, W.-Q.; Liu, W.-T.; Chen, K.-M. Genome-Wide Identification of Malectin/Malectin-Like Domain Containing Protein Family Genes in Rice and Their Expression Regulation Under Various Hormones, Abiotic Stresses, and Heavy Metal Treatments. *J. Plant Growth Regul.* **2019**, *39*, 492–506. [CrossRef]

67. Che, J.; Yokosho, K.; Yamaji, N.; Ma, J.F. A Vacuolar Phytosiderophore Transporter Alters Iron and Zinc Accumulation in Polished Rice Grains. *Plant Physiol.* **2019**, *181*, 276–288. [CrossRef] [PubMed]

68. Wang, T.; Li, Y.; Fu, Y.; Xie, H.; Song, S.; Qiu, M.; Wen, J.; Chen, M.; Chen, G.; Tian, Y.; et al. Mutation at Different Sites of Metal Transporter Gene OsNramp5 Affects Cd Accumulation and Related Agronomic Traits in Rice (*Oryza sativa* L.). *Front. Plant Sci.* **2019**, *10*, 1081. [CrossRef]

69. Chang, J.; Huang, S.; Yamaji, N.; Zhang, W.; Ma, J.F.; Zhao, F. OsNRAMP1 transporter contributes to cadmium and manganese uptake in rice. *Plant Cell Environ.* **2020**, *43*, 2476–2491. [CrossRef]

70. Moulick, D.; Ghosh, D.; Santra, S.C. Evaluation of effectiveness of seed priming with selenium in rice during germination under arsenic stress. *Plant Physiol. Biochem.* **2016**, *109*, 571–578. [CrossRef]

71. Wang, Y.; Xiao, L.; Li, S.; Guo, Y.; Cai, X. Effects of Compound Pollution of Pb and Cd on Soil and Physiological and Biochemistrical Characteristics of Rice Leaves. *Chin. Agr. Sci. Bull.* **2010**, *26*, 369–373.

72. Barzegar, R.; Moghaddam, A.A.; Soltani, S.; Fijani, E.; Tziritis, E.; Kazemian, N. Heavy Metal(loid)s in the Groundwater of Shabestar Area (NW Iran): Source Identification and Health Risk Assessment. *Expo. Health* **2019**, *11*, 251–265. [CrossRef]

73. Mariyanto, M.; Amir, M.F.; Utama, W.; Hamdan, A.M.; Bijaksana, S.; Pratama, A.; Yunginger, R.; Sudarningsih, S. Heavy metal contents and magnetic properties of surface sediments in volcanic and tropical environment from Brantas River, Jawa Timur Province, Indonesia. *Sci. Total Environ.* **2019**, *675*, 632–641. [CrossRef]

74. Zhou, J.; Du, B.; Liu, H.; Cui, H.; Zhang, W.; Fan, X.; Cui, J.; Zhou, J. The bioavailability and contribution of the newly deposited heavy metals (copper and lead) from atmosphere to rice (*Oryza sativa* L.). *J. Hazard. Mater.* **2020**, *384*, 121285. [CrossRef]

75. Luo, L.; Mei, K.; Qu, L.; Zhang, C.; Chen, H.; Wang, S.; Di, D.; Huang, H.; Wang, Z.; Xia, F.; et al. Assessment of the Geographical Detector Method for investigating heavy metal source apportionment in an urban watershed of Eastern China. *Sci. Total Environ.* **2019**, *653*, 714–722. [CrossRef] [PubMed]

76. Atafar, Z.; Mesdaghinia, A.; Nouri, J.; Homaee, M.; Yunesian, M.; Ahmadimoghaddam, M.; Mahvi, A.H. Effect of fertilizer application on soil heavy metal concentration. *Environ. Monit. Assess.* **2010**, *160*, 83–89. [CrossRef] [PubMed]

77. Gassama, U.M.; Bin Puteh, A.; Abd-Halim, M.R.; Kargbo, B. Influence of municipal wastewater on rice seed germination, seedling performance, nutrient uptake, and chlorophyll content. *J. Crop Sci. Biotechnol.* **2015**, *18*, 9–19. [CrossRef]

78. Bussan, D.; Harris, A.; Douvris, C. Monitoring of selected trace elements in sediments of heavily industrialized areas in Calcasieu Parish, Louisiana, United States by inductively coupled plasma-optical emission spectroscopy (ICP-OES). *Microchem. J.* **2019**, *144*, 51–55. [CrossRef]

79. Ning, L.; Liyuan, Y.; Jirui, D.; Xugui, P. Heavy metal pollution in surface water of Linglong gold mining area, China. *Procedia Environ. Sci.* **2011**, *10*, 914–917. [CrossRef]

80. Ignatavičius, G.; Valskys, V.; Bulskaya, I.; Paliulis, D.; Zigmontienė, A.; Satkūnas, J. Heavy metal contamination in surface runoff sediments of the urban area of Vilnius, Lithuania. *Estonian J. Earth Sci.* **2017**, *66*, 13. [CrossRef]

81. Liu, A.; Ma, Y.; Gunawardena, J.M.; Egodawatta, P.; Ayoko, G.A.; Goonetilleke, A. Heavy metals transport pathways: The importance of atmospheric pollution contributing to stormwater pollution. *Ecotoxicol. Environ. Safe* **2018**, *164*, 696–703. [CrossRef]

82. Birch, G.; Taylor, S.; Matthai, C. Small-scale spatial and temporal variance in the concentration of heavy metals in aquatic sediments: A review and some new concepts. *Environ. Pollut.* **2001**, *113*, 357–372. [CrossRef]

83. Liu, M.; Liu, X.; Li, M.; Fang, M.; Chi, W. Neural-network model for estimating leaf chlorophyll concentration in rice under stress from heavy metals using four spectral indices. *Biosyst. Eng.* **2010**, *106*, 223–233. [CrossRef]

84. Liu, M.; Liu, X.; Zhang, B.; Ding, C. Regional heavy metal pollution in crops by integrating physiological function variability with spatio-temporal stability using multi-temporal thermal remote sensing. *Int. J. Appl. Earth Obs. Geoinf.* **2016**, *51*, 91–102. [CrossRef]

85. Zhao, S.; Qian, X.; Liu, X.; Xu, Z. Finding the Key Periods for Assimilating HJ-1A/B CCD Data and the WOFOST Model to Evaluate Heavy Metal Stress in Rice. *Sensors* **2018**, *18*, 1230. [CrossRef] [PubMed]

86. Brus, D.J.; Li, Z.; Song, J.; Koopmans, G.F.; Temminghoff, E.J.M.; Yin, X.; Yao, C.; Zhang, H.; Luo, Y.; Japenga, J. Predictions of Spatially Averaged Cadmium Contents in Rice Grains in the Fuyang Valley, PR China. *J. Environ. Qual.* **2009**, *38*, 1126–1136. [CrossRef] [PubMed]

87. Liu, M.; Liu, X.; Wu, M.; Li, L.; Xiu, L. Integrating spectral indices with environmental parameters for estimating heavy metal concentrations in rice using a dynamic fuzzy neural-network model. *Comput. Geosci.-UK* **2011**, *37*, 1642–1652. [CrossRef]

88. Liu, M.; Liu, X.; Li, J.; Li, T. Estimating regional heavy metal concentrations in rice by scaling up a field-scale heavy metal assessment model. *Int. J. Appl. Earth Obs.* **2012**, *19*, 12–23. [CrossRef]

89. Chen, Q.; Peng, P.-Q.; Long, J.; Li, X.-Y.; Ding, X.; Hou, H.-B.; Liao, B.-H. Cadmium phytoavailability evaluation in rice-soil system using a field capacity-derived soil solution extraction: An entire growth period study in subtropical China. *Soil Tillage Res.* **2019**, *194*, 104315. [CrossRef]

90. Shi, T.; Liu, H.; Wang, J.; Chen, Y.; Fei, T.; Wu, G. Monitoring Arsenic Contamination in Agricultural Soils with Reflectance Spectroscopy of Rice Plants. *Environ. Sci. Technol.* **2014**, *48*, 6264–6272. [CrossRef]

91. Shi, T.; Wang, J.; Chen, Y.; Wu, G. Improving the prediction of arsenic contents in agricultural soils by combining the reflectance spectroscopy of soils and rice plants. *Int. J. Appl. Earth Obs.* **2016**, *52*, 95–103. [CrossRef]

92. Li, X.; Li, L.; Liu, X. Collaborative inversion heavy metal stress in rice by using two-dimensional spectral feature space based on HJ-1 A HSI and radarsat-2 SAR remote sensing data. *Int. J. Appl. Earth Obs. Geoinf.* **2019**, *78*, 39–52. [CrossRef]

93. Abrham, F.; Gholap, A.V. Analysis of heavy metal concentration in some vegetables using atomic absorption spectroscopy. *Pollution* **2021**, *7*, 205–216. [CrossRef]

94. Ay, E.; Tekin, Z.; Özdoğan, N.; Bakırdere, S. Zirconium Nanoparticles Based Vortex Assisted Ligandless Dispersive Solid Phase Extraction for Trace Determination of Lead in Domestic Wastewater using Flame Atomic Absorption Spectrophotometry. *Bull. Environ. Contam. Toxicol.* **2022**, *108*, 324–330. [CrossRef]

95. Goodarzi, L.; Bayatloo, M.R.; Chalavi, S.; Nojavan, S.; Rahmani, T.; Azimi, S.B. Selective extraction and determination of Cr(VI) in food samples based on tandem electromembrane extraction followed by electrothermal atomic absorption spectrometry. *Food Chem.* **2022**, *373*, 131442. [CrossRef] [PubMed]

96. Castiñeira, M.D.M.; Brandt, R.; Jakubowski, N.; Andersson, J.T. Changes of the Metal Composition in German White Wines through the Winemaking Process. A Study of 63 Elements by Inductively Coupled Plasma−Mass Spectrometry. *J. Agric. Food Chem.* **2004**, *52*, 2953–2961. [CrossRef] [PubMed]

97. Fu, L.; Xie, H.; Huang, J.; Chen, L. Determination of the Non-metallic Elements in Herbal Tea by Inductively Coupled Plasma Tandem Mass Spectrometry. *Biol. Trace Element Res.* **2021**, *199*, 769–778. [CrossRef] [PubMed]

98. Bloom, N.; Fitzgerald, W.F. Determination of volatile mercury species at the picogram level by low-temperature gas chromatography with cold-vapour atomic fluorescence detection. *Anal. Chim. Acta* **1988**, *208*, 151–161. [CrossRef]

99. Gómez-Ariza, J.L.; Lorenzo, F.; García-Barrera, T. Comparative study of atomic fluorescence spectroscopy and inductively coupled plasma mass spectrometry for mercury and arsenic multispeciation. *Anal. Bioanal. Chem.* **2005**, *382*, 485–492. [CrossRef]

100. Musil, S.; Matoušek, T.; Currier, J.M.; Stýblo, M.; Dědina, J. Speciation Analysis of Arsenic by Selective Hydride Generation-Cryotrapping-Atomic Fluorescence Spectrometry with Flame-in-Gas-Shield Atomizer: Achieving Extremely Low Detection Limits with Inexpensive Instrumentation. *Anal. Chem.* **2014**, *86*, 10422–10428. [CrossRef]

101. Tang, X.; Qian, Y.; Guo, Y.; Wei, N.; Li, Y.; Yao, J.; Wang, G.; Ma, J.; Liu, W. Analysis of atmospheric pollutant metals by laser ablation inductively coupled plasma mass spectrometry with a radial line-scan dried-droplet approach. *Spectrochim. Acta Part B At. Spectrosc.* **2017**, *138*, 18–22. [CrossRef]

102. Peyneau, P.-E. Poisson process modelling of spike occurrence in single particle inductively coupled plasma mass spectrometry time scans for very dilute nanoparticle dispersions. *Spectrochim. Acta Part B At. Spectrosc.* **2021**, *178*, 106126. [CrossRef]

103. Valenta, P.; Sipos, L.; Kramer, I.; Krumpen, P.; Rützel, H. An automatic voltammetric analyzer for the simultaneous determination of toxic trace metals in water. *Z. Anal. Bioanal. Chem.* **1982**, *312*, 101–108. [CrossRef]

104. Clark, B.R.; DePaoli, D.W.; McTaggart, D.R.; Patton, B.D. An on-line voltammetric analyzer for trace metals in wastewater. *Anal. Chim. Acta* **1988**, *215*, 13–20. [CrossRef]

105. Hseu, Z.-Y.; Su, S.-W.; Lai, H.-Y.; Guo, H.-Y.; Chen, T.-C.; Chen, Z.-S. Remediation techniques and heavy metal uptake by different rice varieties in metal-contaminated soils of Taiwan: New aspects for food safety regulation and sustainable agriculture. *Soil Sci. Plant Nutr.* **2010**, *56*, 31–52. [CrossRef]

106. Li, H.; Liu, Y.; Luo, Z.; Zhou, Y.; Hou, D.; Mao, Q.; Zhi, D.; Zhang, J.; Yang, Y.; Luo, L. Effect of RM-based-passivator for the remediation of two kinds of Cd polluted paddy soils and mechanism of Cd(II) adsorption. *Environ. Technol.* **2019**, *42*, 1623–1633. [CrossRef] [PubMed]

107. Yang, X.; Zhang, W.; Qin, J.; Zhang, X.; Li, H. Role of passivators for Cd alleviation in rice-water spinach intercropping system. *Ecotoxicol. Environ. Safe* **2020**, *205*, 111321. [CrossRef] [PubMed]

108. Daulta, R.; Sridevi, T.; Garg, V.K. Spatial distribution of heavy metals in rice grains, rice husk, and arable soil, their bioaccumulation and associated health risks in Haryana, India. *Toxin Rev.* **2020**, *40*, 859–871. [CrossRef]

109. Yildirim, A.; Baran, M.F.; Acay, H. Kinetic and isotherm investigation into the removal of heavy metals using a fungal-extract-based bio-nanosorbent. *Environ. Technol. Innov.* **2020**, *20*, 101076. [CrossRef]

110. Fan, M.; Luo, L.; Liao, Y.; Tian, J.; Hu, B. Effect of Application on Different Amount of Red Mud on Rice Yield and Soil Bio-logical Properties in Cd-contaminated Paddy Soil. *J. Soil. Water Conserv.* **2011**, *6*, 39. [CrossRef]

111. Trijatmiko, K.; Dueñas, C.; Tsakirpaloglou, N.; Torrizo, L.; Arines, F.M.; Adeva, C.; Balindong, J.; Oliva, N.; Sapasap, M.V.; Borrero, J.; et al. Biofortified indica rice attains iron and zinc nutrition dietary targets in the field. *Sci. Rep.* **2016**, *6*, 19792. [CrossRef]

112. Shao, G.; Chen, M.; Wang, D.; Xu, C.; Mou, R.; Cao, Z.; Zhang, X. Using iron fertilizer to control Cd accumulation in rice plants: A new promising technology. *Sci. China Ser. C Life Sci.* **2008**, *51*, 245–253. [CrossRef]

113. Makino, T.; Nakamura, K.; Katou, H.; Ishikawa, S.; Ito, M.; Honma, T.; Miyazaki, N.; Takehisa, K.; Sano, S.; Matsumoto, S.; et al. Simultaneous decrease of arsenic and cadmium in rice (*Oryza sativa* L.) plants cultivated under submerged field conditions by the application of iron-bearing materials. *Soil Sci. Plant Nutr.* **2016**, *62*, 340–348. [CrossRef]

114. Yu, H.-Y.; Wang, X.; Li, F.; Li, B.; Liu, C.; Wang, Q.; Lei, J. Arsenic mobility and bioavailability in paddy soil under iron compound amendments at different growth stages of rice. *Environ. Pollut.* **2017**, *224*, 136–147. [CrossRef]

115. Chen, C.C.; Dixon, J.B.; Turner, F.T. Iron Coatings on Rice Roots: Mineralogy and Quantity Influencing Factors. *Soil Sci. Soc. Am. J.* **1980**, *44*, 635–639. [CrossRef]

116. Rajendran, M.; Shi, L.; Wu, C.; Li, W.C.; An, W.; Liu, Z.; Xue, S. Effect of sulfur and sulfur-iron modified biochar on cadmium availability and transfer in the soil–rice system. *Chemosphere* **2019**, *222*, 314–322. [CrossRef] [PubMed]

117. Cuong, T.X.; Ullah, H.; Datta, A.; Hanh, T.C. Effects of Silicon-Based Fertilizer on Growth, Yield and Nutrient Uptake of Rice in Tropical Zone of Vietnam. *Rice Sci.* **2017**, *24*, 283–290. [CrossRef]

118. Lu, K.; Yang, X.; Gielen, G.; Bolan, N.; Ok, Y.S.; Niazi, N.K.; Xu, S.; Yuan, G.; Chen, X.; Zhang, X.; et al. Effect of bamboo and rice straw biochars on the mobility and redistribution of heavy metals (Cd, Cu, Pb and Zn) in contaminated soil. *J. Environ. Manag.* **2017**, *186*, 285–292. [CrossRef]

119. Yin, D.; Wang, X.; Peng, B.; Tan, C.; Ma, L.Q. Effect of biochar and Fe-biochar on Cd and As mobility and transfer in soil-rice system. *Chemosphere* **2017**, *186*, 928–937. [CrossRef]

120. Xing, Y.; Wang, J.; Shaheen, S.M.; Feng, X.; Chen, Z.; Zhang, H.; Rinklebe, J. Mitigation of mercury accumulation in rice using rice hull-derived biochar as soil amendment: A field investigation. *J. Hazard. Mater.* **2020**, *388*, 121747. [CrossRef]

121. Huang, D.; Liu, L.; Zeng, G.; Xu, P.; Huang, C.; Deng, L.; Wang, R.; Wan, J. The effects of rice straw biochar on indigenous microbial community and enzymes activity in heavy metal-contaminated sediment. *Chemosphere* **2017**, *174*, 545–553. [CrossRef]

122. Yue, L.; Lian, F.; Han, Y.; Bao, Q.; Wang, Z.; Xing, B. The effect of biochar nanoparticles on rice plant growth and the uptake of heavy metals: Implications for agronomic benefits and potential risk. *Sci. Total Environ.* **2019**, *656*, 9–18. [CrossRef]

123. Farooq, M.U.; Tang, Z.; Zheng, T.; Asghar, M.A.; Zeng, R.; Su, Y.; Ei, H.H.; Liang, Y.; Zhang, Y.; Ye, X.; et al. Cross-Talk between Cadmium and Selenium at Elevated Cadmium Stress Determines the Fate of Selenium Uptake in Rice. *Biomolecules* **2019**, *9*, 247. [CrossRef]

124. Wang, S.; Wang, F.; Gao, S.; Wang, X. Heavy Metal Accumulation in Different Rice Cultivars as Influenced by Foliar Application of Nano-silicon. *Water Air Soil Pollut.* **2016**, *227*, 228. [CrossRef]

125. Liu, J.; Simms, M.; Song, S.; King, R.S.; Cobb, G.P. Physiological Effects of Copper Oxide Nanoparticles and Arsenic on the Growth and Life Cycle of Rice (*Oryza sativa japonica* 'Koshihikari'). *Environ. Sci. Technol.* **2018**, *52*, 13728–13737. [CrossRef] [PubMed]

126. Peng, C.; Xu, C.; Liu, Q.; Sun, L.; Luo, Y.; Shi, J. Fate and Transformation of CuO Nanoparticles in the Soil-Rice System during the Life Cycle of Rice Plants. *Environ. Sci. Technol.* **2017**, *51*, 4907–4917. [CrossRef] [PubMed]

127. Zhang, W.; Long, J.; Li, J.; Zhang, M.; Xiao, G.; Ye, X.; Chang, W.; Zeng, H. Impact of ZnO nanoparticles on Cd toxicity and bioaccumulation in rice (*Oryza sativa* L.). *Environ. Sci. Pollut. Res.* **2019**, *26*, 23119–23128. [CrossRef] [PubMed]

128. Zhang, Y.; Wang, X.; Ji, X.; Liu, Y.; Lin, Z.; Lin, Z.; Xiao, S.; Peng, B.; Tan, C.; Zhang, X. Effect of a novel Ca-Si composite mineral on Cd bioavailability, transport and accumulation in paddy soil-rice system. *J. Environ. Manag.* **2019**, *233*, 802–811. [CrossRef] [PubMed]

129. Li, H.; Xu, H.; Zhou, S.; Yu, Y.; Li, H.; Zhou, C.; Chen, Y.; Li, Y.; Wang, M.; Wang, G. Distribution and transformation of lead in rice plants grown in contaminated soil amended with biochar and lime. *Ecotoxicol. Environ. Safe* **2018**, *165*, 589–596. [CrossRef] [PubMed]

130. Zhai, W.; Zhao, W.; Yuan, H.; Guo, T.; Hashmi, M.Z.; Liu, X.; Tang, X. Reduced Cd, Pb, and As accumulation in rice (*Oryza sativa* L.) by a combined amendment of calcium sulfate and ferric oxide. *Environ. Sci. Pollut. Res.* **2020**, *27*, 1348–1358. [CrossRef]

131. Honma, T.; Ohba, H.; Kaneko, A.; Nakamura, K.; Makino, T.; Katou, H. Effects of soil amendments on arsenic and cadmium uptake by rice plants (*Oryza sativa* L. cv. Koshihikari) under different water management practices. *Soil Sci. Plant Nutr.* **2016**, *62*, 349–356. [CrossRef]

132. Zhou, Y.; Jia, Z.; Wang, J.; Chen, L.; Zou, M.; Li, Y.; Zhou, S. Heavy metal distribution, relationship and prediction in a wheat-rice rotation system. *Geoderma* **2019**, *354*, 113886. [CrossRef]

133. Huang, S.; Rao, G.; Ashraf, U.; He, L.; Zhang, Z.; Zhang, H.; Mo, Z.; Pan, S.; Tang, X. Application of inorganic passivators reduced Cd contents in brown rice in oilseed rape-rice rotation under Cd contaminated soil. *Chemosphere* **2020**, *259*, 127404. [CrossRef]
134. Zhou, J.; Li, P.; Meng, D.; Gu, Y.; Zheng, Z.; Yin, H.; Zhou, Q.; Li, J. Isolation, characterization and inoculation of Cd tolerant rice endophytes and their impacts on rice under Cd contaminated environment. *Environ. Pollut.* **2020**, *260*, 113990. [CrossRef]
135. Lin, X.; Mou, R.; Cao, Z.; Xu, P.; Wu, X.; Zhu, Z.; Chen, M. Characterization of cadmium-resistant bacteria and their potential for reducing accumulation of cadmium in rice grains. *Sci. Total Environ.* **2016**, *569-570*, 97–104. [CrossRef] [PubMed]
136. Tang, X.; Ni, Y. Review of Remediation Technologies for Cadmium in soil. *E3S Web Conf.* **2021**, *233*, 01037. [CrossRef]
137. Zhang, Q.; Chen, H.; Huang, D.; Xu, C.; Zhu, H.; Zhu, Q. Water managements limit heavy metal accumulation in rice: Dual effects of iron-plaque formation and microbial communities. *Sci. Total Environ.* **2019**, *687*, 790–799. [CrossRef] [PubMed]
138. Nakanishi, H.; Ogawa, I.; Ishimaru, Y.; Mori, S.; Nishizawa, N.K. Iron deficiency enhances cadmium uptake and translocation mediated by the Fe^{2+} transporters OsIRT1 and OsIRT2 in rice. *Soil Sci. Plant Nutr.* **2006**, *52*, 464–469. [CrossRef]
139. Arao, T.; Kawasaki, A.; Baba, K.; Mori, S.; Matsumoto, S. Effects of Water Management on Cadmium and Arsenic Accumulation and Dimethylarsinic Acid Concentrations in Japanese Rice. *Environ. Sci. Technol.* **2009**, *43*, 9361–9367. [CrossRef]
140. Pan, Y.; Koopmans, G.F.; Bonten, L.T.C.; Song, J.; Luo, Y.; Temminghoff, E.J.M.; Comans, R.N.J. Temporal variability in trace metal solubility in a paddy soil not reflected in uptake by rice (*Oryza sativa* L.). *Environ. Geochem. Health* **2016**, *38*, 1355–1372. [CrossRef]
141. Honma, T.; Ohba, H.; Kaneko-Kadokura, A.; Makino, T.; Nakamura, K.; Katou, H. Optimal Soil Eh, pH, and Water Management for Simultaneously Minimizing Arsenic and Cadmium Concentrations in Rice Grains. *Environ. Sci. Technol.* **2016**, *50*, 4178–4185. [CrossRef]
142. Ishfaq, M.; Farooq, M.; Zulfiqar, U.; Hussain, S.; Akbar, N.; Nawaz, A.; Anjum, S.A. Alternate wetting and drying: A water-saving and ecofriendly rice production system. *Agric. Water Manag.* **2020**, *241*, 106363. [CrossRef]
143. Deng, X.; Yang, Y.; Zeng, H.; Chen, Y.; Zeng, Q. Variations in iron plaque, root morphology and metal bioavailability response to seedling establishment methods and their impacts on Cd and Pb accumulation and translocation in rice (*Oryza sativa* L.). *J. Hazard. Mater.* **2020**, *384*, 121343. [CrossRef]
144. Farooq, M.U.; Zhu, J. The paradox in accumulation behavior of cadmium and selenium at different planting times in rice. *Environ. Sci. Pollut. Res.* **2019**, *26*, 22421–22430. [CrossRef]
145. Yi, Y.K.; Zhou, Z.B.; Chen, G.H. Effects of soil pH on growth and grain cadmium content in rice. *J. Agro-Environ. Sci.* **2017**, *36*, 428–436.
146. Chen, H.; Yang, Y.; Ye, Y.; Tao, L.; Fu, X.; Liu, B.; Wu, Y. Differences in cadmium accumulation between indica and japonica rice cultivars in the reproductive stage. *Ecotoxicol. Environ. Safe* **2019**, *186*, 109795. [CrossRef] [PubMed]
147. Abedin, M.J.; Meharg, A.A. Relative toxicity of arsenite and arsenate on germination and early seedling growth of rice (*Oryza sativa* L.). *Plant Soil* **2002**, *243*, 57–66. [CrossRef]
148. Songmei, L.; Jie, J.; Yang, L.; Jun, M.; Shouling, X.; Yuanyuan, T.; Youfa, L.; Qingyao, S.; Jianzhong, H. Characterization and Evaluation of OsLCT1 and OsNramp5 Mutants Generated Through CRISPR/Cas9-Mediated Mutagenesis for Breeding Low Cd Rice. *Rice Sci.* **2019**, *26*, 88–97. [CrossRef]
149. Du, Z.; Su, D. Bibliometric analysis of the effects of heavy metal pollution remediation technologies on paddy fields. *J. Agro-Environ. Sci.* **2018**, *37*, 2409–2417.

toxics

Article

Toxicity Effects of Combined Mixtures of BDE-47 and Nickel on the Microalgae *Phaeodactylum tricornutum* (Bacillariophyceae)

Xiaolai Shi, Ruoyu Guo, Douding Lu, Pengbin Wang * and Xinfeng Dai *

Key Laboratory of Marine Ecosystem Dynamics, Second Institute of Oceanography,
Ministry of Natural Resources, 36 Baochubei Road, Hangzhou 310012, China; shxl7509@sio.org.cn (X.S.);
dinoflagellate@sio.org.cn (R.G.); doudinglu@sio.org.cn (D.L.)
* Correspondence: algae@sio.org.cn (P.W.); microalgae@sio.org.cn (X.D.);
  Tel.: +86-182-6886-1647 (P.W.); +86-137-3546-6556 (X.D.)

Abstract: Nickel and 2,2′,4,4′-tetrabromodiphenyl ether (BDE-47) are two environmental pollutants commonly and simultaneously present in aquatic systems. Nickel and BDE-47 are individually toxic to various aquatic organisms. However, their toxicity mechanisms are species-dependent, and the toxic effects of combined mixtures of BDE-47 and nickel have not yet been investigated. The present study investigated the toxic effects of combined mixtures of BDE-47 and nickel in the diatom *Phaeodactylum tricornutum*. BDE-47 and nickel mixtures significantly decreased cell abundance and photosynthetic efficiency, while these cells' reactive oxygen species (ROS) production significantly increased. The EC_{50}-72 h for BDE-47 and mixtures of BDE-47 and nickel were 16.46 ± 0.93 and 1.35 ± 0.06 mg/L, respectively. Thus, combined mixtures of the two pollutants enhance their toxic effects. Interactions between BDE-47 and nickel were evaluated, revealing synergistic interactions that contributed to toxicity in *P. tricornutum*. Moreover, transcriptomic analyses revealed photosynthesis, nitrogen metabolism, the biosynthesis of amino acids, the biosynthesis of secondary metabolites, oxoacid metabolism, organic acid metabolism, carboxylic acid metabolism, and oxidation-reduction processes were considerably affected by the mixtures. This study provides evidence for the mechanisms of toxicity from combined BDE-47 and nickel exposure while also improving our understanding of the ecological risks of toxic chemicals on microalgae.

Keywords: toxicity; physiological; transcriptome; the combination of BDE-47 and nickel; *Phaeodactylum tricornutum*

Citation: Shi, X.; Guo, R.; Lu, D.; Wang, P.; Dai, X. Toxicity Effects of Combined Mixtures of BDE-47 and Nickel on the Microalgae *Phaeodactylum tricornutum* (Bacillariophyceae). *Toxics* **2022**, *10*, 211. https://doi.org/10.3390/toxics10050211

Academic Editors: François Gagné, Stefano Magni and Valerio Matozzo

Received: 27 March 2022
Accepted: 16 April 2022
Published: 22 April 2022

Publisher's Note: MDPI stays neutral with regard to jurisdictional claims in published maps and institutional affiliations.

1. Introduction

Increasing concentrations of chemical pollutants, including heavy metals and organic pollutants from anthropogenic activities, have significantly affected aquatic ecosystems by inhibiting cellular function and growth of aquatic organisms and even killing organisms [1–3]. Numerous ecological risk assessments have been conducted for various individual chemical pollutants [4,5]. However, the need to assess risks for combinations of chemicals has become apparent since organisms are exposed to mixtures of chemical pollutants in natural environments rather than individual compounds, especially in aquatic ecosystems [6–8].

Microalgae are major primary producers in aquatic ecosystems and biospheres. Microalgae can accumulate chemical pollutants in aquatic ecosystems and transfer them to higher trophic levels [9]. In addition, they are also sensitive to certain chemical pollutants, with many pollutants altering microalgal morphology, physiology, and genetics, thereby affecting their functions in aquatic ecosystems [9]. Consequently, microalgae are considered important models for assessing ecotoxicity, with numerous toxicity assessments previously published for microalgae [10–12].

Heavy metals and organic pollutants are two of the major toxic pollutants present. Nickel is a heavy metal used in diverse metallurgical, electronic, chemical, and food industries. The prevalence of nickel-based products has led to the inevitable pollution of the environment, with concentrations reaching 74.56 mg/kg in coastal surface sediments [13]. Nickel is also a micronutrient essential for some microalgal enzymes, such as urease in the diatom *Phaeodactylum tricornutum* [14]. Nevertheless, excess nickel can inhibit photosynthesis and decrease protein, carbohydrate, and lipid concentrations in microalgae [10,15]. The toxic effects of nickel on *P. tricornutum* have recently been investigated, revealing that nickel can inhibit *P. tricornutum* photosynthesis and induce reactive oxygen species (ROS) production [16]. Polybrominated diphenyl ethers (PBDEs) are persistent organic pollutants and flame retardants that have been used in commercial products since the 1960s [17]. PBDE pollutants are globally distributed in aquatic ecosystems and have been extensively detected in biological samples [2,18–20]. Indeed, PBDEs in coastal waters have been detected at a concentration of 65.5 ng/L in the Bohai Sea of China [21]. However, their concentrations are generally higher in coastal sediments than in open waters, as high as 4212 (ng/g dw) [22]. In most areas, decabromodiphenyl ether (BDE-209) and 2,2′,4,4′-tetrabromodiphenyl ether (BDE-47) are the predominant types of identified PBDEs [21,22]. The toxicity of BDE-47 toward microalgae has been evaluated in the algae *Skeletonema costatum*, *Thalassiosira pseudonana*, *Phaeodactylum tricornutum*, *Platymonas subcordiformis*, *Alexandrium minutum*, and *Dunaliella salina* [23–25]. These studies have shown that BDE-47 can inhibit photosynthetic efficiency, arrest cell division, and induce H_2O_2 production in microalgal cells [25–28]. The toxic effects of nickel and BDE-47 have been identified in several microalgae species, although the toxicity mechanisms vary [10,26–29]. In *P. tricornutum*, BDE-47 can damage chloroplasts, reduce the oxygen evolution rate, alter the performance of photosystems, and stimulate ROS production [24]. Nevertheless, studies of the combined toxicity of nickel and BDE-47 in microalgae remain limited.

P. tricornutum is a cosmopolitan marine diatom species widely used to study ecotoxicity [30], and its complete genome sequence has been published [31]. Moreover, the toxicity mechanisms of nickel on *P. tricornutum* have been recently reported [16]. Furthermore, the toxic effects of BDE-47 have also been recently evaluated for *P. tricornutum* based on photosynthesis-related parameters [24]. This study investigated the combined toxicity effects of BDE-47 and nickel on *P. tricornutum* via physiological and transcriptomic responses. At the same time, the interaction patterns of BDE-47 and nickel in *P. tricornutum* were also measured. This study provides additional understanding of the toxic effects of BDE-47 and nickel mixtures on *P. tricornutum*.

2. Materials and Methods

2.1. Cell Cultures and Chemical Treatments

Axenic algal cultures of the diatom *P. tricornutum* CCMP2561 were obtained from the National Center of Marine Algae and Microbiota (USA) and cultured with an f/2 medium at 20 °C (12:12 h light-dark cycles with 65 μmol photons/m^2/s). The chemical toxicity studies used exponential growth phase cultures with an initial cell density of 5.2×10^5 cells. To evaluate the influence of chemical pollutants on cultures, a gradient of individual BDE-47 (Dr. Ehrenstorfer GmbH, Germany), individual nickel (NiCl$_2$·6H$_2$O, Sigma-Aldrich, Louis, MO, USA), and mixtures of the compounds were added to *P. tricornutum* cultures (concentrations shown in Table S1), followed by harvesting cells at 24, 48, 72, and 96 h to assess cell abundance and photosynthetic efficiency. Each 200 mL of *P. tricornutum* cultures was maintained in glass flasks, and all the flasks were maintained in the same incubator.

The actual concentrations of nickel and BDE-47 were verified with ICP-MS (inductively coupled plasma-mass spectrometry) and GC-MS (gas chromatograph-mass spectrometry), respectively. Moreover, relationships among the actual nickel concentrations and BDE-47 in mixtures were assessed with regression analysis (Table S1). BDE-47 was dissolved in dimethyl sulfoxide (DMSO, CNW, China), and untreated DMSO-amended cultures were used as controls. All treatments were performed in triplicate.

2.2. Physiological Parameter Analysis

Cell density, photosynthetic efficiency, and reactive oxidative species (ROS) levels were measured to evaluate the physiological responses of *P. tricornutum* to chemical exposure. *P. tricornutum* cells were fixed with glutaraldehyde (Sigma, Louis, MO, USA) and counted using a Countstar automated algae counter (Countstar algae, Shanghai, China). Photosynthetic efficiency (Fv/Fm) was directly determined using a pulse amplitude-modulated fluorometer (Water-PAM fluorometer, Walz, Germany) after adapting cultures to the dark. ROS production was determined by measuring fluorescence density. *P. tricornutum* cells were stained with 2′,7′-Dichlorodihydrofluorescein diacetate (DCFH-DA) according to the manufacturer's instructions (Nanjing Jiancheng Bioengineering Institute, Nanjing, China). At the same time, ROS fluorescence density was measured using a fluorospectrophotometer (LS45 Fluorescence Spectrometer, PerkinElmer, Llantrisant, UK) with an excitation and emission wavelength of 515 nm and 550 nm, respectively. A sampling of the microalgae was performed in triplicate throughout the experiments.

2.3. Median Effective Concentrations (EC_{50})

The concentrations of BDE-47 and nickel/BDE-47 mixtures that induced a 50% reduction in *P. tricornutum* biomass after 72 h of exposure were calculated according to the Organization for Economic Cooperation and Development (OECD) guidelines [32] by estimating cell numbers. EC_{50}-72 h values were estimated with a sigmoidal curve and the OriginPro 2018 software program (OriginLab Corporation, Northampton, MA, USA).

2.4. Statistical Analyses

One-way analysis of variance (ANOVA) tests with a subsequent Games–Howell test evaluated differences in parameter distributions. Levene's tests were used to investigate homoscedasticity (Table S2). Data are presented as means $\pm$ standard deviations (SD), and statistical significance was determined at $p < 0.05$. The statistical results are shown in Tables S3–S5.

To assess the effects of the mixed treatment, EC_{50}-72 h values were obtained from each compound, and the mixture treatments were used. Concentration addition (CA) and independent addition (IA) models were used to evaluate the interaction patterns of nickel and BED-47 mixtures, followed by an analysis with the MIXTOX tool [33]. Calculation details were described in a previous study [33]. These models evaluate putative interaction patterns for combinations of chemicals. CA models are generally used to evaluate combinations of chemicals that share the same mechanism of action. In contrast, IA models are generally used for those that exhibit different mechanisms of action. Nevertheless, it is difficult to confirm the model that should be used in many cases. The MIXTOX model can be used with the additional parameter 'a' to evaluate potential interaction patterns [34,35]. Briefly, the 'a' parameter was used to evaluate the synergism or antagonism of two chemical interactions, wherein $a < 0$ indicates synergism and $a > 0$ indicates antagonism.

2.5. Transcriptomic Response Analysis

An expected median concentration of 2.4 mg/L for pollutant mixtures was selected for transcriptomic analyses according to the EC_{50}-48 and EC_{50}-72 h values. The *P. tricornutum* cells were harvested at 48 and 72 h for subsequent transcriptome analysis. Specifically, 200 mL of *P. tricornutum* cultures were harvested, centrifuged, and stored at $-80\ ^\circ$C until subsequent RNA extraction. Total RNA was isolated using the TRIzol reagent (Invitrogen, Carlsbad, CA, USA) according to the manufacturer's instructions, followed by an evaluation of RNA integrity and quality with the RNA Nano 6000 Assay Kit for the Bioanalyzer 2100 system (Agilent Technologies, Santa Clara, CA, USA). mRNAs were purified to construct cDNA libraries for sequencing conducted at Novogene (Beijing, China) on the Illumina NovaSeq platform (Illumina Inc., San Diego, CA, USA). According to the manufacturer's instructions, cDNA libraries were constructed with the NEBNext Ultra RNA Library Prep Kit for Illumina (New England BioLabs, USA).

Raw sequence reads were processed to obtain clean reads by removing adapters, N bases, and low-quality reads. Clean data were mapped to the *P. tricornutum* CCAP 1055/1 reference genome retrieved from the NCBI database. The reference genome index was constructed using Hisat2 v2.0.5, which was also used to align the paired-end clean reads to the reference genome. Genes were assigned to the Kyoto Encyclopedia of Genes and Genomes (KEGG) and the Gene Ontology (GO) databases for pathway and functional analyses. Raw reads were submitted to the Sequence Read Archive database under accession number PRJNA744053.

Differential expression analysis was performed using the DESeq2 R package (1.20.0). Thresholds including |fold changes| >2 and $p < 0.05$ were used to identify significantly differentially expressed genes (DEGs). DEG heatmaps were plotted using an online tool (https://www.omicshare.com/tools/, accessed on 29 September 2021).

3. Results

3.1. P. tricornutum Cell Abundance in Response to Toxicity Tests

No significant differences were observed in cell abundance in the BDE-47 toxicity tests at up to 8 mg/L of BDE-47 exposure over 72 h ($p > 0.05$) (Figure 1A). The cell abundance of BDE-47-treated cultures at concentrations of 16-65 mg/L was significantly lower than controls with the same exposure time ($p < 0.001$). After exposure to 65 mg/L BDE-47, cell abundance decreased at 24 h. Overall, the EC_{50}-72 h of BDE-47 for *P. tricornutum* was 16.46 ± 0.93 mg/L.

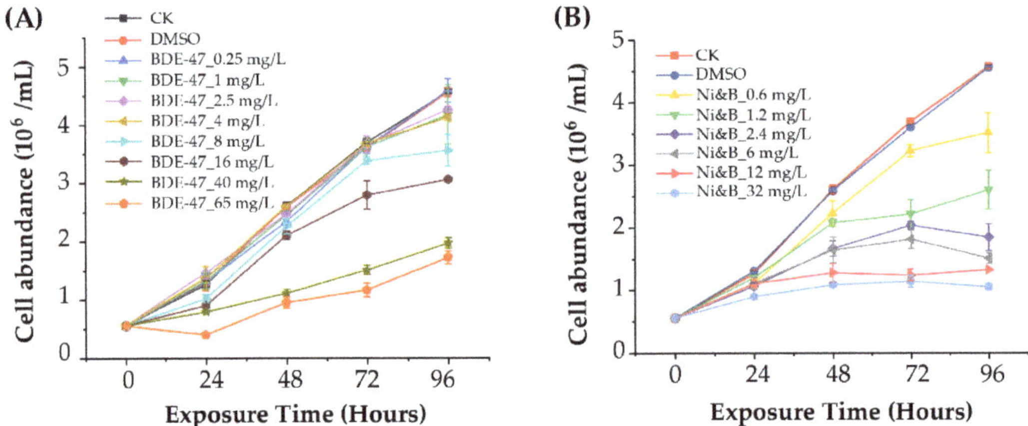

Figure 1. Cell abundance of *Phaeodactylum tricornutum* after individual BDE-47 treatment (**A**) and treatment with mixtures of BDE-47/nickel (**B**). Error bars indicate $\pm$ standard deviations (SD) of means. *p*-values are shown in Table S3.

P. tricornutum growth was inhibited when exposed to BDE-47 and nickel. Further, the inhibitory effect on cell abundances gradually increased with increasing exposure concentration and time (Figure 1B). Significant inhibition of cell abundance was detected at 48, 72, and 96 h when using mixtures at a concentration > 0.6 mg/L ($p < 0.01$). The EC_{50}-48 h and EC_{50}-72 h of the BDE-47 and nickel mixtures for *P. tricornutum* were 3.68 ± 0.11 and 1.35 ± 0.06 mg/L, respectively.

3.2. Photosynthetic Efficiency Response

The photosynthetic efficiency of *P. tricornutum* under different toxin treatments is shown in Figure 2A. BDE-47 exposure caused slight decreases in *P. tricornutum* photosynthetic efficiency. However, significant inhibitory effects were only detected at very high concentrations of BDE-47 (>8 mg/L). Photosynthetic efficiency was still relatively high in the 65 mg/L BDE-47 treatment.

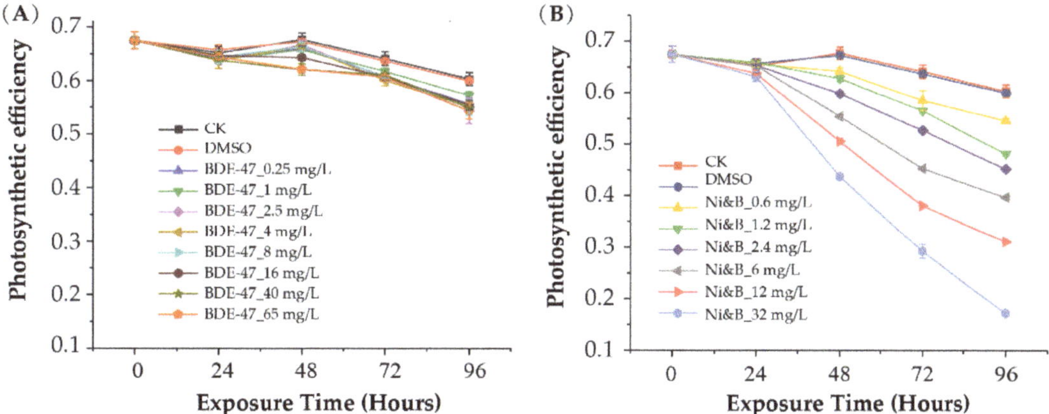

Figure 2. Photosynthetic efficiency of *P. tricornutum* after individual BDE-47 treatment (**A**) and treatment with mixtures of BDE-47/nickel (**B**). Error bars indicate ± SD. *p*-values are shown in Table S4.

Significant differences were not observed after exposure to mixtures of BDE-47 and nickel over 24 h. However, exposure to the mixtures strongly inhibited photosynthesis after 48 h of exposure. Photosynthetic efficiency gradually decreased with increasing exposure time and mixture concentrations (Figure 2B).

3.3. Reactive Oxidative Species (ROS) Production

ROS production levels over 72 h are shown in Figure 3. ROS production was induced by combinations of BDE-47 and nickel, although increases in ROS levels were not detected in the BDE-47-treated samples. The individual nickel treatments also demonstrated significantly increased ROS production [16]. Exposure to the BDE-47 and nickel mixtures significantly increased ROS production over 24 h at an exposure level of 12 mg/L (Figure 3). The highest ROS production was observed with 12 mg/L of mixtures after 72 h of exposure (Figure 3).

3.4. Interactions of BDE-47 and Nickel in P. tricornutum

The toxicity mechanisms of nickel and BDE-47 on *P. tricornutum* cells are not yet fully understood. Consequently, the potential interactions of nickel and BDE-47 were evaluated using concentration addition (CA) and independent addition (IA) models with the MIXTOX tool [33]. The two pollutants' EC50-72h values for nickel, BDE-47, and mixtures were used to evaluate the interaction patterns. The *a* value was identified as −4.5, indicating that the interaction between nickel and BDE-47 was synergistic rather than antagonistic concerning *P. tricornutum* toxicity.

3.5. P. tricornutum Genes Differentially Expressed after Exposure to Mixtures of Nickel and BDE-47

A total of 565 and 325 genes were upregulated and downregulated after 48 h of exposure to nickel and BDE-47, respectively (i.e., those exhibiting |fold changes| > 2 and *p* < 0.05). After 72 h of exposure, these DEG numbers increased to 841 and 435 upregulated and downregulated genes, respectively (Figures S1 and S2).

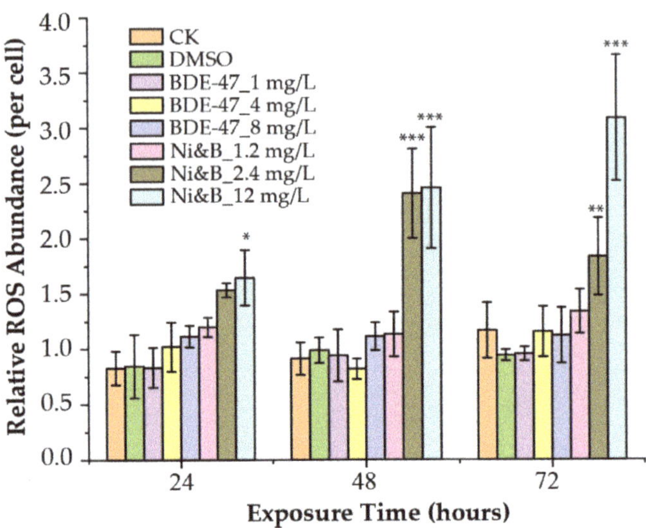

Figure 3. Relative ROS production of *P. tricornutum* after individual BDE-47 treatment and treatment with mixtures of BDE-47/nickel. *** $p < 0.001$; ** $p < 0.01$, and * $p < 0.05$. Error bars indicate $\pm$ SD. *p*-values are shown in Table S5.

Among the DEGs, 532 genes defined as core genes were upregulated or downregulated (I fold changes I >2, $p < 0.05$) after exposure to the mixtures for 48 h and 72 h (Figure S3). KEGG pathway analysis (Figure 4) of the core genes revealed that many of the DEGs were involved in the biosynthesis of secondary metabolites, carbon metabolism, biosynthesis of amino acids, glyoxylate and dicarboxylate metabolism, ribosome biogenesis in eukaryotes, fatty acid metabolism, fatty acid biosynthesis, and porphyrin and chlorophyll metabolism. Further, enrichment GO analysis of the core genes revealed many were involved in oxoacid metabolism, organic acid metabolism, carboxylic acid metabolism, oxidation-reduction reactions, cellular amino acid metabolism, and small molecule metabolism. In addition, oxidoreductase activity and ligase activity were significantly enriched GO terms involved in molecular functions.

3.6. DEGs Related to Photosynthesis and Nitrogen Metabolism

At least 31 genes related to photosynthetic metabolism were responsive to exposure to mixtures of BDE-47 and nickel (Figure 5A, Table S6). Among these, *HemA*, *UbiA*, *Aminotran_3*, *PHATRDRAFT_50740* (magnesium chelates), *hemE*, *GluRS_2*, and *CHLH*, which are involved in porphyrin and chlorophyll metabolism, were downregulated due to exposure to a mixture of BDE-47 and nickel. The photosystem II gene *PsbP* did not exhibit significant changes in expression after 48 h of exposure but decreased expression after 72 h. Most differentially expressed photosynthesis-related genes involved in carbon fixation were upregulated, including *Fba4*, *Fba3*, *FbaC5*, *TIM_1*, *PEPCase_2*, *TPI/GapC3*, *GapC4*, and *GapC1*. The antenna protein-encoding genes *Lhcf15* and *Lhcf11* exhibited increased expression due to exposure to the mixtures, while *Lhcx1* was downregulated after exposure to mixtures between 48 and 72 h. These results indicate that photosynthetic processes display different responses due to exposures to mixtures of BDE-47 and nickel, although the overall photosynthetic balance was disturbed.

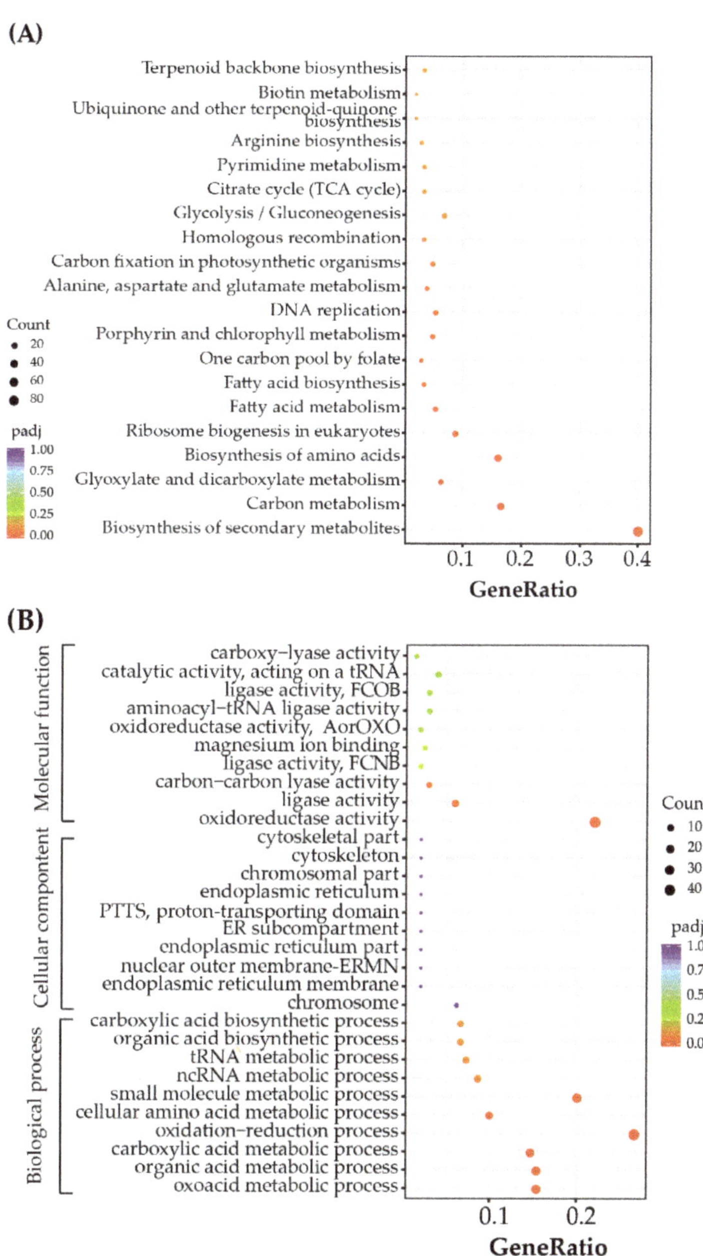

Figure 4. The 20 most abundant Kyoto Encyclopedia of Genes and Genomes pathways (**A**) and 10 most abundant Gene Ontology classifications (**B**) of differentially expressed genes after exposure to mixtures of BDE-47 and nickel treatments. Nuclear outer membrane-endoplasmic reticulum membrane network: ERMN; endoplasmic reticulum: ER; proton-transporting two-sector ATPase complex, proton-transporting domain: PTTS; ligase activity, forming carbon-nitrogen bonds: FCNB; ligase activity, forming carbon-oxygen bonds: FCOB; oxidoreductase activity, acting on the aldehyde or oxo group of donors: AorOXO.

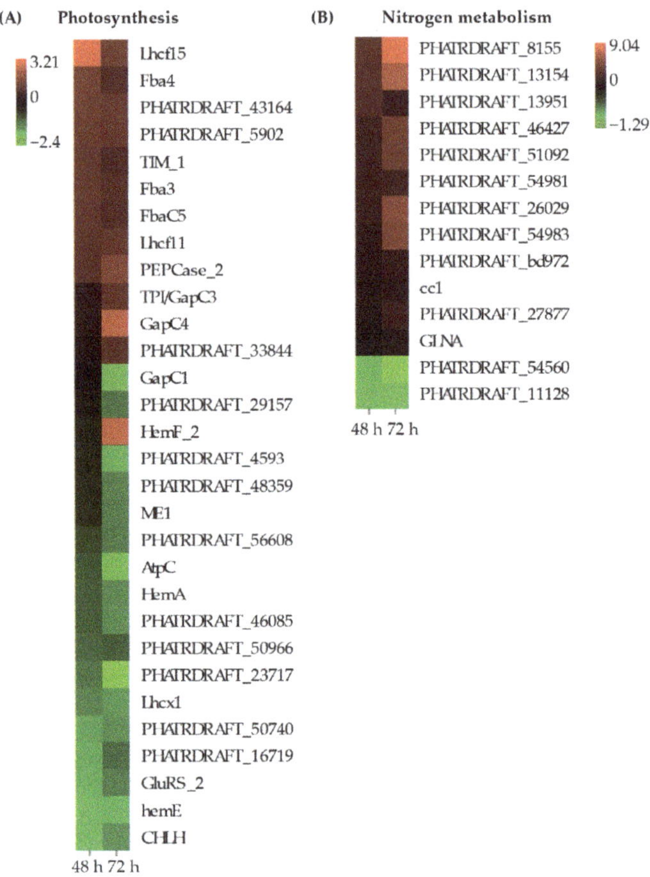

Figure 5. Heatmap of differentially expressed genes involved in photosynthesis and nitrogen metabolism that responded to exposure to mixtures of BDE-47 and nickel. Only genes with |log2 fold| changes > 2 and $p < 0.05$ after 48 h and/or 72 h are shown.

A total of 14 DEGs were involved in nitrogen metabolism (Figure 5B, Table S7). Among these, 12 genes exhibited increased expression after exposure to mixtures of BDE-47 and nickel. *PHATRDRAFT_8155* (nitrite reductase), *PHATRDRAFT_13154* (Pyr_redox_2), *PHATRDRAFT_26029* (MFS_1), and *PHATRDRAFT_54983* (FAD_binding_6) exhibited very high expression levels with log2-fold changes >5 (Figure 5B, Table S7). Thus, nitrogen metabolism might be enhanced by exposure to mixtures of BDE-47 and nickel.

3.7. DEGs Related to Oxidation-Reduction Processes

Many genes involved in oxidation-reduction processes were identified as DEGs (Figure 6, Table S8), indicating that the cellular redox balance was disturbed by exposure to the pollutant mixtures, consistent with the ROS measurements. Among the differentially expressed oxidation-reduction-related genes, some antioxidant genes exhibited increased expression. The most upregulated genes included a thioredoxin-like gene with log2 fold changes of 3.73 and 3.38 upregulated expression after 48 and 72 h, respectively. Glutathione *S*-transferase (*PHATRDRAFT_37658*, *GST*) was also upregulated due to exposure to the mixtures, with upregulation by 1.47 and 2.59 log2 fold changes after 48 and 72 h exposure, respectively.

Oxidation-reduction process

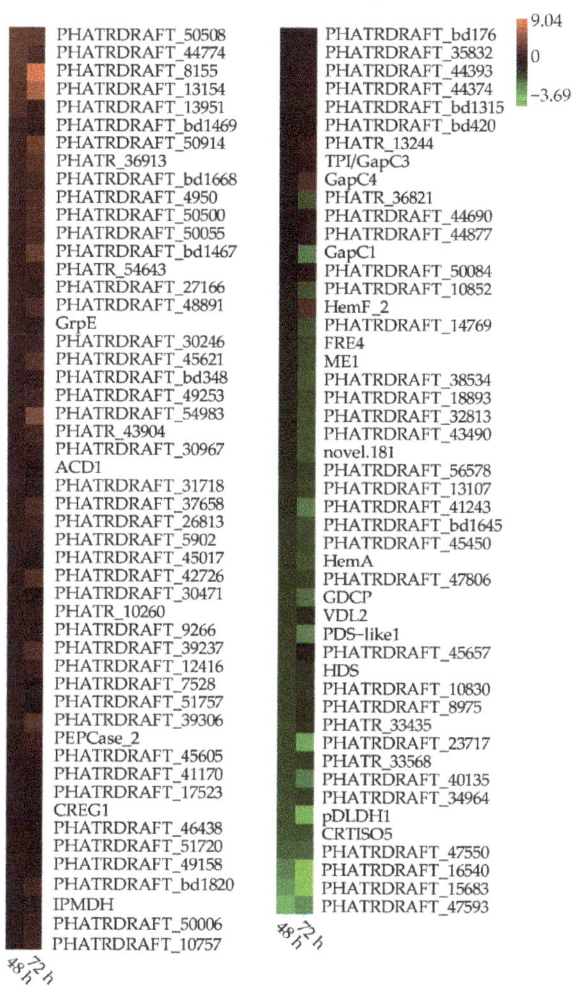

Figure 6. Heatmap of differentially expressed genes involved in oxidation-reduction processes that responded to exposure to mixtures of BDE-47 and nickel. Only genes with |log2 fold| changes > 2 and $p < 0.05$ after 48 h and/or 72 h are shown.

4. Discussion

The toxic effects of nickel by itself on the diatom *P. tricornutum* have been evaluated in our previous studies. Specifically, nickel exposure exhibits an EC_{50}-72 h value of 2.48 ± 0.33 mg/L and leads to ROS production and decreased photosynthesis [16]. In addition, the toxic effects of BDE-47 on *P. tricornutum* have also been reported [24]. Previous studies have indicated that nickel and BDE-47 can cause toxic effects on the diatom *P. tricornutum* by inhibiting cell growth and affecting photosynthesis [24]. However, the present study is the first to investigate the combined toxicity of BDE-47 and nickel on microalgae.

BDE-47 and nickel exhibited synergistic interactions, with enhanced toxicity when mixed. Photosynthetic efficiency was strongly inhibited in *P. tricornutum* due to exposure to mixtures of BDE-47 and nickel, while inhibitory effects increased with increasing exposure time. However, BDE-47 by itself only slightly inhibited the photosynthetic efficiency of

P. tricornutum, while nickel has been shown to strongly inhibit the photosynthetic efficiency of *P. tricornutum* [16]. Other studies of microalgae have shown that BDE-47 inhibits the photosynthetic efficiency of *Skeletonema costatum* and *Alexandrium minutum*. However, BDE-47 also reduced photosynthetic efficiency within 24 h of exposure in *Dunaliella salina* and *Thalassiosira pseudonana*, followed by gradual recovery to normal levels within 120 h [26–28]. Several differentially expressed photosynthetic genes at the transcriptional level, including those involved in chlorophyll synthesis, antenna proteins, oxygen evolution, electron transport, and downstream carbon fixation, were identified as DEGs of *P. tricornutum* after individual BDE-47 exposure, with most being downregulated [24]. Further, BDE-47 also suppressed the expression of some genes involved in photosystem II and antenna proteins in the diatom *T. pseudonana* [25]. In the present study, several genes involved in photosynthesis components such as carbon fixation, chlorophyll synthesis, photosystem II, and antenna proteins exhibited differential expression after exposure to mixtures of BDE-47 and nickel. In addition, nickel might impair the photosystem's oxygen-evolving complex (OEC) in *P. tricornutum*, leading to decreased photosynthetic efficiency [16]. These data indicate that photosynthetic complexes may be one of the toxic effect targets of BDE-47 in microalgae and that its toxic effects on *P. tricornutum* photosynthesis could be enhanced with additional nickel supplementation.

Increased ROS is an indicator that cells were exposed to oxidative stress. Exposure to mixtures of BDE-47 and nickel significantly induced oxidative stress in *P. tricornutum*. Transcriptomic analysis identified DEGs involved in oxidation-reduction processes, with some involved in nitrogen metabolism, photosynthesis, and amino acid metabolism. Among these genes, antioxidant-associated genes, including those encoding thioredoxin (*Thx*) and glutathione *S*-transferase (*GST*), exhibited increased expression, implying that they might be important for oxidative stress resistance after exposure to mixtures of BDE-47 and nickel. Specifically, Thx could be involved in oxidative damage mitigation [36]. Thx possesses a conserved redox site that supports intracellular redox homeostasis, reduces protein thiols, and is one of the most important components of the thioredoxin system [37,38]. Increased expression of Thx has also been detected in individual nickel treatments of *P. tricornutum* [16]. In addition, *GST* plays an important role in the detoxification and reduction of ROS in multiple stress processes [39]. BDE-47-induced *GST* expression has been detected in *T. pseudonana* [25], while *GST* has been identified as a target of Thx in *Arabidopsis* plants [40]. Increased expression of *Thx* and *GST* indicates that detoxification activities were activated to protect cells from oxidative stress due to exposures to mixtures of BDE-47 and nickel. Thus, *Thx* and *GST* might simultaneously protect *P. tricornutum* cells from oxidative stress.

Transcriptomic analysis revealed that exposure to mixtures of BDE-47 and nickel could affect genes involved in oxoacid metabolism, organic acid metabolism, carboxylic acid metabolism, and organic acid biosynthesis (Figure 4B). Organic acids, such as citric acid and malic acid, play important roles in detoxifying chemical stresses, including those from BDE-47 and nickel [41–43]. The regulated organic acid BDE-47 has been detected in *Oryza sativa* [42,43], while nickel is also known to modulate organic acid compositions in *Alyssum murale* [41]. Isocitrate lyase (encoded by *PHATRDRAFT_51088*) is an enzyme that produces succinic acid and exhibited increased expression with 1.95 and 1.65 log2-fold change values after 48 and 72 h exposure to nickel and BDE-47 mixtures in this study. Succinic acid is involved in abiotic stress resistance in some plants [41,44,45]. For example, additional succinic acid concentrations can alleviate aluminum toxicity in alfalfa [44], while exogenous succinic acid concentrations lead to the enhanced tolerance of lead by *Larix olgensis* [45]. These data suggest that some organic/oxo-/carboxylic acids might be involved in the stress responses of *P. tricornutum* to mixtures of BDE-47 and nickel. In addition, organic/oxo-/carboxylic acids could serve as electron donors via their involvement in the dehalogenation process. For example, exogenous organic/oxo-/carboxylic acid-related compounds can be used as carbon sources (e.g., formate, acetate, and lactate) that could enhance the degradation of BDE-47 by microorganisms under anaerobic conditions [46,47]. Some microalgae, such as *Chlorella*, have been shown to transform BDE-47 into the debrominated product of BDE-47 [48]. We consequently hypothesized that some organic/oxo-/carboxylic acids produced by microalgae could be electron donors and are involved in transforming BDE-47 within algal cells.

The biosynthesis of secondary metabolites (Table S9), amino acids, and fatty acids were also extensively affected by exposure to mixtures of BDE-47 and nickel. Many organic acids are precursors for fatty acids, amino acids, and secondary metabolites [49]. Similarly, some secondary metabolites, amino acids, and fatty acids exhibited increased concentrations after BDE-47 treatment of rice [42,43]. Moreover, the biosynthesis of amino acids and fatty acid synthesis pathways were also affected by the BDE-47 treatment of the diatom *T. pseudonana*. Likewise, individual nickel treatment of *P. tricornutum* led to differentially expressed biosynthesis of amino acid and fatty acid synthesis pathways [16]. These results indicate that exposure to mixtures of BDE-47 and nickel extensively affected the primary and secondary metabolite profiles of *P. tricornutum*.

Nitrogen metabolism-associated genes were also detected among the DEGs of this study following exposure to the mixture of pollutants. The most expressed nitrogen metabolism-associated gene was *PHATRDRAFT_8155* (nitrite reductase), which is involved in nitrate assimilation, although other nitrogen metabolism-associated genes were also highly up-expressed. The enhanced expression of nitrogen metabolism-related genes was most prevalent in the individual BDE-47-treated *T. pseudonana* and nickel-treated *P. tricornutum*. These results suggest that the nitrogen metabolism of microalgae is highly sensitive to BDE-47 and/or nickel pollution. In addition, mixtures of BDE-47 and nickel can also enhance toxicity effects on nitrogen metabolism relative to the effects from individual chemicals.

5. Conclusions

In conclusion, the results from this study indicate that combined mixtures of BDE-47 and nickel pollutants were more toxic to *P. tricornutum* cells than either of the chemicals individually. Modeling analysis indicated a synergistic interaction between the pollutants, resulting in increased toxicity to *P. tricornutum*. Specifically, mixtures of BDE-47 and nickel decreased cell abundances, inhibited photosynthetic efficiency, and induced ROS production. Transcriptomic analysis further revealed that the mixtures affected photosynthesis, nitrogen metabolism, primary metabolism, and secondary metabolism processes, ultimately leading to cell growth inhibition in the diatom *P. tricornutum*.

Supplementary Materials: The following supporting information can be downloaded at: https://www.mdpi.com/article/10.3390/toxics10050211/s1, Figure S1: Volcano plot of differentially expressed genes under 2.5 mg/L mixture treatment for 48 h; Figure S2: Volcano plot of differentially expressed genes under 2.5 mg/L mixture treatment for 72 h; Figure S3: Venn diagram of differentially expressed genes under 2.5 mg/L mixture treatment for 48 and 72 h; Table S1: Chemical concentrations tested in this study; Table S2: Results of Levene statistic test to determine homoscedasticity in data and statistic method used in ANOVA; Table S3: The P-value of comparisons with Games–Howell of cell abundance; Table S4: The P-value of comparisons with Games–Howell of photosynthesis; Table S5: The P-value of comparisons with Games–Howell of ROS; Table S6: DEGs related to photosynthesis; Table S7: DEGs involved in nitrogen metabolism; Table S8: DEGs involved in oxidation-reduction process; Table S9: DEGs involved in biosynthesis of secondary metabolites.

Author Contributions: X.S. and X.D. designed the research; R.G. performed the experiments; X.S. and R.G. wrote the manuscript; D.L., P.W. and X.D. reviewed and revised all manuscript versions. All authors have read and agreed to the published version of the manuscript.

Funding: Please add: This work is supported by the following Projects: The National Key Research and Development Program of China (2016YFF0201104); The Zhejiang Provincial Natural Science Foundation of China (LY19D060007); Scientific Research Fund of the Second Institute of Oceanography, MNR (JG2108).

Institutional Review Board Statement: Not applicable.

Informed Consent Statement: Not applicable.

Data Availability Statement: Not applicable.

Conflicts of Interest: The authors declare that they have no known competing financial interests or personal relationships that could have appeared to influence the work reported in this paper.

References

1. Zhang, M.; Sun, X.; Xu, J. Heavy metal pollution in the East China Sea: A review. *Mar. Pollut. Bull.* **2020**, *159*, 111473. [CrossRef] [PubMed]
2. Yogui, G.T.; Sericano, J.L. Polybrominated diphenyl ether flame retardants in the U.S. marine environment: A review. *Environ. Int.* **2009**, *35*, 655–666. [CrossRef] [PubMed]
3. Mearns, A.J.; Morrison, A.M.; Arthur, C.; Rutherford, N.; Bissell, M.; Rempel-Hester, M.A. Effects of pollution on marine organisms. *Water Environ. Res.* **2020**, *92*, 1510–1532. [CrossRef]
4. O'Brien, A.L.; Keough, M.J. Ecological responses to contamination: A meta-analysis of experimental marine studies. *Environ. Pollut.* **2014**, *195*, 185–191. [CrossRef] [PubMed]
5. Schuijt, L.M.; Peng, F.J.; Van den Berg, S.J.P.; Dingemans, M.M.L.; Van den Brink, P.J. (Eco)toxicological tests for assessing impacts of chemical stress to aquatic ecosystems: Facts, challenges, and future. *Sci. Total Environ.* **2021**, *795*, 148776. [CrossRef]
6. Aronzon, C.M.; Peluso, J.; Coll, C.P. Mixture toxicity of copper and nonylphenol on the embryo-larval development of *Rhinella arenarum*. *Environ. Sci. Pollut. Res. Int.* **2020**, *27*, 13985–13994. [CrossRef]
7. Kovalakova, P.; Cizmas, L.; McDonald, T.J.; Marsalek, B.; Feng, M.; Sharma, V.K. Occurrence and toxicity of antibiotics in the aquatic environment: A review. *Chemosphere* **2020**, *251*, 126351. [CrossRef]
8. Topaz, T.; Egozi, R.; Suari, Y.; Ben-Ari, J.; Sade, T.; Chefetz, B.; Yahel, G. Environmental risk dynamics of pesticides toxicity in a Mediterranean micro-estuary. *Environ. Pollut.* **2020**, *265*, 114941. [CrossRef]
9. Torres, M.A.; Barros, M.P.; Campos, S.C.; Pinto, E.; Rajamani, S.; Sayre, R.T.; Colepicolo, P. Biochemical biomarkers in algae and marine pollution: A review. *Ecotoxicol. Environ. Saf.* **2008**, *71*, 1–15. [CrossRef]
10. Martínez-Ruiz, E.B.; Martínez-Jerónimo, F. Nickel has biochemical, physiological, and structural effects on the green microalga *Ankistrodesmus falcatus*: An integrative study. *Aquat. Toxicol.* **2015**, *169*, 27–36. [CrossRef]
11. Nguyen, M.K.; Moon, J.Y.; Lee, Y.C. Microalgal ecotoxicity of nanoparticles: An updated review. *Ecotoxicol. Environ. Saf.* **2020**, *201*, 110781. [CrossRef] [PubMed]
12. Nunes, B.W.; Hedlund, K.F.; Oliveira, M.A.; Carissimi, E. Ecotoxicological analysis of acrylamide using a microalga as an Indicator organism. *Water Environ. Res.* **2018**, *90*, 442–451. [CrossRef] [PubMed]
13. Yunus, K.; Zuraidah, M.A.; John, A. A review on the accumulation of heavy metals in coastal sediment of Peninsular Malaysia. *Ecofeminism Clim. Change* **2020**, *1*, 21–35. [CrossRef]
14. Bekheet, T. The role of nickel in urea assimilation by algae. *Planta* **1982**, *156*, 385–387.
15. Agawany, N.E.; Kaamoush, M.; El-Zeiny, A.; Ahmed, M. Effect of heavy metals on protein content of marine unicellular green alga Dunaliella tertiolecta. *Environ. Monit. Assess.* **2021**, *193*, 1–14. [CrossRef] [PubMed]
16. Guo, R.; Lu, D.; Liu, C.; Hu, J.; Wang, P.; Dai, X. Toxic effect of nickel on microalgae *Phaeodactylum tricornutum* (Bacillariophyceae). *Ecotoxicology* **2022**. [CrossRef]
17. Vonderheide, A.P.; Mueller, K.E.; Meija, J.; Welsh, G.L. Polybrominated diphenyl ethers: Causes for concern and knowledge gaps regarding environmental distribution, fate and toxicity. *Sci. Total Environ.* **2008**, *400*, 425–436. [CrossRef]
18. Kelly, B.C.; Ikonomou, M.G.; Blair, J.D.; Gobas, F.A. Hydroxylated and methoxylated polybrominated diphenyl ethers in a Canadian Arctic marine food web. *Environ. Sci. Technol.* **2008**, *42*, 7069–7077. [CrossRef]
19. Salvadó, J.A.; Sobek, A.; Carrizo, D.; Gustafsson, Ö. Observation-based assessment of PBDE loads in Arctic Ocean Waters. *Environ. Sci. Technol.* **2016**, *50*, 2236–2245. [CrossRef]
20. Xie, Z.; Möller, A.; Ahrens, L.; Sturm, R.; Ebinghaus, R. Brominated flame retardants in seawater and atmosphere of the Atlantic and the Southern Ocean. *Environ. Sci. Technol.* **2011**, *45*, 1820–1826. [CrossRef]
21. Wang, Y.; Wu, X.; Zhao, H.; Xie, Q.; Hou, M.; Zhang, Q.; Du, J.; Chen, J. Characterization of PBDEs and novel brominated flame retardants in seawater near a coastal mariculture area of the Bohai Sea, China. *Sci. Total Environ.* **2017**, *580*, 1446–1452. [CrossRef] [PubMed]
22. Yin, H.; Tang, Z.; Meng, T.; Zhang, M. Concentration profile, spatial distributions and temporal trends of polybrominated diphenyl ethers in sediments across China: Implications for risk assessment. *Ecotoxicol. Environ. Saf.* **2020**, *206*, 111205. [CrossRef] [PubMed]
23. Li, Z.; Fenghua, J.; Xiangfeng, K.; Yang, W.; Jingru, W.; Tianpeng, Z.; Zhaoyu, W.; Yingying, Z. Toxic effect of BDE-47 on the marine alga *Skeletonema costatum*: Population dynamics, photosynthesis, antioxidation and morphological changes. *Chemosphere* **2022**, *286*, 131674. [CrossRef]
24. Liu, Q.; Tang, X.; Zhang, X.; Yang, Y.; Sun, Z.; Jian, X.; Zhao, Y.; Zhang, X. Evaluation of the toxic response induced by BDE-47 in a marine alga, *Phaeodactylum tricornutum*, based on photosynthesis-related parameters. *Aquat. Toxicol.* **2020**, *227*, 105588. [CrossRef] [PubMed]

25. Zhao, Y.; Tang, X.; Lv, M.; Liu, Q.; Li, J.; Zhang, B.; Li, L.; Zhang, X.; Zhao, Y. The molecular response mechanisms of a diatom *Thalassiosira pseudonana* to the toxicity of BDE-47 based on whole transcriptome analysis. *Aquat. Toxicol.* **2020**, *229*, 105669. [CrossRef] [PubMed]

26. Zhao, X.; Lin, X.; Qu, K.; Xia, B.; Sun, X.; Chen, B. Toxicity of BDE-47, BDE-99 and BDE-153 on swimming behavior of the unicellular marine microalgae *Platymonas subcordiformis* and implications for seawater quality assessment. *Ecotoxicol. Environ. Saf.* **2019**, *174*, 408–416. [CrossRef]

27. Zhao, Y.; Tang, X.; Quigg, A.; Lv, M.; Zhao, Y. The toxic mechanisms of BDE-47 to the marine diatom *Thalassiosira pseudonana*-a study based on multiple physiological processes. *Aquat. Toxicol.* **2019**, *212*, 20–27. [CrossRef]

28. Zhao, Y.; Wang, Y.; Li, Y.; Santschi, P.H.; Quigg, A. Response of photosynthesis and the antioxidant defense system of two microalgal species (*Alexandrium minutum* and *Dunaliella salina*) to the toxicity of BDE-47. *Mar. Pollut. Bull.* **2017**, *124*, 459–469. [CrossRef]

29. Boisvert, S.; Joly, D.; Leclerc, S.; Govindachary, S.; Harnois, J.; Carpentier, R. Inhibition of the oxygen-evolving complex of photosystem II and depletion of extrinsic polypeptides by nickel. *Biometals Int. J. Role Metal Ions Biol. Biochem. Med.* **2007**, *20*, 879–889. [CrossRef]

30. Fastelli, P.; Renzi, M. Exposure of key marine species to sunscreens: Changing ecotoxicity as a possible indirect effect of global warming. *Mar. Pollut. Bull.* **2019**, *149*, 110517. [CrossRef]

31. Bowler, C.; Allen, A.; Badger, J.H.; Grimwood, J.; Jabbari, K.; Al, E. The *Phaeodactylum* genome reveals the evolutionary history of diatom genomes. *Nature* **2008**, *456*, 239–244. [CrossRef] [PubMed]

32. OECD. *Test No. 201: Freshwater Alga and Cyanobacteria, Growth Inhibition Test*; Organisation for Economic Cooperation and Development (OECD): Paris, France, 2011.

33. Jonker, M.J.; Svendsen, C.; Bedaux, J.J.; Bongers, M.; Kammenga, J.E. Significance testing of synergistic/antagonistic, dose level-dependent, or dose ratio-dependent effects in mixture dose-response analysis. *Environ. Toxicol. Chem.* **2005**, *24*, 2701–2713. [CrossRef] [PubMed]

34. Greco, W.; Unkelbach, H.D.; Pöch, G.; Sühnel, J.; Kundi, M.; Bödeker, W. Consensus on concepts and terminology for combined-action assessment: The Saariselkä agreement. *Arch. Complex Environ. Stud.* **1992**, *4*, 65–69.

35. Greco, W.R.; Bravo, G.; Parsons, J.C. The search for synergy: A critical review from a response surface perspective. *Pharmacol. Rev.* **1995**, *47*, 331–385. [PubMed]

36. Vieira Dos Santos, C.; Rey, P. Plant thioredoxins are key actors in the oxidative stress response. *Trends Plant Sci.* **2006**, *11*, 329–334. [CrossRef]

37. Nordberg, J.; Arnér, E.S. Reactive oxygen species, antioxidants, and the mammalian thioredoxin system. *Free Radic. Biol. Med.* **2001**, *31*, 1287–1312. [CrossRef]

38. Patwari, P.; Lee, R.T. Thioredoxins, mitochondria, and hypertension. *Am. J. Pathol.* **2007**, *170*, 805–808. [CrossRef]

39. Kumar, S.; Trivedi, P.K. Glutathione *S*-Transferases: Role in combating abiotic stresses including arsenic detoxification in plants. *Front. Plant Sci.* **2018**, *9*, 751. [CrossRef]

40. Marchand, C.; Le Maréchal, P.; Meyer, Y.; Miginiac-Maslow, M.; Issakidis-Bourguet, E.; Decottignies, P. New targets of *Arabidopsis* thioredoxins revealed by proteomic analysis. *Proteomics* **2004**, *4*, 2696–2706. [CrossRef]

41. Agrawal, B.; Czymmek, K.J.; Sparks, D.L.; Bais, H.P. Transient Influx of nickel in root mitochondria modulates organic acid and reactive oxygen species production in nickel hyperaccumulator *Alyssum murale*. *J. Biol. Chem.* **2013**, *288*, 7351–7362. [CrossRef]

42. Chen, J.; Le, X.C.; Zhu, L. Metabolomics and transcriptomics reveal defense mechanism of rice (*Oryza sativa*) grains under stress of 2,2′,4,4′-tetrabromodiphenyl ether. *Environ. Int.* **2019**, *133*, 105154. [CrossRef] [PubMed]

43. Chen, J.; Li, K.; Le, X.C.; Zhu, L. Metabolomic analysis of two rice (*Oryza sativa*) varieties exposed to 2, 2′, 4, 4′-tetrabromodiphenyl ether. *Environ. Pollut.* **2018**, *237*, 308–317. [CrossRef] [PubMed]

44. An, Y.; Zhou, P.; Xiao, Q.; Shi, D. Effects of foliar application of organic acids on alleviation of aluminum toxicity in alfalfa. *J. Plant Nutr. Soil Sci.* **2014**, *177*, 421–430. [CrossRef]

45. Song, J.; Zhang, H.; Duan, C.; Cui, X. Exogenous application of succinic acid enhances tolerance of *Larix olgensis* seedling to lead stress. *J. For. Res.* **2018**, *29*, 1497–1505. [CrossRef]

46. Pan, Y.; Chen, J.; Zhou, H.; Cheung, S.G.; Tam, N.F.Y. Degradation of BDE-47 in mangrove sediments with amendment of extra carbon sources. *Mar. Pollut. Bull.* **2020**, *153*, 110972. [CrossRef] [PubMed]

47. Zhao, S.; Rogers, M.J.; Ding, C.; He, J. Reductive Debromination of Polybrominated Diphenyl Ethers—Microbes, Processes and Dehalogenases. *Front. Microbiol.* **2018**, *9*, 1292. [CrossRef] [PubMed]

48. Deng, D.; Chen, H.X.; Wong, Y.S.; Tam, N. Physiological response and oxidative transformation of 2,2′,4,4′-tetrabromodiphenyl ether (BDE-47) by a *Chlorella* isolate. *Sci. Total Environ.* **2020**, *744*, 140869. [CrossRef]

49. Singh, S.K.; Barnaby, J.Y.; Reddy, V.R.; Sicher, R.C. Varying response of the concentration and yield of soybean seed mineral elements, carbohydrates, organic acids, amino acids, protein, and oil tophosphorus starvation and CO(2) enrichment. *Front. Plant Sci.* **2016**, *7*, 1967. [CrossRef]

Article

Genotoxicity Set Up in *Artemia franciscana* Nauplii and Adults Exposed to Phenanthrene, Naphthalene, Fluoranthene, and Benzo(k)fluoranthene

Luisa Albarano [1,2,*], Sara Serafini [1], Maria Toscanesi [3], Marco Trifuoggi [3], Valerio Zupo [4], Maria Costantini [2], Davide A. L. Vignati [5], Marco Guida [1] and Giovanni Libralato [1,2]

1   Department of Biology, University of Naples Federico II, Complesso Universitario di Monte Sant'Angelo, Via Vicinale Cupa Cintia 26, 80126 Naples, Italy; sara.serafini@unina.it (S.S.); marco.guida@unina.it (M.G.); giovanni.libralato@unina.it (G.L.)
2   Stazione Zoologica Anton Dohrn, Department of Marine Biotechnology, Villa Comunale, 80121 Naples, Italy; maria.costantini@szn.it
3   Department of Chemical Sciences, University of Naples Federico II, Complesso Universitario di Monte Sant'Angelo, Via Vicinale Cupa Cintia 26, 80126 Naples, Italy; maria.toscanesi@unina.it (M.T.); marco.trifuoggi@unina.it (M.T.)
4   Stazione Zoologica Anton Dohrn, Department of Marine Biotechnology, Villa Dohrn, Punta San Pietro, 80077 Naples, Italy; valerio.zupo@szn.it
5   Université de Lorraine, CNRS, LIEC, F-57000 Metz, France; david-anselmo.vignati@univ-lorraine.fr
*   Correspondence: luisa.albarano@unina.it

Citation: Albarano, L.; Serafini, S.; Toscanesi, M.; Trifuoggi, M.; Zupo, V.; Costantini, M.; Vignati, D.A.L.; Guida, M.; Libralato, G. Genotoxicity Set Up in *Artemia franciscana* Nauplii and Adults Exposed to Phenanthrene, Naphthalene, Fluoranthene, and Benzo(k)fluoranthene. *Water* **2022**, *14*, 1594. https://doi.org/10.3390/w14101594

Academic Editors: François Gagné, Stefano Magni and Valerio Matozzo

Received: 2 April 2022
Accepted: 13 May 2022
Published: 16 May 2022

Publisher's Note: MDPI stays neutral with regard to jurisdictional claims in published maps and institutional affiliations.

Abstract: Polycyclic aromatic hydrocarbons (PAHs) consist of a group of over 100 different organic compounds mainly generated by anthropogenic activities. Because of their low water solubility, they tend to be accumulated in sediment, where their degradation rate is very low. Few studies have been carried out so far to investigate the effects of PAHs on *Artemia franciscana*. *Artemia* is easy to manage at laboratory scale, but it is not a sensitive biological model considering the traditional endpoints (i.e., mortality). In addition to evaluating the lethality on nauplii and adults of *A. franciscana* after 24 and 48 h, we focused on the genotoxicity to investigate the potential effects of phenanthrene (PHE), naphthalene (NAP), fluoranthene (FLT), and benzo(k)fluoranthene (BkF). Results showed that FLT was the most toxic both for nauplii and adults after 48 h of exposure. Real-time qPCR showed that all toxicants, including BkF, which had no negative effects on the survival of the crustacean, were able to switch the gene expression of all nine genes. This work has important ecological implications, especially on contaminated sediment assessment considering that PAHs represent the most abundant organic group of compounds in marine environment, opening new perspectives in understanding the molecular pathways activated by crustaceans.

Keywords: polycyclic aromatic hydrocarbons; crustacean; short-term effects; toxicity; genotoxicity

1. Introduction

Polycyclic aromatic hydrocarbons (PAHs) are a group of over 100 different organic compounds generated by natural events or anthropogenic activities. PAHs predominantly originate from anthropogenic processes, especially from incomplete combustions of organic fuels. Certain naturally occurring processes, such as volcanic eruptions and forest fires, contribute to the increase of these organic compounds in the environment. In Santos Bay and Estuary, the anthropogenic contributions to PAHs in sediments resulted of about 99% (i.e., concentrations varied from 79.6 for uninhabited area to 15,389.1 ng/g for area located in the proximity of industries [1]).

PAHs are formed by two or more fused benzene rings, and their toxicity depends on the number of benzene rings [2–4]. Because of their low water solubility and hydrophobicity, in the water column, PAHs tend to associate with suspended particulate matter and are eventually deposited in sediments, where their degradation is very slow [5]. The level of PAHs in the water column is closely linked to the level of PAHs in sediments [6,7]. In fact, PAHs concentration in the water column increase with increased concentration in surface sediment [6,7]. Although they are not very soluble in water, their concentration in the water column remains stable for a long time, thus representing a great problem for biota and consequently for human health [8–10]. Specifically, the half-lives of low-molecular-weight PAHs (naphthalene, acenaphthene, fluorene, and phenanthrene) ranged from approximately 3 to 8 days, whereas half-lives of high-molecular-weight PAHs (pyrene, chrysene, benzo[a]pyrene, dibenz[a,h]anthracene) ranged from 73 to 1780 days [11–13].

For these reasons, the European Water Framework Directive 2000/60/EC (WFD) was developed aiming to achieve and ensure good ecological and chemical water status [14]. The list of monitored pollutants has recently been updated with a new daughter Directive (2013/39/EU) to identify a number of emerging chemicals of concern, including non-polar organic substances (e.g., PAHs and PCB) and polar compounds (e.g., pharmaceuticals and pesticides) [15].

In aquatic environments, PAHs can have several toxic effects, such as immunotoxicity, embriotoxicity, and cardiotoxicity, especially impacting fish, benthic organisms, and other marine vertebrates [16–19]. Five different PAHs (naphthalene (NAP), phenanthrene (PHE), fluoranthene (FLT), fluorene (FLR), pyrene (PYR), and hydroxypyrene), known to be potentially toxic, inhibited and reduced the larval development and growth of both *Mytilus galloprovincialis* and *Paracentrotus lividus*, whereas NAP was able to impact the embryos and larval stages of *Ciona intestinalis* [16]. The benzo(a)anthracene (BaA), one of the most toxic PAHs, showed higher toxicity on crustaceans *Daphnia magna* (LC50 = 4.3 µg/L) and *Ceriodaphnia reticulata* (LC50 = 4.7 µg/L) than that displayed *Artemia salina* at conceivable concentrations in the environment (from 1 to 32 µg/L) [20,21]. At the same manner, PHE and FLT were able to impact the survival of *D. magna* (LC50 = 50 and 10 mg/kg, respectively), *Hyalella azteca* (LC50 = 15 and 5 mg/kg, respectively), and *Chiromonus riparius* (LC50 = 20 and 15 mg/kg, respectively) [22]. After PYR, FLT, and anthracene (ANT) exposure, *D. magna* and *Artemia salina* crustaceans displayed higher sensibility than those registered for the mosquito *Aedes aegypti*, the amphibian *Rana pipiens*, and the fish *Pimephales promelas* [23].

To the best of our knowledge, few studies have been carried out so far to investigate direct toxic effects of individual PAHs on *A. franciscana*, but no work has been conducted to establish the possible changes in expression levels of genes after organic compounds exposure. Rojo-Nieto et al. [24] established that mixtures of ten PAHs (naphthalene, acenaphthene, phenanthrene, fluoranthene, fluorene, pyrene, anthracene, benzo(a)pyrene, benzo(a)anthracene, and chrysene) found in sediment samples from the Bay of Algeciras did not have impact on survival of *A. franciscana* using passive dosing. Similarly, the passive dosing of three PAHs (toluene, 1-methylnaphthalene, and phenanthrene) did not impact the hutching of cysts [25].

The crustacean *A. franciscana* has been considered as a model species to investigate the ecotoxicological response of marine invertebrates to environmental pollutants [26–28]. The main advantage of this species is that nauplii can be hatched as needed from commercially available durable cysts to avoid the maintenance of laboratory cultures as required for many model species used in ecotoxicity tests. In any case, these tests (namely "Toxkit") employing dormant stages ("cryptobiotic eggs") have the same efficacy and sensitivity as tests with cultured animals [29]. Moreover, the embryo hatches and grows rapidly in laboratory conditions (the nauplius stage is reached in 24 h), and the small body size permits to conduct tests in small beakers or even plates. In addition, *Artemia* is a eurialine organism with large adaptability to a range of salinities (5–300 PSU) and temperatures (6–40 °C) [30]. However, *Artemia* models revealed several disadvantages due to a limited sensitivity towards a wide range of substances in comparison to other species so that the possibility to underestimate potential effects may occur [26,31]. In fact, in the recent years, the use of this crustacean in ecotoxicology has become increasingly rare [31]. For these reasons, in this work, we are interested in giving a new life to this model organism by proposing genotoxicity as a new endpoint. Since changes of gene expression induced by some toxicants may be very subtle and differences of animal reactivity between experimental groups may not be noticed by simple observations, the genotoxicity could be considered a good approach providing more detailed toxicological information. Therefore, the use of *A. franciscana* for evaluating the molecular aspects that are on the base of toxicological effects could confirm this branchiopod crustacean as a good biological model.

Thus far, few studies investigated the stress response of *Artemia* spp. through the evaluation of key genes involved in larval growth, molting, stress, and detoxification processes [32–37]. In this work, as well as evaluating PHE, NAP, FLT, and BkF acute (24 h–48 h-LC50) toxicity on nauplii and adults by measuring survival, we defined for the first time the molecular response of PAHs toxicity. In particular, after 48 h under sublethal exposure for both tested life stages, the effect on several key genes involved in stress response (*hsp26*, *hsp60*, *hsp70*, *COXI*, and *COXIII*) was assessed. In addition, the impact on developmental genes (*HAD-like*, *tcp*, *UCP2*, and *CDC48*) was also evaluated for nauplii.

Sediment can be the final main sink and source of PAHs and genotoxicity can represent an easy and fast screening method for their ranking [38–40]. Prior to direct PAHs contaminated sediment investigation, we decided to highlight the sensitivity of genotoxicity endpoint in *A. franciscana* from spiked saltwater solutions.

In this study, we tested the PHE, NAP, FLT, and benzo(k)fluoranthene (BkF) toxicity on embryos and adults of the branchiopod crustacean *A. franciscana* Kellog 1906, using environmental concentrations (from 0.025 to 10 mg/L, from 0.36 to 2.3×10^2 mg/L, from 0.41 to 3.9×10^2 mg/L, and from 0.025 to 9.4×10^1 mg/L for NAP, PHE, FLT, and BkF, respectively) detected in polluted sediments subjected to various pollution sources [38].

2. Materials and Methods

2.1. Ecotoxicity Test

Acute toxicity test using both *A. franciscana* nauplii and adults were performed according to standard methods [41] using lethality as an endpoint. Effects were measured after 24 and 48 h of exposure for both adults and nauplii up to the third instar (corresponding to 48 h old specimen that are considered as the most sensitive stage). Certified dehydrated cysts of brine shrimp *A. franciscana* (AF/F2005) were purchased from the company ECOTOX LDS (Gallarate, Italy). Hatching of the cysts was obtained by incubating 100 mg of cysts in glass Petri dishes containing seawater prepared by dissolving 36 g of Instant Ocean® salt in deionized water, stirred for 24 h under aeration and then filtered through 0.45 μm Millipore cellulose filters. Newly hatched brine shrimp larvae (Instar I nauplius stage) were separated from unhatched cysts and transferred, taking advantage of phototactic movements, into new glass Petri dishes with synthetic seawater (SSW) prepared according to ISO 10253/16 [42].

2.2. Chemicals

The naphthalene, phenanthrene, fluoranthene, and benzo(k)fluoranthene (Sigma-Aldrich, Saint Louis, USA) were used in the toxicity tests. The purity was greater than 97%. Stock solutions of NAP, PHE, FLT, and BkF were prepared by dissolving the above indicated chemicals in dimethyl sulfoxide (DMSO) [43–45]. Maximum DMSO in test solutions did not exceed 1% v/v, which is not toxic to *A. franciscana* [46]. The solubility of PAHs in seawater depends on temperature, salinity, and the analytical method used for the determination. Details about were reported in Supplementary Materials (Supplementary Table S1). Autoclaved glass tubes containing stock solutions (XX mg/L) of each PAH in DMSO were kept in the dark at room temperature. The solutions were sampled and analyzed for four PAHs according to Carotenuto et al. [47].

2.3. Acute Toxicity Test

Acute toxicity test on nauplii was performed by adding 10 nauplii to each well of 24-well plates, containing 2 mL of solutions at increasing concentrations of NAP

(0, 0.025, 0.05, 0.1, 0.2, 0.4, 0.5, 1, 2.5, 5, 10 mg/L), PHE (0, 0.36, 1, 2, 3, 4, 5, 10, 57.5, 115, 230 mg/L), FLT (0, 0.41, 1, 2.5, 5, 12.5, 25, 50, 97.5, 195, 390 mg/L), and BkF (0, 0.025, 0.5, 1, 1.5, 3, 6, 12, 23.5, 47, 94 mg/L) tested in SSW.

Toxicity on adults was evaluated by adding 5 crustaceans to each well of 6-well plates, containing 10 mL with solutions of various concentrations of NAP, PHE, FLT, and BkF mentioned above. The plates were kept at 25 ± 1 °C with salinity 35 ppm for 48 h in a light regime of 16:8 h light:dark, without providing food. At 24 and 48 h, the number of nauplii and adults (which were motionless for 10 s) was counted under a stereomicroscope (Leica EZ4 HD) to calculate the mortality. Tests were considered as valid when mortality in control (organisms exposed to 0 mg/L of organic compounds) was <10% after 48 h. All the experiments were performed in triplicates.

2.4. Organisms Exposures for RNA Extraction

Two hundred nauplii of *A. franciscana* were exposed to NAP, PHE, FLT, and BkF at 0.26 mg/L, 1.15 mg/L, 0.81 mg/L, and 84.6 mg/L, respectively, whereas 10 *A. franciscana* adults were exposed to NAP, PHE, FLT, and BkF at 1.45 mg/L, 1.15 mg/L, 0.81 mg/L, and 84.6 mg/L, respectively. These concentrations were chosen because they did not cause mortality in acute tests. All the experiments were performed in triplicates.

Samples were collected after 48 h of exposure by centrifugation at $4000 \times g$ for 15 min in a swing-out rotor at 4 °C in a 2 mL tube, kept on ice, and were further homogenized in TRIzol (Invitrogen, Paisley, UK) using a TissueLyser II (Qiagen, Valencia, CA, USA) and steal beads of 7 mm diameter (Qiagen, Valencia, CA, USA). Total RNA was extracted and purified using Direct-zolTM RNA Miniprep Plus Kit (ZYMO RESEARCH). The amount of total RNA extracted was estimated by the absorbance at 260 nm and the purity by 260/280 and 260/230 nm ratios, using a NanoDrop spectrophotometer 2000 (Thermo Scientific Inc., Waltham, MA, USA), to exclude the presence of proteins, phenol, and other contaminants [48].

2.5. cDNA Synthesis and Real Time q-PCR

For each sample, 1000 ng of total RNA was retrotranscribed with an iScript™ cDNA Synthesis kit (Bio-Rad, Milan, Italy), following the manufacturer's instructions. The variations in the expression of five genes involved in stress response (*hsp26, hsp60, hsp70, COXI,* and *COXIII* [49]; see Supplementary Figure S1) were evaluated for adults. For nauplii, the variations in the expression of four other genes involved in developmental and differentiation processes (*HAD-like, tcp, UCP2,* and *CDC48,* [49]) were also tested (Supplementary Figure S1).

Undiluted cDNA was used as a template in a reaction containing a final concentration of 0.3 mM for each primer and $1 \times$ SensiFAST™ SYBR Green master mix (total volume of 10 µL) (Meridiana Bioline). PCR amplifications were performed in AriaMx Real-Time PCR instrument (Agilent Technologies, Inc., Santa Clara, CA, USA), according to the manufacturer's instructions, using the following thermal profile: 95 °C for 10 min, one cycle for cDNA denaturation; 95 °C for 15 s and 60 °C for 1 min, 40 cycles for amplification; 95 °C for 15 s, one cycle for final elongation; and one cycle for melting curve analysis (from 60 °C to 95 °C) to verify the presence of a single product. Each assay included a no-template control for each primer pair. To capture intra-assay variability, all real-time qPCR reactions were carried out in triplicate. Fluorescence was measured using Agilent Aria 1.7 software (Agilent Technologies, Inc.). The relative expression ratios were calculated according to [50,51] using REST software (Version No., Relative Expression Software Tool, Weihenstephan, Germany; Equation (1)):

$$R = \frac{E_{target}^{\Delta Cq\ target\ (Mean\ Control - Mean\ Sample)}}{E_{reference}^{\Delta Cq\ reference\ (Mean\ Control - Mean\ Sample)}} \tag{1}$$

The expression of each gene was analyzed and internally normalized against *GAPDH* [49] using REST software (Relative Expression Software Tool, Weihenstephan, Germany) based on the Pfaffl method [50,51]. Relative expression ratios above 1.5 were considered as significant. The nonparametric Mann–Whitney test was applied to ΔCq (Cq gene of interest—Cq reference) values between treated and control samples (n = 3). *p*-Values < 0.05 were considered significant. Statistical analyses were performed using GraphPad Prism Software (version 8.02 for Windows, GraphPad Software, La Jolla, CA, USA, www.graphpad.com, accessed on 1 February 2021). Fold-change values were represented through a Heatmap generated by GraphPad Prism Software.

2.6. LC50 Calculation and Statistical Analyses

Toxicity data were reported as mean ± standard deviation (SD). Data were checked for normality using the Shapiro–Wilk's (S–W) test (*p*-value < 0.05). The significance of differences among treatments and the control was checked by two-way ANOVA followed by post hoc Tukey's test for multiple comparisons (GraphPad Prism Software version 8.02 for Windows, GraphPad Software, La Jolla, CA, USA, www.graphpad.com, accessed on 1 February 2021). *p*-Values < 0.05 were considered statistically significant. The calculation of LC50 values was done by GraphPad Software through four parameters of the logistic equation, which corresponds to the dose-response curve with the slope of the variable slope.

3. Results

3.1. PAHs Analysis

Samples were analyzed to verify NAP, PHE, FLT, and BkF nominal concentrations ranging from 0.025 to 10 mg/L, 0.36 to 2.3×10^2 mg/L, 0.41 to 3.9×10^2 mg/L, and 0.025 to 94 mg/L, respectively (shown in Table 1). The gas chromatography-mass spectrometry (GC-MS) (2010plus-TQ8030, Shimadzu, Japan) determinations showed a good agreement between nominal vs. analytical concentrations, whose ratios were less than 1.5 in most cases (Table 1).

3.2. Naphthalene, Phenanthrene, Fluoranthene, and Benzo(k)fluoranthene Toxicity on Nauplii

After 24 h of exposure to NAP, a statistically significant increase in toxicity (*p* < 0.0001; see Supplementary Table S2) was observed only at the two highest tested concentrations of 4.23 mg/L (38 % mortality) and 10.1 mg/L (43% mortality) (Figure 1A).

After 48 h, mortality became statistically significant (*p* < 0.01) already at 0.11 and 0.26 mg/L and reached 80% (*p* < 0.0001) at 4.23 and 10.1 mg/L (Figure 1A and Supplementary Table S2).

After 24 h of exposure (Figure 1B) PHE induced an increase of the percentage of dead (about 26%) with respect to control already from 3.45, 4.23, and 7.48 mg/L. The data reported at these concentrations were statistically significant as compared to the two lowest (0 and 0.21 mg/L; *p* < 0.01) and highest (98.6 and 223.4 mg/L; *p* < 0.01; see Supplementary Table S2) concentrations. At 48.7, 98.6, and 223.4 mg/L, significant increase of toxicity (about 43.3%, 50%, and 50%, respectively) respect lower tested concentrations (0, 0.21, 0.71, 1.15, and 2.26 mg/L; *p* < 0.0001; see also Supplementary Table S2) has been shown. The scenario after 48 h of exposure was similar to the one described after 24 h. About 26% of dead nauplii was registered at 1.15 mg/L (*p* < 0.05; Supplementary Table S2). At 2.26 and 3.45 mg/L, the toxicity (about 34%) was statistically significant compared to 0, 0.21, 7.48, 48.7, 98.6, and 223.4 mg/L (*p* < 0.0001; Supplementary Table S2). At 7.48, 48.7, and 98.6 mg/L, about 70% of mortality was registered, whereas a percentage of about 90% of dead was displayed at 223.4 mg/L.

Table 1. Comparisons of nominal vs. analytical concentrations of naphthalene (NAP), phenanthrene (PHE), fluoranthene (FLT), and benzo(k)fluoranthene (BkF) in seawater. The data were reported in mg/L.

Compounds	Nominal Concentration	Analytical Concentration	Nominal/Analytical Concentration Ratio
NAP	0.025	0.015	1.67
	0.05	0.032	1.56
	0.1	0.078	1.28
	0.2	0.11	1.82
	0.4	0.26	1.54
	0.5	0.41	1.22
	1	0.76	1.32
	2.5	1.45	1.72
	5	4.23	1.18
	10	10.1	0.99
PHE	0.36	0.21	1.71
	1	0.71	1.41
	2	1.15	1.74
	3	2.26	1.33
	4	3.45	1.16
	5	4.23	1.18
	10	7.48	1.34
	57.5	48.7	1.18
	115	98.6	1.17
	230	223.4	1.03
FLT	0.41	0.29	1.41
	1	0.81	1.23
	2.5	2.14	1.17
	5	4.41	1.13
	12.5	9.91	1.26
	25	20.4	1.23
	50	45.6	1.10
	97.5	91.6	1.06
	195	179	1.09
	390	325	1.20
BkF	0.025	0.016	1.56
	0.5	0.41	1.22
	1	0.78	1.28
	1.5	0.98	1.53
	3	2.4	1.25
	6	5.3	1.13
	12	10.4	1.15
	23.5	19.5	1.21
	47	41.7	1.13
	94	84.6	1.11

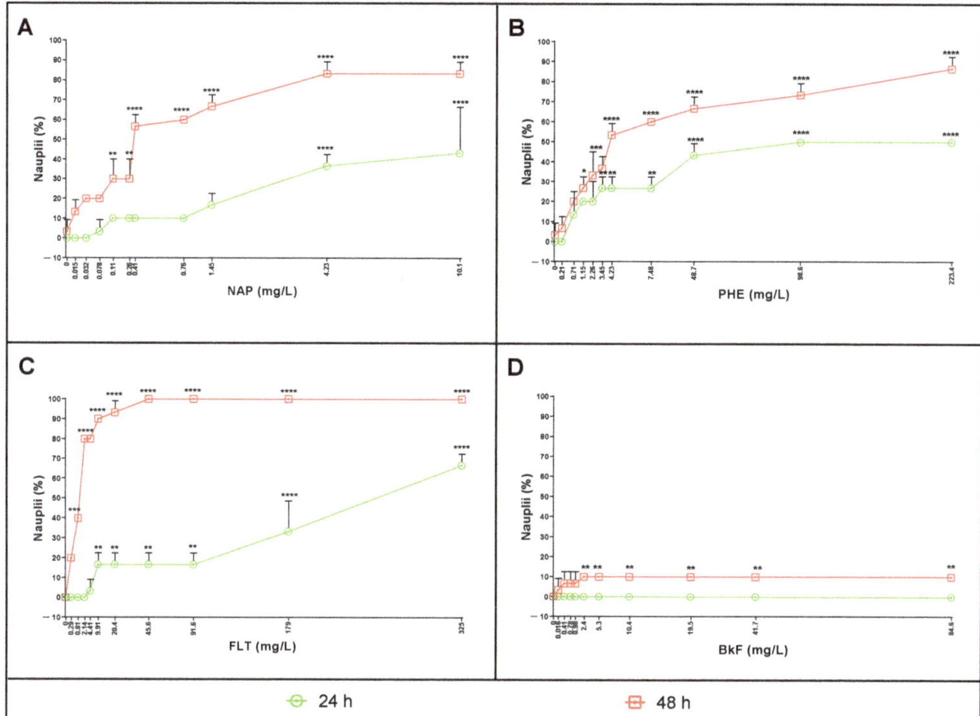

Figure 1. After 24 h and 48 h, the percentage of dead nauplii in control (0 mg/L) and treated samples with (**A**) NAP at the concentrations of 0.025, 0.05, 0.1, 0.2, 0.4, 0.5, 1, 2.5, 5, and 10 mg/L; (**B**) PHE at the concentrations of 0.36, 1, 2, 3, 4, 5, 10, 57.5, 115, and 230 mg/L; (**C**) FLT at the concentrations of 0.41, 1, 2.5, 5, 12.5, 25, 50, 97.5, 195, and 390 mg/L; and (**D**) BkF at the concentrations of 0.025, 0.5, 1, 1.5, 3, 6, 12, 23.5, 47, and 94 mg/L was regarded. Data are reported as mean $\pm$ standard deviation two-way ANOVA by Tukey's test (* $p < 0.05$, ** $p < 0.01$, *** $p < 0.001$, **** $p < 0.0001$).

Taking into the consideration FLT exposure (Figure 1C), an increase of the percentage of dead (about 18%) with respect to control was already recorded from 9.91, 20.4, 45.6, and 91.6 mg/L. The data reported at these concentrations were statistically significant compared to the four lowest (0, 0.29, 0.81, and 2.14 mg/L; $p < 0.01$) and the two highest (179 and 325 mg/L; $p < 0.0001$; see Supplementary Table S2) concentrations. At 179 and 325 mg/L, significant increase of toxicity (about 35% and 70%, respectively) with respect to lower tested concentrations (4.41, 9.91, 20.4, and 45.6 mg/L; $p < 0.0001$; Supplementary Table S2) has been shown. After 48 h, at 0.81 mg/L, a significant increase of toxicity (about 40%) has been displayed with respect to 0 ($p < 0.0001$) and 0.29 mg/L ($p < 0.001$; see also Supplementary Table S2). However, the toxicity increases of about 80% (2.14 and 4.41 mg/L). These data were statistically significant respect to compared to the three lowest (0, 0.29, and 0.81 mg/L; $p < 0.0001$) and the highest (20.4, 45.6, 91.6, 179, and 325 mg/L; $p < 0.001$; Supplementary Table S2).

When we considered BkF toxicity (Figure 1D) after 24 h, no effect has been recorded. Only after 48 h, at 2.4, 5.3, 10.4, 19.5, 41.7, and 84.6 mg/L, a significant increase in toxicity (about 10%) was shown with respect to the control ($p < 0.01$; Supplementary Table S2).

3.3. Naphthalene, Phenanthrene, Fluoranthene, and Benzo(k)fluoranthene Toxicity on Adults

After 24 h of exposure, NAP and PHE and BkF did not affect the survival of *A. franciscana* at all tested concentrations (Figure 2).

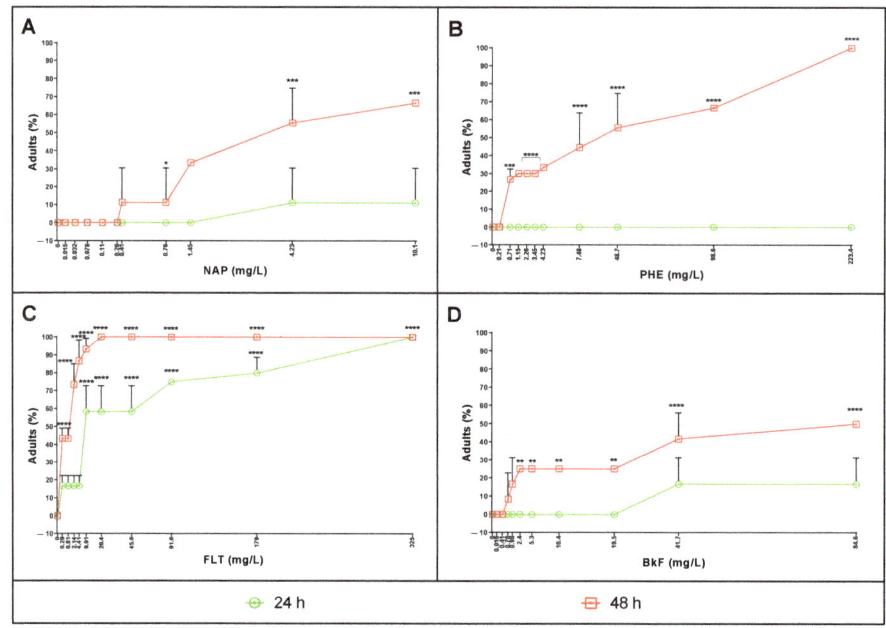

Figure 2. After 24 h and 48 h, the percentage of dead adults in control (0 mg/L) and treated samples with (**A**) NAP at the concentrations of 0.025, 0.05, 0.1, 0.2, 0.4, 0.5, 1, 2.5, 5, and 10 mg/L; (**B**) PHE at the concentrations of 0.36, 1, 2, 3, 4, 5, 10, 57.5, 115, and 230 mg/L; (**C**) FLT at the concentrations of 0.41, 1, 2.5, 5, 12.5, 25, 50, 97.5, 195, and 390 mg/L; and (**D**) BkF at the concentrations of 0.025, 0.5, 1, 1.5, 3, 6, 12, 23.5, 47, and 94 mg/L was regarded. Data are reported as mean $\pm$ standard deviation two-way ANOVA by Tukey's test (* $p < 0.05$, ** $p < 0.01$, *** $p < 0.001$, **** $p < 0.0001$).

However, only FLT showed toxic effects already after 24 h of exposure (Figure 2C). An increase of the percentage of dead (about 60%) with respect to control was already recorded from 9.91, 20.4, 45.6, and 91.6 mg/L. The data reported at these concentrations were statistically significant compared to the four lowest (0, 0.29, 0.81, and 2.14 mg/L; $p < 0.0001$; Supplementary Table S3) and the two highest (179 and 325 mg/L; $p < 0.0001$; Supplementary Table S3) concentrations. At 179 and 325 mg/L, significant increase of toxicity (about 95% and 100%, respectively) with respect to all other concentrations ($p < 0.0001$; Supplementary Table S3) was shown. After 48 h, at 0.29 and 0.81 mg/L, a significant increase of toxicity (about 45%) was displayed with respect to 0 ($p < 0.0001$). However, toxicity increases of about 66% (2.14 and 4.41 mg/L) were shown. These data were statistically significant compared to the three lowest (0, 0.29 and 0.81 mg/L; $p < 0.0001$) and the highest (20.4, 45.6, 91.6, 179, and 325 mg/L; $p < 0.05$; Supplementary Table S3), where a percentage of about 100% was registered.

After 48 h, NAP caused a statistically significant mortality starting from 1.45 mg/L (30%), with a maximum effect of 60% at 10.1 mg/L (Figure 2A).

As shown in Figure 2B, PHE induced an increase of the percentage of dead (about 26.6%) with respect to control already from 0.71 mg/L. The data reported at this concentration were statistically significant compared to the two lowest (0 and 0.21 mg/L; $p < 0.001$) and other concentrations ($p < 0.0001$; Supplementary Table S3). At 1.15, 2.26, 3.45, 4.23, and 7.48 mg/L, significant increase of toxicity (about 30%, 30%, 30%, 33%, and 45%, respectively) with respect to lower (0, 0.21, and 0.71 mg/L; $p < 0.0001$) and higher (48.7, 98.6, and 223.4 mg/L; $p < 0.0001$; see also Supplementary Table S3) tested concentrations was shown. At 48.7 and 98.6 mg/L, the toxicity (about 66%) was statistically significant compared to other used concentrations ($p < 0.0001$). At 223.4 mg/L, 100% of mortality was registered ($p < 0.0001$; Supplementary Table S3).

When we considered BkF toxicity (Figure 2D) after 24 h, no effect was recorded. Only after 48 h, at 2.4, 5.3, 10.4, and 19.5 mg/L, a significant increase in toxicity (about 25%) was shown with respect to the three lowest concentrations (0, 0.016, 0.41 mg/L; $p < 0.01$) and the highest concentration (84.6 mg/L; $p < 0.01$; Supplementary Table S3), where a percentage of about 50% was registered.

3.4. Lethal Concentrations after 24 and 48 h of Exposure

Considering nauplii exposure, the NAP solution has a LC50 value of 1.73 mg/L (1.52–46.28 mg/L) and 0.60 mg/L (0.21–90.38 mg/L) after 24 h and 48 h, respectively; PHE solution has a LC50 value of 4.44 (3.66–56.76 mg/L) and 3.07 mg/L (1.32–81.01 mg/L) after 24 h and 48 h, respectively; FLT solution has a LC50 value of 1.30 (0.45–107.5 mg/L) and 0.09 mg/L (0.01–99.1 mg/L) after 24 h and 48 h, respectively (Supplementary Table S4). When considering adults, the NAP solution has a LC50 value of 0.11 mg/L (0.02–12.06 mg/L) and 44.31 mg/L (5.81–268.12 mg/L) after 24 h and 48 h, respectively; PHE solution has a LC50 value of 1.68 mg/L (1.35–234.09 mg/L) after 48 h; FLT solution has a LC50 value of 32.03 mg/L (0.10–120.08 mg/L) and 0.77 mg/L (0.10–103.67 mg/L) after 24 h and 48 h, respectively; and BkF solution has a LC50 value of 28.67 mg/L (0.5–36.67 mg/L) and 6.12 mg/L (0.05–48.72 mg/L) after 24 h and 48 h, respectively (see also Supplementary Table S4).

3.5. Gene Response to NAP, PHE, FLT, and BkF Exposure

Five genes were analyzed for adults, and all were targeted by four PAHs with the exception of *hsp70*, *COXI*, and *COXIII* (Figure 3; see also Supplementary Table S5 for the values).

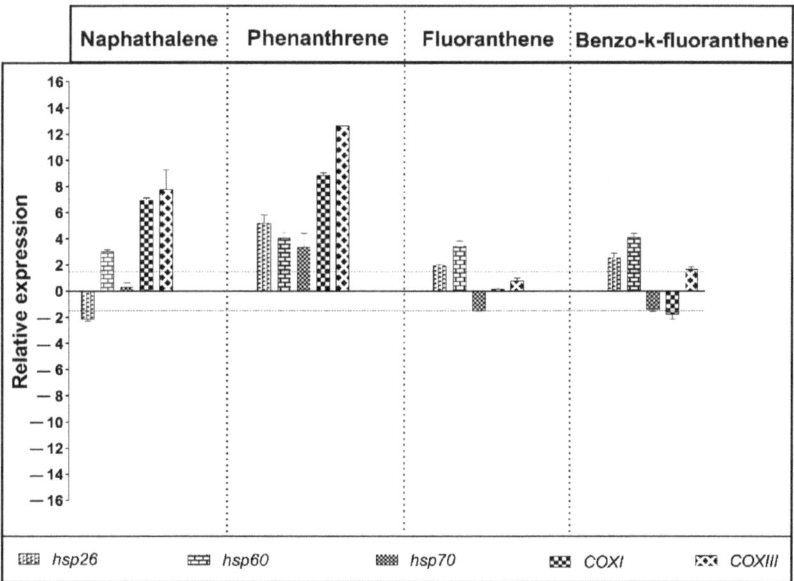

Figure 3. Histograms show the differences in expression levels of five genes involved in stress response. *A. franciscana* adults were exposed to naphthalene, phenanthrene, fluoranthene, and benzo(k)fluoranthene at 1.45 mg/L, 1.15 mg/L, 0.81 mg/L, and 84.6 mg/L, respectively. Fold differences greater than ±1.5 (see red dotted horizontal guidelines at values of +1.5 and −1.5) were considered significant (see Supplementary Table S2 for the values). Real-time qPCR reactions were carried out in triplicate. Statistical differences were evaluated by nonparametric Mann–Whitney test. *p*-Values < 0.05 were considered significant.

In fact, *hsp70* was targeted only by PHE and FLT, whereas *COXI* and *COXIII* were not targeted only by FLT. NAP, PHE, and BkF, increased the expression levels of three genes (*hsp60*, *COXI*, and *COXIII*). Moreover, treatment with NAP also down-regulated *hsp26*; the exposure to PHE up-regulated *hsp26* and *hsp70*; FLT is able to up-regulate *hsp26* and *hsp60*, and down-regulate *hsp70*, whereas the exposure to BkF up-regulated *hsp26* and down-regulated *hsp70* (see also Supplementary Table S5).

As shown in Figure 4, among the nine genes analyzed, only one gene (*hsp70*) was not targeted by NAP, PHE, and FLT. In fact, *hsp70* was target only of BkF. Common molecular targets for four contaminants were *HAD-like*, *tcp*, *UCP2*, and *CDC48*, of which only *UCP2* was up-regulated by all treatment, whereas *tcp* and *CDC48* were up-regulated by NAP, PHE, and BkF and down-regulated by FLT; and *HAD-like* was up-regulated by PHE, FLT, and BkF and down-regulated by NAP.

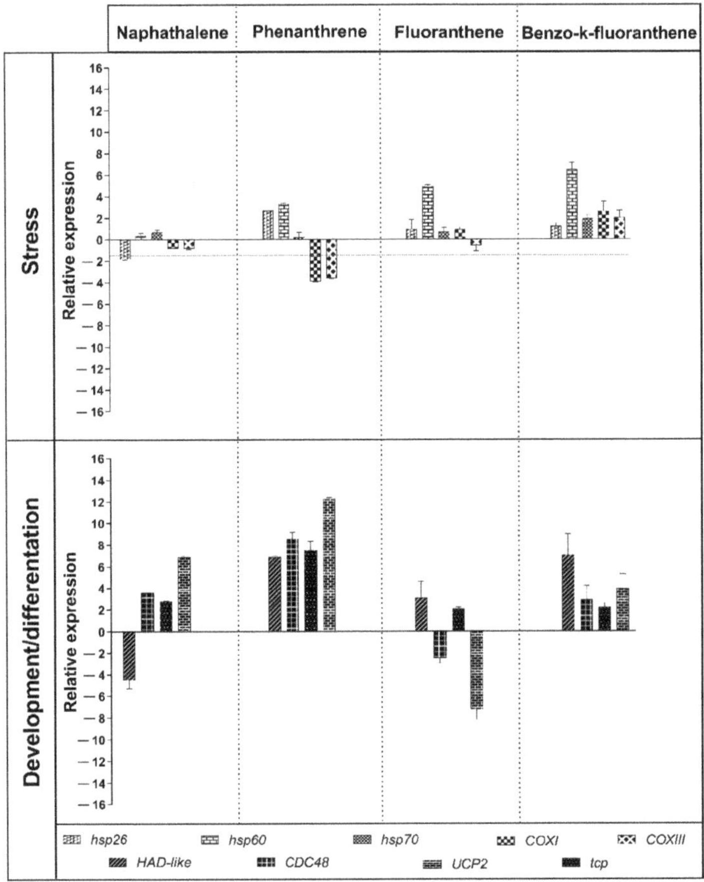

Figure 4. Histograms show the differences in expression levels of five genes involved in stress response. *A. franciscana* nauplii were exposed to naphthalene, phenanthrene, fluoranthene, and benzo(k)fluoranthene at 0.26 mg/L, 1.15 mg/L, 0.81 mg/L, and 84.6 mg/L, respectively. Fold differences greater than ±1.5 (see red dotted horizontal guidelines at values of +1.5 and −1.5) were considered significant (see Supplementary Table S3 for the values). Real-time qPCR reactions were carried out in triplicate. Statistical differences were evaluated by nonparametric Mann–Whitney test. *p*-Values < 0.05 were considered significant.

Moreover, *hsp60* was up-regulated by all PAHs with exception of NAP; *hsp26* resulted up-regulated and down-regulated only after PHE and NAP treatment, respectively; and *COXI* and *COXIII* were down-regulated by PHE and up-regulated by BkF (see also Supplementary Table S6 for the values).

4. Discussion

Acute toxicity tests of PAHs showed similar negative impact of single four pollutants (NAP, PHE, FLT, and BkF) on both adults and nauplii. The NAP, PHE, and FLT were able to induce an increase of nauplii death already after 24 h of exposure, whereas the survival of *A. franciscana* was unaffected by exposure to BkF both after 24 h and 48 h. On the basis of lethal concentrations, the FLT (1.30 mg/L) and NAP (1.73 mg/L) appeared to be more toxic than PHE (4.44 mg/L) at 24 h. As shown in Table 2, Bellas et al. [16] showed similar results in both *C. intestinalis* and *P. lividus* embryos. In fact, exposing these two crustaceans to five PAHs for 24 h, they revealed that FLT and NAP were two and six times more toxic than NAP for *C. intestinalis* and *P. lividus*, respectively.

Table 2. Lethal concentration 50% (LC50) values (mg/L) of PAHs.

	PAHs				References
	NAP	**PHE**	**FLT**	**BkF**	
D. magna	7.924 (24 h)	0.458 (24 h); 0.8 (48 h)			[17]
M. galloprovincialis	0.009 (24 h)	0.0002 (24 h)	0.036 (24 h)		[16]
P. lividus	0.012 (24 h)	0.069 (24 h)	0.036 (24 h)		[16]
C. intestinalis	0.001 (24 h)	0.069 (24 h)	0.036 (24 h)		[16]
C. elegans		4.7 (48 h)			[17]
E. fetidas		0.1 (48 h)			[17]
C. tentans	2.81 (48 h)	0.49 (48 h)			[2]
S. capricornutum	2.96 (48 h)	0.94 (48 h)			[2]
N. palea	2.82 (48 h)	0.87 (48 h)			[2]
P. gyrina	5.02 (48 h)				[2]
G. minus	3.93 (48 h)	0.46 (48 h)			[2]
P. promelas	1.99 (48 h)				[2]
S. gairdneri	0.12 (48 h)	0.03 (48 h)			[2]
M. salmonid	0.68 (48 h)	0.25 (48 h)			[2]
A. franciscana	1.73 (24 h); 0.40 (48 h)	4.44 (24 h); 3.07 (48 h)	1.30 (24 h); 0.09 (48 h)		This study

Instead, after 48 h of exposure, the NAP (0.40 mg/L) and FLT (0.09 mg/L) showed a toxicity 7- and 34-times higher than that established for PHE (3.07 mg/L). In comparison with PHE results from this study, the 48 h LC50 was similar to that of *C. elegans* (4.7 mg/L) but lower than those of *D. magna* (0.8 mg/L), *Chironomus tentans* (0.4 mg/L), and *Eisenia fetida* (0.1 mg/L) reported in a previous study (see also Table 2) [17]. When considering adults exposure, after 48 h, the FLT and PHE showed higher toxicity than these of NAP and BkF. Our results suggest that there is a direct relationship between toxicity and aromatic ring number of the tested compounds. Millemann et al. [2] showed the same relationship for the number of aromatic rings and their toxicity. In fact, they revealed that the PAHs with three or four aromatic rings was always more toxic than those with two aromatic rings for each of the nine exposed species (*Selenastrum capricornutum, Nitzschia palea, Physa gyrina, D. magna, Chironomus tentans, Gaintaurus minus, Pimephales promelas, Salmo gairdneri,* and *Micropterus salmonid*; see Table 2).

Our study also provides new information on the large-scale genotoxicity information of PAHs on *A. franciscana* adults and embryos. Firstly, when considering the genotoxicity on adults, all five genes were molecular targets of four pollutants, with only the exceptions of *hsp70, COXI,* and *COXIII* (Figure 3). When considering the real-time qPCR experiments on nauplii, all nine genes were molecular targets of four pollutants, with the only exception of *hsp26, hsp60, hsp70, COXI,* and *COXIII* (Figure 4). These data suggest that the nauplii treated with NAP, PHE, FLT, and BkF were very similar, as also shown in the heatmap reported in Figure 5.

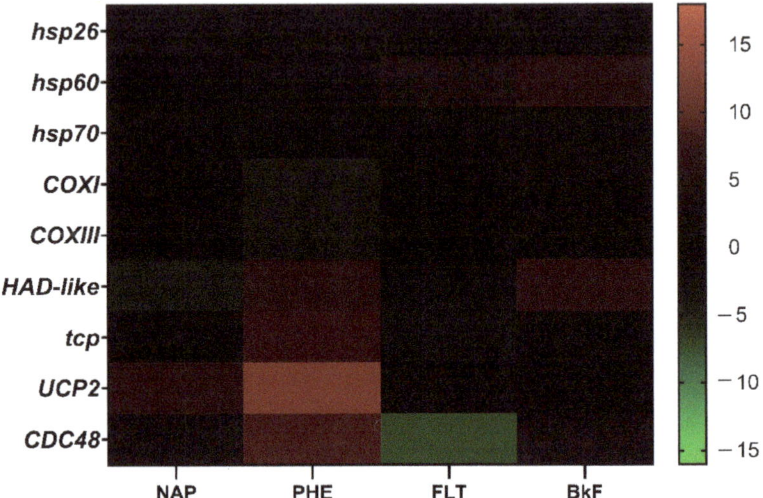

Figure 5. Heatmaps showing the expression profiles and hierarchical clustering of nine genes analyzed through real-time qPCR in nauplii treated with naphthalene (NAP), phenanthrene (PHE), fluoranthene (FLT), and benzo(k)fluoranthene (BkF). Color code: red, up-regulated genes with respect to the control; green, down-regulated genes with respect to the control; black, genes for which there was no variation in expression with respect to the control.

All together, these molecular results revealed that the majority of affected genes in *A. franciscana* were involved in the development processes. In fact, all genes belonging to these classes were affected by the four toxicants. *HAD-like, tcp, UCP2,* and *CDC48* in *A. franciscana* are involved into molecular mechanisms underlying post-diapause, a common event in diverse taxa from plants to animals [49,52–54]. These data could indicate that PAHs affect some common molecular pathways by changing the normal biological mechanisms, which, in turn, generate death in nauplii and adults. Interestingly, several genes followed by RT-qPCR in the present study were previously found to be functionally interconnected [35,49,52]. In particular, Varó et al. [35] showed that nanoparticles (PS NPs) altered the expression of all genes belonging to the network except for *tcp*, whose relative expression was not significant (Supplementary Figure S1). It is important to underline how the evaluation of the changes in gene expression induced by these toxicants has given the opportunity to uncover some key results that are not easily noticed through observations (i.e., mortality). In fact, the BkF was not able to impact the survival of both nauplii and adults but contemporarily altered the expression level of all tested genes.

It has been widely demonstrated that PAHs are mutagenic, carcinogenic, and teratogenic compounds with long-term effect, especially on human health [55,56]. For these reasons, embryos and larvae of marine invertebrates can be considered as suitable indicators in understanding the toxicological response induced by organic compounds since they are also capable to accumulate high levels of them in their tissues [9,16,19]. Moreover, invertebrate species have a key trophic position in benthic food web, playing the role of intermediate consumers [57]. As a result, the toxicological risk is addressed not only to marine species but also to human beings, which could be exposed to such contamination through the food chain [58,59]. Thus, there is the need to develop early warning systems on consolidated biological models supporting sensitive sub-lethal endpoints. An increase of knowledge on changes of *A. franciscana* genes expression can provide great added values in toxicity assessment. In fact, despite its widespread past use, few studies have been conducted on the change of gene expression of *A. franciscana* in response to environmental contamination. The identification of molecular pathways in which the targeted genes were involved represents a key step in understanding how crustacean *A. franciscana* protects itself from the stress caused by toxic substances.

In conclusion, genotoxicity may be considered as a possible new biomarker to detect the presence and effects of key environmental pollutants impacting the survival of marine invertebrates. The great simplicity of handling *A. franciscana* in laboratory conditions together with the high sensitiveness of the molecular endpoints could support future applications of this model organism.

Supplementary Materials: The following supporting information can be downloaded at: https://www.mdpi.com/article/10.3390/w14101594/s1, Figure S1: Gene network; Table S1: Solubility of PAHs [60–62]; Table S2: Tukey's test on nauplii survival data after 24 h and 48 h of PAHs exposure; Table S3: Tukey's test on adults survival data after 24 h and 48 h of PAHs exposure; Table S4: LC50 and 95% confidence intervals calculated after 24 h and 48 h of PAHs exposure; Table S5: Data of gene expression levels in adults; Table S6: Data of gene expression levels in nauplii.

Author Contributions: Conceptualization, L.A. and G.L.; methodology, L.A. and S.S.; software, L.A. and V.Z.; validation, L.A., G.L., M.C. and M.G.; formal analysis, L.A. and M.T. (Maria Toscanesi); investigation, L.A. and S.S.; resources, G.L., M.G. and M.C.; data curation, L.A.; writing—original draft preparation, L.A. and G.L.; writing—review and editing, G.L., D.A.L.V., M.T. (Marco Trifuoggi) and M.G.; visualization, D.A.L.V.; supervision, G.L. and M.G.; project administration, L.A. and G.L.; funding acquisition, G.L. and M.G. All authors have read and agreed to the published version of the manuscript.

Funding: This research received no external funding.

Institutional Review Board Statement: Not applicable.

Informed Consent Statement: Not applicable.

Data Availability Statement: Not applicable.

Acknowledgments: Luisa Albarano was supported by a Ph.D. (Ph.D. in Biology, University of Naples Federico II) fellowship co-funded by the Stazione Zoologica Anton Dohrn and University of Naples Federico II.

Conflicts of Interest: The authors declare no conflict of interest.

References

1. Medeiros, P.M.; Caruso Bícego, M. Investigation of natural and anthropogenic hydrocarbon inputs in sediments using geochemical markers. I. Santos, SP—Brazil. *Mar. Pollut. Bull.* **2004**, *49*, 761–769. [CrossRef] [PubMed]
2. Millemann, R.E.; Birge, W.J.; Black, J.A.; Cushman, R.M.; Daniels, K.L.; Franco, P.J.; Giddings, J.M.; McCarthy, J.F.; Stewart, A.J. Comparative Acute Toxicity to Aquatic Organisms of Components of Coal-Derived Synthetic Fuels. *Trans. Am. Fish. Soc.* **2014**, *113*, 37–41. [CrossRef]
3. Giddings, J.M. Acute toxicity to Selenastrum capricornutum of aromatic compounds from coal conversion. *Bull. Environ. Contam. Toxicol.* **1979**, *23*, 360–364. [CrossRef]
4. Patel, A.B.; Shaikh, S.; Jain, K.R.; Desai, C.; Madamwar, D. Polycyclic Aromatic Hydrocarbons: Sources, Toxicity, and Remediation Approaches. *Front. Microbiol.* **2020**, *11*, 2675. [CrossRef]
5. Ghosal, D.; Ghosh, S.; Dutta, T.K.; Ahn, Y. Current state of knowledge in microbial degradation of polycyclic aromatic hydrocarbons (PAHs): A review. *Front. Microbiol.* **2016**, *7*, 1837. [CrossRef] [PubMed]
6. Qiu, Y.W.; Zhang, G.; Liu, G.Q.; Guo, L.L.; Li, X.D.; Wai, O. Polycyclic aromatic hydrocarbons (PAHs) in the water column and sediment core of Deep Bay, South China. *Estuar. Coast. Shelf Sci.* **2009**, *83*, 60–66. [CrossRef]
7. Castro-Jiménez, J.; Berrojalbiz, N.; Wollgast, J.; Dachs, J. Polycyclic aromatic hydrocarbons (PAHs) in the Mediterranean Sea: Atmospheric occurrence, deposition and decoupling with settling fluxes in the water column. *Environ. Pollut.* **2012**, *166*, 40–47. [CrossRef]
8. Liu, M.; Chen, L.; He, Y.; Baumann, Z.; Mason, R.P.; Shen, H.; Yu, C.; Zhang, W.; Zhang, Q.; Wang, X. Impacts of farmed fish consumption and food trade on methylmercury exposure in China. *Environ. Int.* **2018**, *120*, 333–344. [CrossRef]
9. Gregorin, C.; Albarano, L.; Somma, E.; Costantini, M.; Zupo, V. Assessing the ecotoxicity of copper and polycyclic aromatic hydrocarbons: Comparison of effects on Paracentrotus lividus and Botryllus schlosseri, as alternative bioassay methods. *Water* **2021**, *13*, 711. [CrossRef]

10. Albarano, L.; Zupo, V.; Guida, M.; Libralato, G.; Caramiello, D.; Ruocco, N.; Costantini, M. PAHs and PCBs affect functionally intercorrelated genes in the sea urchin paracentrotus lividus embryos. *Int. J. Mol. Sci.* **2021**, *22*, 12498. [CrossRef]

11. MacRae, J.D.; Hall, K.J. Biodegradation of polycyclic aromatic hydrocarbons (PAH) in marine sediment under denitrifying conditions. *Water Sci. Technol.* **1998**, *38*, 177–185. [CrossRef]

12. Shi, D.; Bera, G.; Knap, A.H.; Quigg, A.; Al Atwah, I.; Gold-Bouchot, G.; Wade, T.L. A mesocosm experiment to determine half-lives of individual hydrocarbons in simulated oil spill scenarios with and without the dispersant, Corexit. *Mar. Pollut. Bull.* **2020**, *151*, 110804. [CrossRef] [PubMed]

13. Tansel, B.; Fuentes, C.; Sanchez, M.; Predoi, K.; Acevedo, M. Persistence profile of polyaromatic hydrocarbons in shallow and deep Gulf waters and sediments: Effect of water temperature and sediment-water partitioning characteristics. *Mar. Pollut. Bull.* **2011**, *62*, 2659–2665. [CrossRef] [PubMed]

14. Rice, J.; Arvanitidis, C.; Borja, A.; Frid, C.; Hiddink, J.G.; Krause, J.; Lorance, P.; Ragnarsson, S.Á.; Sköld, M.; Trabucco, B.; et al. Indicators for sea-floor integrity under the european marine strategy framework directive. *Ecol. Indic.* **2012**, *12*, 174–184. [CrossRef]

15. Munné, A.; Ginebreda, A.; Prat, N. (Eds.) *Experiences from Surface Water Quality Monitoring: The EU Water Framework Directive Implementation in the Catalan River Basin District (Part I)*; Springer: Cham, Switzerland, 2015.

16. Bellas, J.; Saco-Álvarez, L.; Nieto, Ó.; Beiras, R. Ecotoxicological evaluation of polycyclic aromatic hydrocarbons using marine invertebrate embryo-larval bioassays. *Mar. Pollut. Bull.* **2008**, *57*, 493–502. [CrossRef] [PubMed]

17. Honda, M.; Suzuki, N. Toxicities of polycyclic aromatic hydrocarbons for aquatic animals. *Int. J. Environ. Res. Public Health* **2020**, *17*, 1363. [CrossRef] [PubMed]

18. Ruocco, N.; Bertocci, I.; Munari, M.; Musco, L.; Caramiello, D.; Danovaro, R.; Zupo, V.; Costantini, M. Morphological and molecular responses of the sea urchin Paracentrotus lividus to highly contaminated marine sediments: The case study of Bagnoli-Coroglio brownfield (Mediterranean Sea). *Mar. Environ. Res.* **2020**, *154*, 104865. [CrossRef]

19. Albarano, L.; Zupo, V.; Caramiello, D.; Toscanesi, M.; Trifuoggi, M.; Guida, M.; Libralato, G.; Costantini, M. Sub-chronic effects of slight pah-and pcb-contaminated mesocosms in Paracentrotus lividus lmk: A multi-endpoint approach and de novo transcriptomic. *Int. J. Mol. Sci.* **2021**, *22*, 6674. [CrossRef]

20. Ikenaka, Y.; Sakamoto, M.; Nagata, T. Effects of polycyclic aromatic hydrocarbons (PAHs) on an aquatic ecosystem: Acute toxicity and community-level toxic impact tests of benzo[a]pyrene using lake zooplankton community. *J. Toxicol. Sci.* **2013**, *38*, 131–136. [CrossRef]

21. Sese, B.T.; Grant, A.; Reid, B.J. Toxicity of polycyclic aromatic hydrocarbons to the nematode caenorhabditis elegans. *J. Toxicol. Environ. Health—Part A Curr. Issues* **2009**, *72*, 1168–1180. [CrossRef]

22. Verrhiest, G.; Clément, B.; Blake, G. Single and combined effects of sediment-associated PAHs on three species of freshwater macroinvertebrates. *Ecotoxicology* **2001**, *10*, 363–372. [CrossRef] [PubMed]

23. Kagan, J.; Kagan, E.D.; Kagan, I.A.; Kagan, P.A.; Quigley, S. The phototoxicity of non-carcinogenic polycyclic aromatic hydrocarbons in aquatic organisms. *Chemosphere* **1985**, *14*, 1829–1834. [CrossRef]

24. Rojo-Nieto, E.; Smith, K.E.C.; Perales, J.A.; Mayer, P. Recreating the seawater mixture composition of HOCs in toxicity tests with Artemia franciscana by passive dosing. *Aquat. Toxicol.* **2012**, *120–121*, 27–34. [CrossRef]

25. Colvin, K.A.; Parkerton, T.F.; Redman, A.D.; Lewis, C.; Galloway, T.S. Miniaturised marine tests as indicators of aromatic hydrocarbon toxicity: Potential applicability to oil spill assessment. *Mar. Pollut. Bull.* **2021**, *165*, 112151. [CrossRef] [PubMed]

26. Libralato, G. The case of Artemia spp. in nanoecotoxicology. *Mar. Environ. Res.* **2014**, *101*, 38–43. [CrossRef]

27. Libralato, G.; Losso, C.; Ghirardini, A.V. Toxicity of untreated wood leachates towards two saltwater organisms (Crassostrea gigas and Artemia franciscana). *J. Hazard. Mater.* **2007**, *144*, 590–593. [CrossRef]

28. Migliore, L.; Brambilla, G.; Grassitellis, A.; Delupis, G.D.; Vergata, U.T.; Rm, S.L.; Pubblica, S.I. Toxicity and bioaccumulation of sulphadimethoxine in *Artemia* (Crustacea, Anostraca). *Int. J. Salt Lake Res.* **1993**, *2*, 141–152. [CrossRef]

29. Persoone, G.; Baudo, R.; Cotman, M.; Blaise, C.; Thompson, K.C.; Moreira-Santos, M.; Vollat, B.; Törökne, A.; Han, T. Review on the acute Daphnia magna toxicity test ? Evaluation of the sensitivity and the precision of assays performed with organisms from laboratory cultures or hatched from dormant eggs. *Knowl. Manag. Aquat. Ecosyst.* **2009**, *393*, 1–29. [CrossRef]

30. Manfra, L.; Canepa, S.; Piazza, V.; Faimali, M. Lethal and sublethal endpoints observed for Artemia exposed to two reference toxicants and an ecotoxicological concern organic compound. *Ecotoxicol. Environ. Saf.* **2016**, *123*, 60–64. [CrossRef]

31. Libralato, G.; Prato, E.; Migliore, L.; Cicero, A.M.; Manfra, L. A review of toxicity testing protocols and endpoints with *Artemia* spp. *Ecol. Indic.* **2016**, *69*, 35–49. [CrossRef]

32. Bergami, E.; Pugnalini, S.; Vannuccini, M.L.; Manfra, L.; Faleri, C.; Savorelli, F.; Dawson, K.A.; Corsi, I. Long-term toxicity of surface-charged polystyrene nanoplastics to marine planktonic species *Dunaliella tertiolecta* and *Artemia franciscana*. *Aquat. Toxicol.* **2017**, *189*, 159–169. [CrossRef] [PubMed]

33. Comeche, A.; Martín-Villamil, M.; Picó, Y.; Varó, I. Effect of methylparaben in Artemia franciscana. *Comp. Biochem. Physiol. Part—C Toxicol. Pharmacol.* **2017**, *199*, 98–105. [CrossRef] [PubMed]

34. De Vos, S.; Van Stappen, G.; Sorgeloos, P.; Vuylsteke, M.; Rombauts, S.; Bossier, P. Identification of salt stress response genes using the Artemia transcriptome. *Aquaculture* **2019**, *500*, 305–314. [CrossRef]

35. Varó, I.; Perini, A.; Torreblanca, A.; Garcia, Y.; Bergami, E.; Vannuccini, M.L.; Corsi, I. Time-dependent effects of polystyrene nanoparticles in brine shrimp Artemia franciscana at physiological, biochemical and molecular levels. *Sci. Total Environ.* **2019**, *675*, 570–580. [CrossRef]

36. Yi, X.; Zhang, K.; Liu, R.; Giesy, J.P.; Li, Z.; Li, W.; Zhan, J.; Liu, L.; Gong, Y. Transcriptomic responses of Artemia salina exposed to an environmentally relevant dose of Alexandrium minutum cells or Gonyautoxin2/3. *Chemosphere* **2020**, *238*, 124661. [CrossRef]

37. Lee, J.M.; Cho, B.C.; Park, J.S. Transcriptomic analysis of brine shrimp Artemia franciscana across a wide range of salinities. *Mar. Genomics* **2022**, *61*, 100919. [CrossRef]

38. Arienzo, M.; Donadio, C.; Mangoni, O.; Bolinesi, F.; Stanislao, C.; Trifuoggi, M.; Toscanesi, M.; Di Natale, G.; Ferrara, L. Characterization and source apportionment of polycyclic aromatic hydrocarbons (pahs) in the sediments of gulf of Pozzuoli (Campania, Italy). *Mar. Pollut. Bull.* **2017**, *124*, 480–487. [CrossRef] [PubMed]

39. Mora, M.; Walker, T.R.; Willis, R. Spatiotemporal characterization of petroleum hydrocarbons and polychlorinated biphenyls in small craft harbour sediments in Nova Scotia, Canada. *Mar. Pollut. Bull.* **2022**, *177*, 113524. [CrossRef]

40. Armiento, G.; Caprioli, R.; Cerbone, A.; Chiavarini, S.; Crovato, C.; De Cassan, M.; De Rosa, L.; Montereali, M.R.; Nardi, E.; Nardi, L.; et al. Current status of coastal sediments contamination in the former industrial area of Bagnoli-Coroglio (Naples, Italy). *Chem. Ecol.* **2020**, *36*, 579–597. [CrossRef]

41. APAT; IRSA; CNR. *Metodo 8060 di Valutazione della Tossicità Acuta con Artemia sp. In Manuali e Linee Guida-Metodi Analitici per le Acque*; APAT IRSA CNR: Rome, Italy, 2003; Volume 3, pp. 1043–1049.

42. *ISO 10253*; Water Quality—Marine Algal Growth Inhibition Test with *Skeletonema* sp. and *Phaeodactylum tricornutum*. ISO: London, UK, 2016.

43. Kwon, H.C.; Kwon, J.H. Measuring aqueous solubility in the presence of small cosolvent volume fractions by passive dosing. *Environ. Sci. Technol.* **2012**, *46*, 12550–12556. [CrossRef]

44. Vemula, V.R.; Lagishetty, V.; Lingala, S. Solubility enhancement techniques. *Int. J. Pharm. Sci. Rev. Res.* **2010**, *5*, 41–51.

45. Miyako, Y.; Khalef, N.; Matsuzaki, K.; Pinal, R. Solubility enhancement of hydrophobic compounds by cosolvents: Role of solute hydrophobicity on the solubilization effect. *Int. J. Pharm.* **2010**, *393*, 48–54. [CrossRef] [PubMed]

46. Barahona-Gomariz, M.V.; Sanz-Barrera, F.; Sánchez-Fortún, S. Acute toxicity of organic solvents on Artemia salina. *Bull. Environ. Contam. Toxicol.* **1994**, *52*, 766–771. [CrossRef] [PubMed]

47. Carotenuto, Y.; Vitiello, V.; Gallo, A.; Libralato, G.; Trifuoggi, M.; Toscanesi, M.; Lofrano, G.; Esposito, F.; Buttino, I. Assessment of the relative sensitivity of the copepods *Acartia tonsa* and *Acartia clausi* exposed to sediment-derived elutriates from the Bagnoli-Coroglio industrial area: Sensitivity of Acartia tonsa and Acartia clausi to sediment elutriates. *Mar. Environ. Res.* **2020**, *155*, 104878. [CrossRef]

48. Riesgo, A.; Pérez-Porro, A.R.; Carmona, S.; Leys, S.P.; Giribet, G. Optimization of preservation and storage time of sponge tissues to obtain quality mRNA for next-generation sequencing. *Mol. Ecol. Resour.* **2012**, *12*, 312–322. [CrossRef]

49. Chen, W.H.; Ge, X.; Wang, W.; Yu, J.; Hu, S. A gene catalogue for post-diapause development of an anhydrobiotic arthropod Artemia franciscana. *BMC Genom.* **2009**, *10*, 1–9. [CrossRef]

50. Pfaffl, M.W. A new mathematical model for relative quantification in real-time RT–PCR. *Nucleic Acids Res.* **2001**, *29*, e45. [CrossRef]

51. Pfaffl, M.W.; Horgan, G.W.; Dempfle, L. Relative expression software tool (REST©) for group-wise comparison and statistical analysis of relative expression results in real-time PCR. *Nucleic Acids Res.* **2002**, *30*, e36. [CrossRef]

52. Wang, X.; Shi, G.X.; Xu, Q.S.; Xu, B.J.; Zhao, J. Lanthanum- and cerium-induced oxidative stress in submerged *Hydrilla verticillata* plants. *Russ. J. Plant Physiol.* **2007**, *54*, 693–697. [CrossRef]

53. Dambroski, H.R.; Feder, J.L. Host plant and latitude-related diapause variation in Rhagoletis pomonella: A test for multifaceted life history adaptation on different stages of diapause development. *J. Evol. Biol.* **2007**, *20*, 2101–2112. [CrossRef]

54. Ragland, G.J.; Fuller, J.; Feder, J.L.; Hahn, D.A. Biphasic metabolic rate trajectory of pupal diapause termination and post-diapause development in a tephritid fly. *J. Insect Physiol.* **2009**, *55*, 344–350. [CrossRef] [PubMed]

55. Chiarelli, R.; Roccheri, M.C. Marine Invertebrates as Bioindicators of Heavy Metal Pollution. *Open J. Met.* **2014**, *4*, 93–106. [CrossRef]

56. Ferrarese, E.; Andreottola, G.; Oprea, I.A. Remediation of PAH-contaminated sediments by chemical oxidation. *J. Hazard. Mater.* **2008**, *152*, 128–139. [CrossRef] [PubMed]

57. Depledge, M.H.; Fossi, M.C. The role of biomarkers in environmental assessment (2). Invertebrates. *Ecotoxicology* **1994**, *3*, 161–172. [CrossRef]

58. Balcıoğlu, E.B. Potential effects of polycyclic aromatic hydrocarbons (PAHs) in marine foods on human health: A critical review. *Toxin Rev.* **2016**, *35*, 98–105. [CrossRef]

59. Tongo, I.; Ogbeide, O.; Ezemonye, L. Human health risk assessment of polycyclic aromatic hydrocarbons (PAHs) in smoked fish species from markets in Southern Nigeria. *Toxicol. Rep.* **2017**, *4*, 55–61. [CrossRef]

60. Eganhouse, R.P.; Calder, J.A. The solubility of medium molecular weight aromatic hydrocarbons and the effects of hydrocarbon co-solutes and salinity. *Geochim. Cosmochim. Acta* **1976**, *40*, 555–561. [CrossRef]

61. May, W.E.; Wasik, S.P.; Freeman, D.H. Determination of the solubility behavior of some polycyclic aromatic hydrocarbons in water. *Anal. Chem.* **1978**, *50*, 997–1000. [CrossRef]

62. NOAA Cameo Chemicals. Available online: https://cameochemicals.noaa.gov/ (accessed on 1 February 2021).

Article

Triflumizole Induces Developmental Toxicity, Liver Damage, Oxidative Stress, Heat Shock Response, Inflammation, and Lipid Synthesis in Zebrafish

Lina Bai [1], Peng Shi [1], Kun Jia [1], Hua Yin [2,*], Jilin Xu [1], Xiaojun Yan [1] and Kai Liao [1,*]

[1] School of Marine Sciences, Ningbo University, Ningbo 315211, China
[2] Ningbo No. 2 Hospital, Ningbo 315010, China
* Correspondence: jerry_yin@163.com (H.Y.); liaokai@nbu.edu.cn (K.L.)

Abstract: Triflumizole (TFZ) toxicity must be investigated in the aquatic environment to understand the potential risks to aquatic species. Accordingly, the adverse effects of TFZ exposure in zebrafish were investigated. Results demonstrate that, after TFZ exposure, the lethal concentration 50 (LC$_{50}$) in 3 d post-fertilization (dpf) embryos and 6 dpf larvae were 4.872 and 2.580 mg/L, respectively. The development (including pericardium edema, yolk sac retention, and liver degeneration) was apparently affected in 3 dpf embryos. Furthermore, the alanine aminotransferase (ALT) activity, superoxide dismutase (SOD) activity, catalase (CAT) activity, and malondialdehyde (MDA) content in 6 dpf larvae were significantly increased. Additionally, the expression of heat shock response genes (including *hsp70*, *grp78*, *hsp90*, and *grp94*), inflammatory genes (including *p65-nfκb*, *il-1β*, and *cox2a*), and lipid synthetic genes (including *srebp1*, *fas*, *acc*, and *ppar-γ*) in 3 dpf embryos was significantly increased, which was also partially observed in the intestinal cell line form *Pampus argenteus*. Taken together, TFZ could affect the development of zebrafish, accompanied by disturbances of oxidative stress, heat shock response, inflammation, and lipid synthesis. Our findings provide an original insight into the potential risks of TFZ to the aquatic ecosystem.

Keywords: triflumizole; zebrafish; oxidative stress; heat shock response; inflammation; lipid synthesis

Citation: Bai, L.; Shi, P.; Jia, K.; Yin, H.; Xu, J.; Yan, X.; Liao, K. Triflumizole Induces Developmental Toxicity, Liver Damage, Oxidative Stress, Heat Shock Response, Inflammation, and Lipid Synthesis in Zebrafish. *Toxics* **2022**, *10*, 698. https://doi.org/10.3390/toxics10110698

Academic Editors: François Gagné, Stefano Magni and Valerio Matozzo

Received: 18 October 2022
Accepted: 15 November 2022
Published: 17 November 2022

Publisher's Note: MDPI stays neutral with regard to jurisdictional claims in published maps and institutional affiliations.

1. Introduction

The widespread use of pesticides may endanger aquatic species by contaminating water through surface runoff or leachingOerke [1–3]. Triflumizole (TFZ, C$_{15}$H$_{15}$ClF$_3$N$_3$O), a triazole fungicide (Nippon Soda Co., Ltd., Tokyo, Japan), has not been approved by the European Commission. It is mostly used to prevent powdery mildew and rust in cereals, vegetables, fruit trees, and other crops [4–7]. It has the potential to hinder the biological synthesis of ergosterol by limiting C$_{14}$-demethylation in sterol [8]. As recently as 2009, 56,231 pounds of TFZ were spilled in California, although the actual level of human exposure is unknown [9]. Further, TFZ has been found in pears, apples, and cucumbers [10]. Environmental risk assessments of TFZ to non-target species must be carried out immediately [11].

Several studies have demonstrated the toxicity of TFZ to aquatic species, such as freshwater algae [12] and fish [13]. Xi et al. (2019) have indicated that TFZ is potentially toxic to freshwater algae in hydrophytic ecosystems [12]. Ecological studies have suggested that TFZ be classified as showing medium toxicity to fish compared to other triazoles [5]. It has also been found that the 72-h lethal concentration 50 (LC$_{50}$) of TFZ for rare minnow (*Gobiocypris rarus*) embryos was 7.11 mg/L, and that it induced abnormal development, extensively modified enzyme activities, and genes expression [10]. Therefore, further toxicological investigation of TFZ for aquatic organisms is required.

Many aquatic species can survive in polluted environments because of defense mechanisms such as antioxidant system and stress response [14]. In fish, toxicant bioaccumulation initiates redox reactions that generate reactive oxygen species (ROS) changing metabolism [15]. Organisms scavenge it using antioxidants including superoxide dismutase (SOD), catalase (CAT), and malondialdehyde (MDA), which are used to assess environmental risk [16]. Furthermore, ROS production is deeply involved in inflammatory reactions [17]. ROS might be accumulated or released due to the increase in oxygen uptake caused by the aggregation of immune cells in damaged tissues. Furthermore, it has been documented that oxidative stress leads to cellular proteins being damaged so that they must subsequently be refolded [18]. Heat shock proteins (HSPs) are the most effectively protective mechanism, and their syntheses increase remarkably with stress. HSPs enable cells to accommodate various xenobiotic factors and naturally derived cytotoxic factors, and previous research has demonstrated the crucial role of HSPs in oxidative stress [19,20].

Some studies have reported that triazoles may affect lipid metabolism, harm embryonic development, and alter the expression of lipid synthetic genes [21–23]. In addition to these target mechanisms, TFZ could also be recognized as an obesogen in mice by acting on the peroxisome proliferator-activated receptor-gamma (PPAR-γ) pathway to increase weight [9], this suggests that TFZ could participate in the lipid synthesis of organisms.

Zebrafish have been commonly used to assess the toxicity of pesticides in the environment [24,25]. Our research aim to study the toxicity and lipid synthesis effects of TFZ by exposing 3 d post-fertilization (dpf) zebrafish embryos and 6 dpf zebrafish larvae. Our results will be conducive to an original understanding of the harmful impacts of TFZ on fish and its mechanism.

2. Materials and Methods

2.1. Chemicals

Triflumizole (Aladdin, China, CAS: 99387-89-0, 99% purity) was dissolved in dimethyl sulfoxide (DMSO) to 40 mg/mL as a stock solution, then diluted in $1 \times$ EM medium ($20 \times$ EM medium: NaCl, 17.5 g; KCl, 0.75 g; $CaCl_2$ anhydrous, 2.18 g; KH_2PO_4, 0.41 g; Na_2HPO_4 anhydrous, 0.142 g; $MgSO_4$-$7H_2O$, 4.9 g) to the exposure concentrations ensuring that the final DMSO volume was less than 0.1%.

2.2. Zebrafish Maintenance and Breeding

All zebrafish research procedures were approved by the Ethical Committee of Ningbo University. The AB strain adult zebrafish (*Danio rerio*) (Shanghai FishBio Co., Ltd., Shanghai, China) were maintained in a light/dark cycle of 14/10 h. The zebrafish were fed with freshly hatched brine shrimp twice daily. Water quality was monitored every day to ensure pH 7.3–7.4, conductivity within 450–500 μS, and a temperature of 28 ± 1 °C. Male and female (2:2) zebrafish were placed in isolation in the mating tanks in order to trigger spawning when the next morning lights on. After two hours, the normal embryos were collected and cultivated in $1 \times$ EM medium at 28 °C.

2.3. Zebrafish Exposure

An overview of the exposure procedures is shown in Figure 1A,B. In the control groups, we added 0.1% DMSO in 5 mL $1 \times$ EM medium. We chose 3 dpf zebrafish embryos and 6 dpf zebrafish larvae for TFZ exposure. Firstly, they were exposed to TFZ ranging from 0 to 150 mg/L for 24, 48 and 72 h. the TFZ exposure solutions were replaced per 24 h. Based on the morphological changes and mortality data, 3 dpf zebrafish embryos were exposed to 0, 1, 2, and 3 mg/L TFZ for 48 h. Likewise, 6 dpf zebrafish larvae were exposed to 0, 0.5, 1, and 1.5 mg/L TFZ for 24 h. These were individually dispensed into the 6-well plates, thirty embryos/larvae per well with 5 mL exposure solution, and biological triplicate was collected for each exposure. In order to retain the suit concentrations and water quality, the TFZ exposure solutions were exchanged daily. The 6-well plates were incubated at 28 °C for light/dark cycles of 14/10 h during the exposure period. The survival rates of embryos and larvae were counted in each exposed group every 24 h.

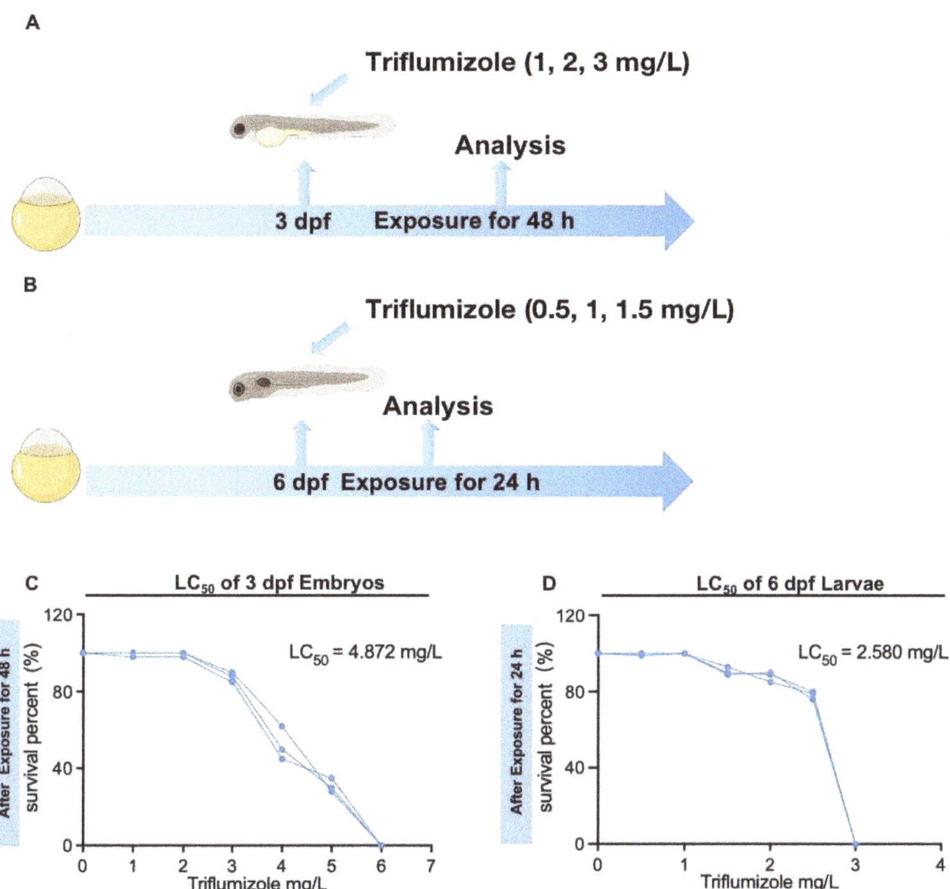

Figure 1. Exposure patterns and the survival percentage of zebrafish after triflumizole exposure. (**A**) Three dpf embryos were exposed to 1, 2, and 3 mg/L triflumizole for 48 h, (**B**) Six dpf larvae were exposed to 0.5, 1, and 1.5 mg/L triflumizole for 24 h. (**C**) The percentage of survival of 3 dpf embryos after triflumizole exposure for 48 h. (**D**) The percentage of survival of 6 dpf larvae after triflumizole exposure for 24 h. dpf: d post-fertilization; LC_{50}: lethal concentration 50.

2.4. Fish Intestinal Cell Line Exposure

The intestinal cell line obtained from a marine fish, silver pomfret (*Pampus argenteus*), was obtained and cultured as our previous method [26]. Cells were plated on 96-well plates, then exposed to 0, 5, 10, 15, 20, 25, 30, 35, and 40 mg/L TFZ in triplicate for 24 h. We used a Cell Counting Kit-8 (Dojindo, Kumamoto, Japan) to assess cell viability. Briefly, cells were incubated with CCK-8 solution for 4 h at 28 °C. Finally, samples were detected in the absorbance at 450 nm by a microplate reader (BioTek, Winooski, VT, USA). Based on cell viability data, the intestinal cell line was exposed to 0, 5, 10, and 20 mg/L TFZ for 24 h to evaluate its toxicity. Then cells were trypsinized at 37 °C for 1 min and collected for the next analysis.

2.5. Morphological Assay

After TFZ exposure, phenotypic changes were evaluated in 3 dpf zebrafish embryos and 6 dpf zebrafish larvae. After anesthetizing with 0.16% tricaine (Sangon Biotech, Shanghai, China), they were fixed on the culture dish in a lateral view using 3% methylcellulose.

A lateral view of the entire zebrafish larvae was observed and photographed with a stereomicroscope (Olympus, Tokyo, Japan).

2.6. Enzyme Activities Assay

Thirty embryos/larvae (each replicate) were homogenized with 300 µL saline solution, followed by centrifuging at 2500 rpm for 15 min at 4 °C to collect the supernatant for use in subsequent experiments. Three replicates were performed. The activities of SOD, CAT, alanine aminotransferase (ALT), and the content of MDA were assayed by commercially available biochemical assay kits (Nanjing Jiancheng Bioengineering Institute, Nanjing, China) in accordance with the manufacturer's instructions. The SOD activity was tested by the xanthine oxidase method. In brief, superoxide anions can oxidize hydroxylamine to nitrite and then turn amaranth purple in the presence of a chromogenic agent. Data were recorded by reading the absorbance at 550 nm and computing the activity of the SOD. One unit of SOD activity (U) was defined as the amount of enzyme required to inhibit the oxidation reaction by 50% and was expressed as U/g protein. The CAT activity was tested by the ammonium molybdate method. The decomposition of hydrogen peroxide is quickly terminated by adding ammonium molybdate. The rest of the hydrogen peroxide reacts with ammonium molybdate to form a pale-yellow complex compound, which was detected at 405 nm and used to compute the CAT activity. One unit of CAT activity was defined as the amount of enzyme required to consume 1 µmol H_2O_2 in 1 min at 25 °C and was expressed as µmol/min/g protein. The products of lipid peroxidation (measured MDA content) were tested by the thiobarbituric acid (TBA) method, and the amount of TBA substance that occurred through lipid peroxidation was detected after incubation at 95 °C with TBA. The pink color generated in these reactions was detected by spectrophotometry at 532 nm, and MDA content was expressed as nmol/g protein. The pyruvate reacts with 2-4-dinitrophenylhydrazine to generate the hydrazone, which acquires maximum staining by the addition of NaOH to measure ALT activity [27]. It was examined at 505 nm and was expressed as µmol/min/g protein. Protein concentration was tested using the bicinchoninic acid (BCA) protein assay kit (Nanjing Jiancheng Bioengineering Institute, Nanjing, China) for Cu^{2+} following protein-mediated reduction of Cu^{2+} by an alkaline environment. All measurements were performed on a microplate reader, using the A590 microwell plate protocol.

2.7. Quantitative Real Time PCR (qPCR)

The total RNA was extracted using Trizol reagent (Omgea, Norcross, GA, USA) according to the manufacturer's instructions. Next, the samples of RNA were reverse transcribed to generate cDNA using the Reverse Transcription Kit (TRAN, Beijing, China). Each 25 µL application contained 4 µL diluted cDNA, 10 µL qPCR PowerUp SYBR Green Master Mix (TRAN, Beijing, China), 1 µL forward and reverse primers (10 µM), and 4 µL RNA-free water. Two tests were performed for each gene in each biological replicate. The qPCR condition was as follows: 94 °C for 30 s (1 cycle), 94 °C for 5 s, 60 °C for 15 s, and 72 °C for 10 s (45 cycles). The results were subjected to relative quantitative analysis using *β-actin* as an endogenous control gene [26]. The primers used for qPCR are listed in the Table S1. The expression of relative genes was calculated by the $2^{-\Delta\Delta CT}$ method [18].

2.8. Statistical Analysis

SPSS version 20.0 (SPSS Inc., Chicago, IL, USA) was used for statistical analysis. LC$_{50}$ values were determined using GraphPad Prism software version 9.0 (GraphPad Software, San Diego, CA, USA). The data were expressed as mean ± standard error of means (SEM). The results were demonstrated through three independent experiments. One-way analysis of variance (ANOVA) with Tukey's multiple range test was used to analyze cell activity differences. Differences among two groups were analyzed by unpaired *t*-test with Welch's correction. * $p < 0.05$ was considered significant, and ** $p < 0.01$ was considered highly significant.

3. Results

3.1. Exposure Concentrations

To determine the TFZ dose which caused harmful effects in zebrafish, different concentrations were used for exposure and survival rates were recorded. For the primary screening of tolerance ranges in 3 dpf zebrafish embryos and 6 dpf zebrafish larvae, we used 0, 30, 60, 90, 120, and 150 mg/L TFZ. Next, we exposed 3 dpf zebrafish embryos and 6 dpf zebrafish larvae to 0, 1, 2, 3 mg/L and 0, 0.5, 1, 1.5 mg/L TFZ, respectively (Figure 1A,B). After TFZ exposure, the 48 h LC_{50} of 3 dpf zebrafish embryos is 4.872 mg/L (Figure 1C), and the 24 h LC_{50} of 6 dpf zebrafish larvae is 2.580 mg/L (Figure 1D).

3.2. Zebrafish Larvae Morphology after TFZ Exposure

To investigate the developmental toxicity of TFZ, morphological changes were scored. After TFZ exposure for 48 h in 3 dpf zebrafish embryos, it was discovered that TFZ induced pericardium edema in the 3 mg/L group, and induced yolk sac retention and liver degeneration in the 2 and 3 mg/L groups (Figure 2A). After TFZ exposure for 24 h in 6 dpf zebrafish larvae, liver degeneration was also noted in 1 and 1.5 mg/L groups, pericardial edema was not observed in 0.5, 1, and 1.5 mg/L groups (Figure 2B).

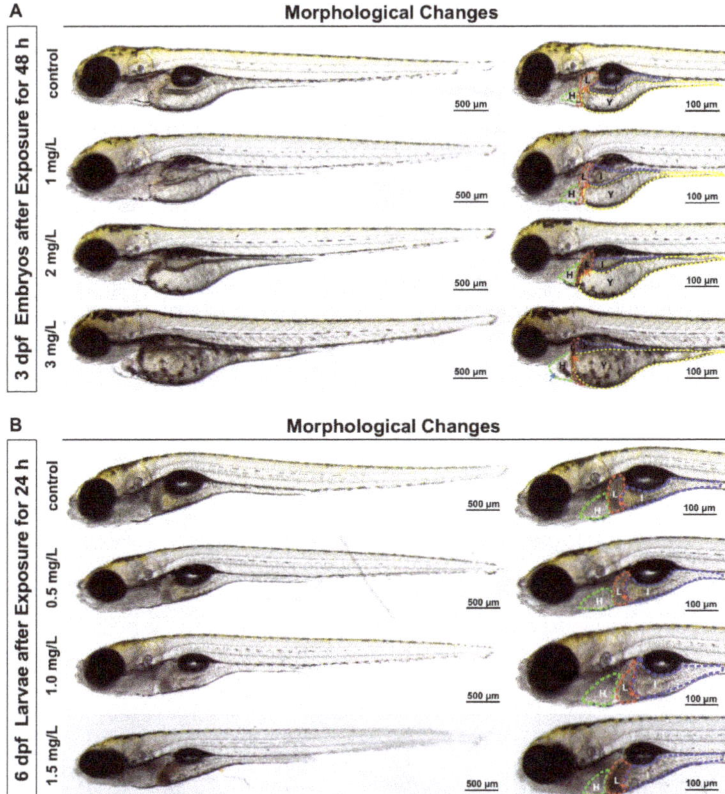

Figure 2. Effects of development after triflumizole exposure in zebrafish. (**A**) 3 dpf zebrafish embryos with morphological changes after triflumizole exposure for 48 h. (**B**) 6 dpf zebrafish larvae with liver degeneration after triflumizole exposure for 24 h. Blue arrows and red arrows indicate pericardial edema and liver degeneration, respectively. H: heart, the green dotted line; L: liver, the red dotted line; I: intestine, the blue dotted line; Y: yolk sac, the yellow dotted line. Scale bars = 500 µm; Scale bars = 100 µm.

3.3. Enzyme Activities after TFZ Exposure

The SOD activity in 3 dpf zebrafish embryos was markedly decreased in the 3 mg/L TFZ group relative to the control group ($p < 0.01$, Figure 3A). Compared to the control group, the CAT activity in 3 dpf zebrafish embryos was of no significance in 3 mg/L TFZ group ($p > 0.05$, Figure 3B). Meanwhile, the MDA content in 3 dpf zebrafish embryos was significantly higher in 3 mg/L TFZ group than the control group ($p < 0.01$, Figure 3C). The ALT activity in 3 dpf zebrafish embryos was significantly decreased in 3 mg/L TFZ group compared to the control group ($p < 0.01$, Figure 3D).

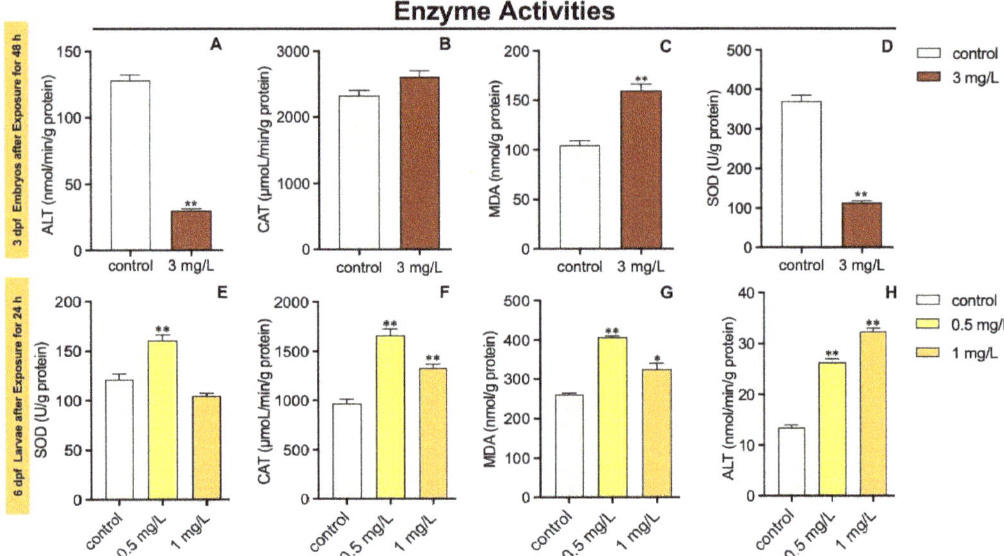

Figure 3. Effects of triflumizole on enzyme activities in zebrafish. For 3 dpf zebrafish embryos, SOD activity (**A**), CAT activity (**B**), MDA content (**C**), and ALT activity (**D**) after triflumizole exposure for 48 h. For 6 dpf zebrafish larvae, SOD activity (**E**), CAT activity (**F**), MDA content (**G**), and ALT activity (**H**) after triflumizole exposure for 24 h. Data are expressed as the mean of three replicates ± standard error (SEM). Asterisks denote significant differences between the control group and TFZ groups (determined by Dunnett post hoc comparison, * $p < 0.05$, ** $p < 0.01$). SOD: superoxide dismutase; CAT: catalase; MDA: malondialdehyde; ALT: alanine transaminase.

The SOD activity in 6 dpf zebrafish larvae was dramatically increased in 0.5 mg/L TFZ group relative to the control group ($p < 0.01$, Figure 3E). However, compared to the control group, the SOD activity in 6 dpf zebrafish larvae was of no significance after 1 mg/L TFZ exposure ($p > 0.05$, Figure 3E). The CAT activity in 6 dpf zebrafish larvae was significantly increased in 0.5 and 1 mg/L TFZ groups compared with the control group ($p < 0.01$, Figure 3F). Meanwhile, the MDA content in 6 dpf larvae was significantly higher in 0.5 ($p < 0.01$) and 1 mg/L ($p < 0.05$) TFZ groups than the control group (Figure 3G). The ALT activity in 6 dpf zebrafish larvae was significantly increased in 0.5 and 1 mg/L TFZ groups compared to the control group ($p < 0.01$, Figure 3H).

3.4. Heat Shock Response in Zebrafish Larvae after TFZ Exposure

The expression of glucose-regulated protein 94 (*grp94*) in 3 dpf zebrafish embryos was markedly increased in 1 mg/L TFZ group relative to the control group ($p < 0.05$, Figure 4A). Furthermore, compared to the control group, the expression of heat shock protein 70 (*hsp70*), *grp78*, *hsp90*, and *grp94* in 3 dpf zebrafish embryos was significantly increased in 2 and 3 mg/L TFZ groups ($p < 0.01$, Figure 4A).

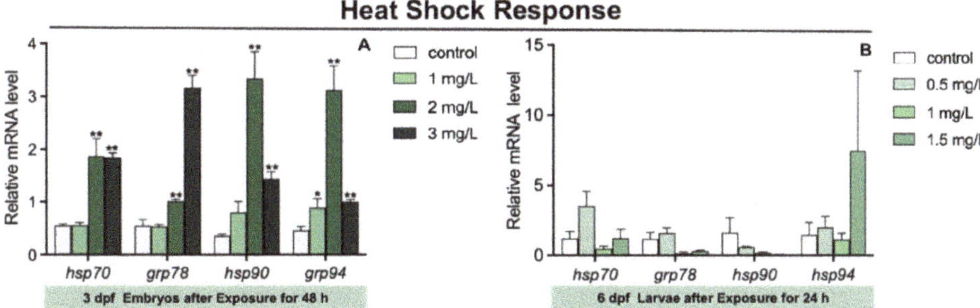

Figure 4. The expression of heat shock response genes (including *hsp70*, *grp78*, *hsp90*, and *grp94*) in zebrafish. (**A**) For 3 dpf zebrafish embryos, the expression of *hsp70*, *grp78*, *hsp90*, and *grp94* after triflumizole exposure for 48 h. (**B**) For 6 dpf zebrafish larvae, the expression of *hsp70*, *grp78*, *hsp90*, and *grp94* after triflumizole exposure for 48 h. Data are expressed as the mean of three replicates ± standard error (SEM). Asterisks denote significant differences between the control group and TFZ groups (determined by Dunnett post hoc comparison, * $p < 0.05$, ** $p < 0.01$). *hsp70*: heat shock protein 70; *grp78*: glucose-regulated protein 78; *hsp90*: heat shock protein 90; *grp94*: glucose-regulated protein 94.

Compared to the control group, the expression of *hsp70*, *grp78*, *hsp90*, and *grp94* in 6 dpf zebrafish larvae was of no significance in 0.5, 1, and 1.5 mg/L TFZ groups ($p > 0.05$, Figure 4B).

3.5. Inflammatory Genes Expression in Zebrafish Larvae after TFZ Exposure

The expression of tumor necrosis factor α (*tnfα*) in 3 dpf zebrafish embryos was significantly decreased in 2 mg/L TFZ group compared to the control group ($p < 0.01$, Figure 5A). Compared to the control group, the expression of p65-nuclear transcription factor κB (*p65-nfκb*), interleukin 1, beta (*il-1β*), and cyclooxygenase type 2 a (*cox2a*) in 3 dpf zebrafish embryos was significantly increased in 2 mg/L TFZ group ($p < 0.01$, Figure 5A). The expression of *p65-nfκb* and *il-1β* in 3 dpf zebrafish embryos was obviously increased in 3 mg/L TFZ group compared to the control group ($p < 0.01$, Figure 5A).

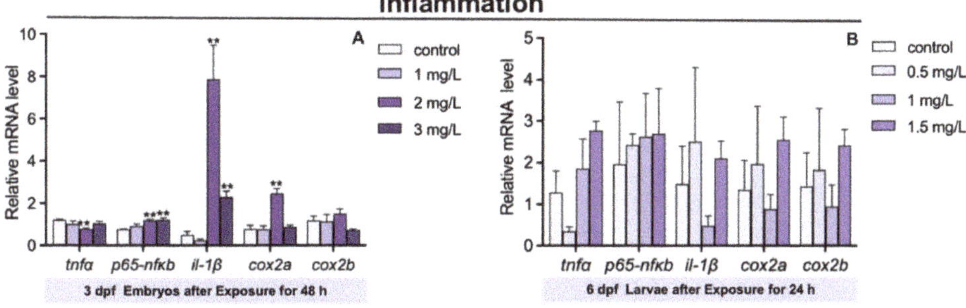

Figure 5. The expression of inflammatory genes (including *tnfα*, *il-1β*, *p65-nfκb*, *cox2a*, and *cox2b*) in zebrafish. (**A**) For 3 dpf zebrafish embryos, the expression of *tnfα*, *il-1β*, *p65-nfκb*, *cox2a*, and *cox2b* after triflumizole exposure for 48 h. (**B**) For 6 dpf zebrafish larvae, the expression of *tnfα*, *il-1β*, *p65-nfκb*, *cox2a*, and *cox2b* after triflumizole exposure for 24 h. Data are expressed as the mean of three replicates ± standard error (SEM). Asterisks denote significant differences between the control group and TFZ groups (determined by Dunnett post hoc comparison, ** $p < 0.01$). *tnfα*: tumor necrosis factor α; *il-1β*: interleukin 1, beta; *p65-nfκb*: p65-nuclear transcription factor κB; *cox2a*: cyclooxygenase type 2 a; *cox2b*: cyclooxygenase type 2 b.

Compared to the control group, the expression of *tnfα*, *p65-nfκb*, *il-1β*, *cox2a*, and *cox2b* in 6 dpf zebrafish larvae was of no significance in the 0.5, 1, and 1.5 mg/L TFZ groups ($p > 0.05$, Figure 5B).

3.6. Lipid Synthesis Gene Expression in Zebrafish Larvae after TFZ Exposure

The expression of sterol regulatory element binding transcription protein 1 (*srebp1*) and *ppar-γ* in 3 dpf zebrafish embryos was markedly increased in 1 mg/L TFZ group relative to the control group ($p < 0.05$, Figure 6A). Compared to the control group, the expression of *srebp1*, fas cell surface death receptor (*fas*), acetyl-coa carboxylase (*acc*), and *ppar-γ* in 3 dpf zebrafish embryos was significantly increased in 2 mg/L TFZ group ($p < 0.01$, Figure 6A). The expression of *srebp1* and *fas* in 3 dpf zebrafish embryos was dramatically increased in 3 mg/L TFZ group compared to the control group ($p < 0.01$, Figure 6A).

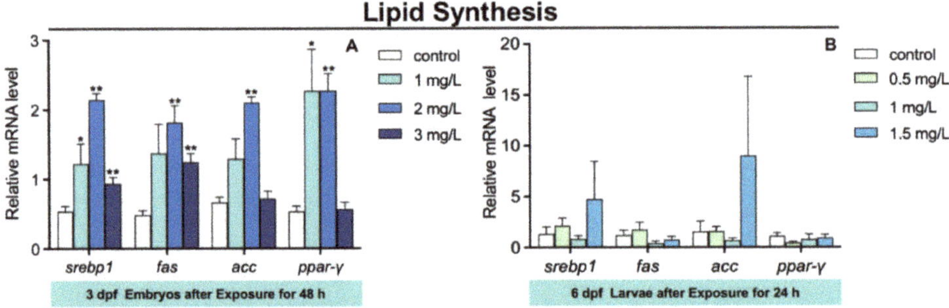

Figure 6. The expression of lipid synthesis genes (including *srebp1*, *fas*, *acc*, and *ppar-γ*) in zebrafish. (**A**) For 3 dpf zebrafish embryos, the expression of *srebp1*, *fas*, *acc*, and *ppar-γ* after triflumizole exposure for 48 h. (**B**) For 6 dpf zebrafish larvae, the expression of *srebp1*, *fas*, *acc*, and *ppar-γ* after triflumizole exposure for 48 h. Data are expressed as the mean of three replicates ± standard error (SEM). Asterisks denote significant differences between the control group and TFZ groups (determined by Dunnett post hoc comparison, * $p < 0.05$, ** $p < 0.01$). *srebp1*: sterol regulatory element binding transcription protein 1; *fas*: fas cell surface death receptor; *acc*: acetyl-coa carboxylase; *ppar-γ*: peroxisome proliferator-activated receptor gamma.

Compared to the control group, the expression of *srebp1*, *fas*, *acc*, and *ppar-γ* in 6 dpf zebrafish larvae was of no significance in the 0.5, 1, and 1.5 mg/L TFZ groups ($p > 0.05$, Figure 6B).

3.7. Effects of TFZ on Fish Intestinal Cell Line

As the TFZ concentration increased in the intestinal cell line, the cell activity generally decreased (Figure 7A). The expression of *hsp90* and *grp94* was increased in the 5 mg/L TFZ group relative to the control group ($p < 0.05$, Figure 7B). Additionally, the expression of *grp94* was significantly increased in the 10 mg/L TFZ group compared to the control group ($p < 0.05$, Figure 7B). Compared to the control group, the expression of *hsp90* was markedly increased in the 20 mg/L TFZ group ($p < 0.01$, Figure 7B). Compared to the control group, the expression of *cox2* was obviously increased in the 10 ($p < 0.01$) and 20 mg/L ($p < 0.05$) TFZ groups (Figure 7C). The expression of *srebp1* was significantly increased in the 20 mg/L TFZ group compared to the control group ($p < 0.05$, Figure 7D). The expression of *ppar-γ* was significantly increased in the 5 ($p < 0.05$), 10 ($p < 0.01$), and 20 mg/L ($p < 0.01$) TFZ groups relative to the control group (Figure 7D).

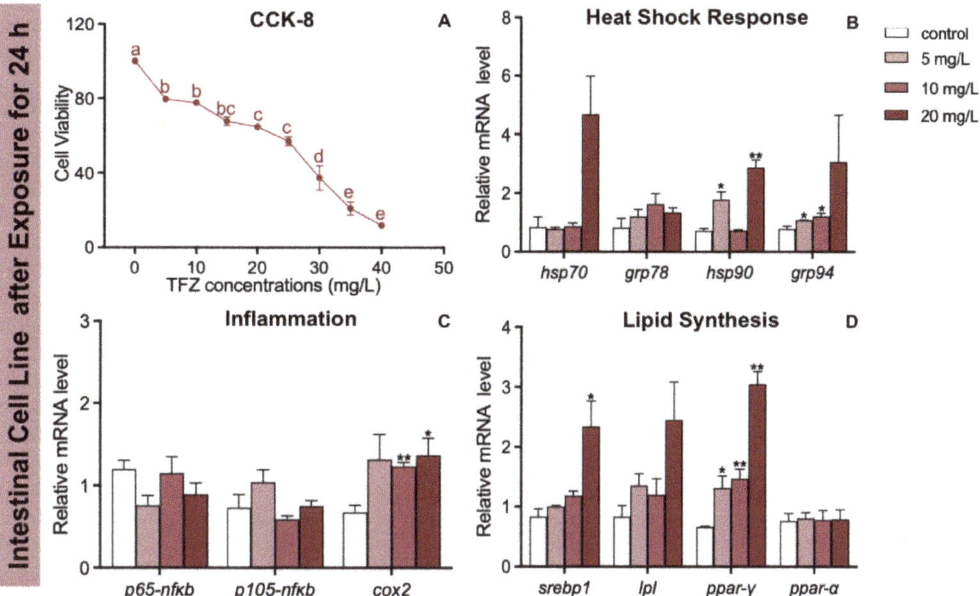

Figure 7. Effects of the fish intestinal cell line after triflumizole exposure for 24 h. (**A**) The cell viability. (**B**) The expression of *hsp70*, *grp78*, *hsp90*, and *grp94* after triflumizole exposure. (**C**) The expression of *p65-nfκb*, *p105-nfκb*, and *cox2* after triflumizole exposure. (**D**) The expression of *srebp1*, *lpl*, *ppar-γ* and *ppar-α* after triflumizole exposure. Data are expressed as the mean of three replicates $\pm$ standard error (SEM). Values without a common superscript letter differ ($p < 0.05$, Tukey's test). Asterisks denote significant differences between the control group and TFZ groups (determined by Dunnett post hoc comparison, * $p < 0.05$, ** $p < 0.01$). Data are expressed as the mean of three replicates $\pm$ standard error (SEM). *p65-nfκb*: p65-nuclear transcription factor κB; *p105-nfκb*: p105-nuclear transcription factor κB; *cox2*: cyclooxygenase 2; *lpl*: lipoprotein lipase; *ppar-α*: peroxisome proliferator-activated receptor alpha.

4. Discussion

From our results, it is clear that TFZ strongly affected the early development of zebrafish in a concentration-dependent manner. After TFZ exposure, the 48 h LC_{50} in 3 dpf zebrafish embryos and the 24 h LC_{50} in 6 dpf zebrafish larvae were 4.872 mg/L and 2.580 mg/L, respectively. Previous studies have suggested that the 72 h LC_{50} of TFZ for rare minnow (*Goboicypris rarus*) embryos was 7.11 (6.69–7.51) mg/L [13], which was to some degree higher than the results in this study. The species tested and exposure time caused these differences in acute toxicity experiments. For *Oncorhynchus mykiss*, the acute 96 h LC_{50} was 0.57 mg/L in Lewis's study [28], Hermsen's study found that the benchmark concertation of six triazoles ranges from 1.5 mg/L to 25 mg/L in zebrafish embryos [29], which was in accordance with our research.

In addition, we discovered obvious morphological changes induced by TFZ, including pericardial edema, yolk sac swelling, liver size reduction and liver color darkening in 3 dpf zebrafish embryos after exposure for 48 h. Zebrafish embryonic toxicity tests found that pericardial edema and yolk sac retention were extensively noted after triazoles exposure [29–32]. According to Jiang's study, difenoconazole can induce liver degeneration, including the retention of yolk sac in zebrafish [33]. It has been reported that a delay in yolk sac uptake is indicative of dysfunction of the liver during zebrafish larval development following tamoxifen exposure [34]. We found that yolk sac retention and liver degeneration occur in a concentration-dependent manner in zebrafish larvae, which was in accordance with the results obtained from a previous report [35]. It is noteworthy that the yolk sac is the only nutrient source in zebrafish larvae during embryonic development, and the retention of the yolk sac may affect nutrient absorption and lipid metabolism [36]. We subsequently observed that ALT activity was obviously increased in the 6 dpf zebrafish larvae after TFZ exposure. In conclusion, morphological changes in the liver and a significant increase in ALT activity indicate liver damage.

The antioxidant defense systems were affected by oxidative stress following TFZ exposure. SOD and CAT are potent enzymes for defending against oxygen toxicity because of their inhibition of effects on oxyradical formation [37]. In addition, the MDA content may indirectly represent the extent of lipid peroxidation. According to our study, SOD activity, CAT activity, and MDA content were significantly higher in 6 dpf zebrafish larvae. The increase in SOD activity, CAT activity, and MDA content is likely to toxicant stress and counteract the damage from ROS [38]. However, SOD activity was strongly inhibited in the 3 mg/L TFZ group in 3 dpf zebrafish embryos, indicating that it might destroy the protective system of zebrafish larvae. We observed that the expression of HSPs (including *hsp70*, *grp78*, *hsp90*, and *grp94*) in 3 dpf zebrafish embryos sharply increased in the 2 mg/L TFZ group, indicating a protective effect against protein misfolding [18]. Furthermore, we found that the expression of *hsp90* in the intestinal cell line was significantly increased in the 5 and 20 mg/L TFZ groups, which was in accordance with 3 dpf zebrafish embryos.

Several studies in aquatic organisms have demonstrated the potential for environmental pollutants to disrupt inflammatory gene expression. Inflammatory genes expressed in zebrafish larvae could be increased by glyphosate exposure [39]. Tricyclazole, a pesticide, also altered the transcription of inflammatory factors such as *tnfα* after exposure [24]. Tissue damage caused by environmental stimuli could lead to inflammation [40]. Our study suggests that TFZ exposure significantly increased the expression of *p65-nfκb*, *il-1β*, and *cox2a* which could be infer that potential tissue damage was caused by TFZ. The expression of *cox2* was markedly upregulated in the 10 and 20 mg/L TFZ groups in the intestinal cell line, which was in accordance with zebrafish larvae.

The increased oxidative stress, heat shock response, and inflammatory response are closely related to lipid synthesis in zebrafish [31,41]. In our research, we found that the expression of *srebp1*, *fas*, *acc*, and *ppar-γ* in 3 dpf zebrafish embryos was markedly increased after TFZ exposure, which was in accordance with the expression of *srebp1* and *ppar-γ* in the intestinal cell line. In reverse, TFZ caused liver damage and lipid synthesis through oxidative stress, heat shock response, and inflammation. Further studies should focus on the mechanism that TFZ contributes to lipid synthesis.

5. Conclusions

In summary, our results show TFZ-induced morphological changes (including pericardium edema, yolk retention, and liver degeneration) as well as death in zebrafish embryos. TFZ exposure could cause oxidative stress in zebrafish larvae. Moreover, the upregulation of *ppar-γ* suggests that the PPAR signaling pathway might participate in the imbalance of lipid metabolism. These findings provide a new insight into the potential mechanisms underlying the lipid metabolism imbalance of TFZ in zebrafish.

Supplementary Materials: The following supporting information can be downloaded at: https://www.mdpi.com/article/10.3390/toxics10110698/s1, Table S1: The sequence of primers for qPCR [42–51].

Author Contributions: L.B.: Methodology, investigation, writing. P.S.: cell culture, software, formal analysis, data curation. K.J.: methodology, manuscript-revised. H.Y.: resources. K.L.: conceptualization, methodology, manuscript-finalized, supervision, project administration, funding acquisition. J.X.: supervision, resources. X.Y.: resources. All authors have read and agreed to the published version of the manuscript.

Funding: This study was funded by the National Natural Science Foundation of China (grant number: 42076118 and 41706159), the Key Research Foundation of Ningbo No.2 Hospital, China (grant number: 2022HMZD08) and partly sponsored by K.C. Wong Magna Fund in Ningbo University.

Institutional Review Board Statement: The study was conducted in accordance with guidelines for the care and use of laboratory animals, as approved by the Ethical Committee of Ningbo University.

Informed Consent Statement: Not applicable.

Data Availability Statement: Data are contained within the article or Supplementary Materials.

Acknowledgments: We sincerely thank G.C.Y. and L.Z. for maintaining the zebrafish in a healthy condition.

Conflicts of Interest: The authors have no conflict of interest.

References

1. Oerke, E. Crop losses to pests. *J. Agric. Sci.* **2006**, *144*, 31–43. [CrossRef]
2. Popp, J.; Pető, K.; Nagy, J. Pesticide productivity and food security. A review. *Agron. Sustain* **2013**, *33*, 243–255. [CrossRef]
3. Konwick, B.J.; Garrison, A.W.; Avants, J.K.; Fisk, A.T. Bioaccumulation and biotransformation of chiral triazole fungicides in rainbow trout (*Oncorhynchus mykiss*). *Aquat. Toxicol.* **2006**, *80*, 372–381. [CrossRef] [PubMed]
4. Nakata, A.; Hashimoto, S.; Ikura, K.; Katsuura, K. Development of a New Fungicide Triflumizole. *J. Pestic. Sci.* **1991**, *16*, 301–313. [CrossRef]
5. İnam, R.; Gülerman, E.; Sarıgül, T. Determination of triflumizole by differential pulse polarography in formulation, soil and natural water samples. *Anal. Chim. Acta* **2006**, *579*, 117–123. [CrossRef]
6. Fajardo, J. Triflumizole (Procure): Control of diseases in tree nut, fruit, vegetable, and vine crops. *Phytopathology* **2006**, 96.
7. Tang, G.; Zhang, W.B.; Tang, J.Y.; Niu, J.F.; Yang, J.L.; Dong, H.Q.; Liang, Y.; Luo, L.X.; Cao, Y.S. Development of triflumizole ionic liquids containing anions of natural origin for improving the utilization and minimizing the adverse impacts on aquatic ecosystems. *Sci. Total Environ.* **2019**, *670*, 606–612. [CrossRef]
8. Brancato, A.; Brocca, D.; De Lentdecker, C.; Erdos, Z.; Ferreira, L.; Greco, L.; Jarrah, S.; Kardassi, D.; Leuschner, R.; Lythgo, C. European Food Safety. *Efsa J.* **2017**, *15*, 47–49.
9. Li, X.; Pham, H.T.; Janesick, A.S.; Blumberg, B. Triflumizole is an obesogen in mice that acts through peroxisome proliferator activated receptor gamma (PPAR γ). *Environ. Health Perspect.* **2012**, *120*, 1720–1726. [CrossRef]
10. Khay, S.; El-Aty, A.M.A.; Choi, J.H.; Shim, J.H. Analysis of residual triflumizole, an imidazole fungicide, in apples, pears and cucumbers using high performance liquid chromatography. *Toxicol. Res.* **2008**, *24*, 87–91. [CrossRef]
11. Maurya, P.K.; Malik, D.S.; Sharma, A. Impacts of pesticide application on aquatic environments and fish diversity. *Contam. Agric. Environ. Health Risks Remediat.* **2019**, *1*, 111.
12. Xi, J.; Shao, J.; Wang, Y.; Wang, X.; Yang, H.; Zhang, X.; Xiong, D. Acute toxicity of triflumizole to freshwater green algae Chlorella vulgaris. *Pestic. Biochem. Physiol.* **2019**, *158*, 135–142. [CrossRef] [PubMed]
13. Zhu, B.; Liu, L.; Gong, Y.X.; Ling, F.; Wang, G.X. Triazole-induced toxicity in developing rare minnow (Gobiocypris rarus) embryos. *Environ. Sci. Pollut. Res.* **2014**, *21*, 13625–13635. [CrossRef] [PubMed]
14. Franco, J.L.; Trivella, D.B.; Trevisan, R.; Dinslaken, D.F.; Marques, M.R.; Bainy, A.C.; Dafre, A.L. Antioxidant status and stress proteins in the gills of the brown mussel Perna perna exposed to zinc. *Chem.-Biol. Interact.* **2006**, *160*, 232–240. [CrossRef] [PubMed]
15. Sayeed, I.; Parvez, S.; Pandey, S.; Bin-Hafeez, B.; Haque, R.; Raisuddin, S. Oxidative stress biomarkers of exposure to deltamethrin in freshwater fish, Channa punctatus Bloch. *Ecotoxicol. Environ. Saf.* **2003**, *56*, 295–301. [CrossRef]
16. Li, Z.H.; Zlabek, V.; Velisek, J.; Grabic, R.; Machova, J.; Randak, T. Modulation of antioxidant defence system in brain of rainbow trout (*Oncorhynchus mykiss*) after chronic carbamazepine treatment. *Comp. Biochem. Physiol. Part C Toxicol. Pharmacol.* **2010**, *151*, 137–141. [CrossRef]
17. Pereira, T.C.B.; Campos, M.M.; Bogo, M.R. Copper toxicology, oxidative stress and inflammation using zebrafish as experimental model. *J. Appl. Toxicol.* **2016**, *36*, 86–885. [CrossRef]
18. Kurashova, N.A.; Madaeva, I.M.; Kolesnikova, L.I. Expression of heat shock proteins HSP70 under oxidative stress. *Adv. Gerontol.=Uspekhi Gerontol.* **2019**, *32*, 502–508. [CrossRef]
19. Madaeva, I.M.; Kurashova, N.A.; Semenova, N.V.; Ukhinov, E.B.; Kolesnikov, S.I.; Kolesnikova, L.I. Heat Shock Protein HSP70 in Oxidative Stress in Apnea Patients. *Bull. Exp. Biol. Med.* **2020**, *169*, 695–697. [CrossRef]
20. Szyller, J.; Bil-Lula, I. Heat shock proteins in oxidative stress and ischemia/reperfusion injury and benefits from physical exercises: A review to the current knowledge. *Oxidative Med. Cell. Longev.* **2021**, *2021*, 6678457. [CrossRef]
21. Wang, Y.; Teng, M.; Wang, D.; Yan, J.; Miao, J.; Zhou, Z.; Zhu, W. Enantioselective bioaccumulation following exposure of adult zebrafish (*Danio rerio*) to epoxiconazole and its effects on metabolomic profile as well as genes expression. *Environ. Pollut.* **2017**, *229*, 264–271. [CrossRef] [PubMed]
22. Wang, Y.; Zhu, W.; Wang, D.; Teng, M.; Yan, J.; Miao, J.; Zhou, Z. 1H NMR-based metabolomics analysis of adult zebrafish (*Danio rerio*) after exposure to diniconazole as well as its bioaccumulation behavior. *Chemosphere* **2017**, *168*, 1571–1577. [CrossRef] [PubMed]

23. Jia, M.; Wang, Y.; Wang, D.; Teng, M.; Yan, J.; Yan, S.; Meng, Z.; Li, R.; Zhou, Z.; Zhu, W. The effects of hexaconazole and epoxiconazole enantiomers on metabolic profile following exposure to zebrafish (*Danio rerio*) as well as the histopathological changes. *Chemosphere* **2019**, *226*, 520–533. [CrossRef] [PubMed]

24. Qiu, L.; Jia, K.; Huang, L.; Liao, X.; Guo, X.; Lu, H. Hepatotoxicity of tricyclazole in zebrafish (Danio rerio). *Chemosphere* **2019**, *232*, 171–179. [CrossRef] [PubMed]

25. Jia, K.; Chen, G.; Zeng, J.; Liu, F.; Liao, X.; Guo, C.; Luo, J.; Xiong, G.; Lu, H. Low trifloxystrobin-tebuconazole concentrations induce cardiac and developmental toxicity in zebrafish by regulating notch mediated-oxidative stress generation. *Ecotoxicol. Environ. Saf.* **2022**, *241*, 113752. [CrossRef]

26. Shi, P.; Liao, K.; Xu, J.; Wang, Y.; Xu, S.; Yan, X. Eicosapentaenoic acid mitigates palmitic acid-induced heat shock response, inflammation and repair processes in fish intestine. *Fish Shellfish Immunol.* **2022**, *124*, 362–371. [CrossRef]

27. Batista-Silva, H.; Dambrós, B.F.; de Moura, K.R.S.; Elie, N.; Delalande, C.; Van Der Kraak, G.; Silva, F.R.M.B. Calcium influx and spermatogenesis in the testis and liver enzyme activities in the zebrafish are rapidly modulated by the calcium content of the water. *Comp. Biochem. Physiol. Part A Mol. Integr. Physiol.* **2022**, *270*, 111227. [CrossRef]

28. Lewis, K.A.; Tzilivakis, J.; Warner, D.J.; Green, A. An international database for pesticide risk assessments and management. *Hum. Ecol. Risk Assess. Int. J.* **2016**, *22*, 1050–1064. [CrossRef]

29. Hermsen, S.A.; van den Brandhof, E.J.; van der Ven, L.T.; Piersma, A.H. Relative embryotoxicity of two classes of chemicals in a modified zebrafish embryotoxicity test and comparison with their in vivo potencies. *Toxicol. Vitr.* **2011**, *25*, 745–753. [CrossRef]

30. Tian, S.; Teng, M.; Meng, Z.; Yan, S.; Jia, M.; Li, R.; Liu, L.; Yan, J.; Zhou, Z.; Zhu, W. Toxicity effects in zebrafish embryos (*Danio rerio*) induced by prothioconazole. *Environ. Pollut.* **2019**, *255*, 113269. [CrossRef]

31. Qin, Z.; Wang, W.; Weng, Y.; Bao, Z.; Yang, G.; Jin, Y. Bromuconazole exposure induces cardiotoxicity and lipid transport disorder in larval zebrafish. *Comp. Biochem. Physiol. Part C Toxicol. Pharmacol.* **2022**, *262*, 109451. [CrossRef]

32. Mu, X.; Pang, S.; Sun, X.; Gao, J.; Chen, J.; Chen, X.; Li, X.; Wang, C. Evaluation of acute and developmental effects of difenoconazole via multiple stage zebrafish assays. *Environ. Pollut.* **2013**, *175*, 147–157. [CrossRef]

33. Jiang, J.; Chen, L.; Wu, S.; Lv, L.; Liu, X.; Wang, Q.; Zhao, X. Effects of difenoconazole on hepatotoxicity, lipid metabolism and gut microbiota in zebrafish (*Danio rerio*). *Environ. Pollut.* **2020**, *265*, 114844. [CrossRef]

34. Yu, Q.; Huo, J.; Zhang, Y.; Liu, K.; Cai, Y.; Xiang, T.; Jiang, Z.; Zhang, L. Tamoxifen-induced hepatotoxicity via lipid accumulation and inflammation in zebrafish. *Chemosphere* **2020**, *239*, 124705. [CrossRef] [PubMed]

35. Jia, K.; Cheng, B.; Huang, L.; Xiao, J.; Bai, Z.; Liao, X.; Cao, Z.; Shen, T.; Zhang, C.; Hu, C. Thiophanate-methyl induces severe hepatotoxicity in zebrafish. *Chemosphere* **2020**, *248*, 125941. [CrossRef] [PubMed]

36. Zoupa, M.; Machera, K. Zebrafish as an alternative vertebrate model for investigating developmental toxicity—The triadimefon example. *Int. J. Mol. Sci.* **2017**, *18*, 817. [CrossRef] [PubMed]

37. Pandey, S.; Parvez, S.; Sayeed, I.; Haque, R.; Bin-Hafeez, B.; Raisuddin, S. Biomarkers of oxidative stress: A comparative study of river Yamuna fish Wallago attu (Bl. & Schn.). *Sci. Total Environ.* **2003**, *309*, 105–115.

38. John, S.; Kale, M.; Rathore, N.; Bhatnagar, D. Protective effect of vitamin E in dimethoate and malathion induced oxidative stress in rat erythrocytes. *J. Nutr. Biochem.* **2001**, *12*, 500–504. [CrossRef]

39. Liu, Z.; Shangguan, Y.; Zhu, P.; Sultan, Y.; Feng, Y.; Li, X.; Ma, J. Developmental toxicity of glyphosate on embryo-larval zebrafish (*Danio rerio*). *Ecotoxicol. Environ. Saf.* **2022**, *236*, 113493. [CrossRef]

40. Lu, K.; Qiao, R.; An, H.; Zhang, Y. Influence of microplastics on the accumulation and chronic toxic effects of cadmium in zebrafish (*Danio rerio*). *Chemosphere* **2018**, *202*, 514–520. [CrossRef]

41. Mu, X.; Chai, T.; Wang, K.; Zhang, J.; Zhu, L.; Li, X.; Wang, C. Occurrence and origin of sensitivity toward difenoconazole in zebrafish (*Danio reio*) during different life stages. *Aquat. Toxicol.* **2015**, *160*, 57–68. [CrossRef] [PubMed]

42. Zhang, C.N.; Zhang, J.L.; Ren, H.T.; Zhou, B.H.; Wu, Q.J.; Sun, P. Effect of tributyltin on antioxidant ability and immune responses of zebrafish (Danio rerio). *Ecotoxicol. Environ. Saf.* **2017**, *138*, 1–8. [CrossRef] [PubMed]

43. Ding, Q.; Zhang, Z.; Ran, C.; He, S.; Yang, Y.; Du, Z.; Zhang, J.; Zhou, Z. The hepatotoxicity of palmitic acid in zebrafish involves the intestinal microbiota. *J. Nutr.* **2018**, *148*, 1217–1228. [CrossRef] [PubMed]

44. Pressley, M.E.; Phelan III, P.E.; Witten, P.E.; Mellon, M.T.; Kim, C.H. Pathogenesis and inflammatory response to Edwardsiella tarda infection in the zebrafish. *Dev. Comp. Immunol.* **2005**, *29*, 501–513. [CrossRef]

45. Ko, E.Y.; Cho, S.H.; Kwon, S.H.; Eom, C.Y.; Jeong, M.S.; Lee, W.; Kim, S.Y.; Heo, S.J.; Ahn, G.; Lee, K.P.; et al. The roles of NF-κB and ROS in regulation of pro-inflammatory mediators of inflammation induction in LPS-stimulated zebrafish embryos. *Fish Shellfish Immunol.* **2017**, *68*, 525–529. [CrossRef]

46. Krishnaraj, C.; Harper, S.L.; Yun, S.I. In Vivo toxicological assessment of biologically synthesized silver nanoparticles in adult Zebrafish (Danio rerio). *J. Hazard. Mater.* **2016**, *301*, 480–491. [CrossRef]

47. Jönsson, M.E.; Kubota, A.; Timme-Laragy, A.R.; Woodin, B.; Stegeman, J.J. Ahr2-dependence of PCB126 effects on the swim bladder in relation to expression of CYP1 and cox-2 genes in developing zebrafish. *Toxicol. Appl. Pharmacol.* **2012**, *265*, 166–174. [CrossRef]

48. Pan, Y.X.; Zhuo, M.Q.; Li, D.D.; Xu, Y.H.; Wu, K.; Luo, Z. SREBP-1 and LXRα pathways mediated Cu-induced hepatic lipid metabolism in zebrafish Danio rerio. *Chemosphere* **2019**, *215*, 370–379. [CrossRef]

49. Zhang, J.; Qian, L.; Teng, M.; Mu, X.; Qi, S.; Chen, X.; Zhou, Y.; Cheng, Y.; Pang, S.; Li, X.; et al. The lipid metabolism alteration of three spirocyclic tetramic acids on zebrafish (Danio rerio) embryos. *Environ. Pollut.* **2019**, *248*, 715–725. [CrossRef]

50. Jin, M.; Zhang, B.; Sun, Y.; Zhang, S.; Li, X.; Sik, A.; Bai, Y.; Zheng, X.; Liu, K. Involvement of peroxisome proliferator-activated receptor γ in anticonvulsant activity of α-asaronol against pentylenetetrazole-induced seizures in zebrafish. *Neuropharmacology* **2020**, *162*, 107760. [CrossRef]
51. Shi, P.; Meng, R.; Liao, K.; Li, S.; Hu, J.; Xu, J.; Zhang, L.; Cao, J.; Ran, Z.; Wang, D.; et al. Cadmium transcriptionally regulates Scd1 expression in silver pomfret. *Environ. Toxicol.* **2020**, *35*, 404–413. [CrossRef] [PubMed]

Article

Study of Ultrastructural Abnormalities in the Renal Cells of *Cyprinus carpio* Induced by Toxicants

Sumayya Nazir [1,*], **Md. Niamat Ali** [2,*], **Javeed Ahmad Tantray** [3], **Irfan Akram Baba** [4], **Arizo Jan** [5], **Simona Mariana Popescu** [6], **Bilal Ahamad Paray** [7] **and Aneela Gulnaz** [8]

[1] Department of Zoology, University of Kashmir, Srinagar 190006, J&K, India
[2] Cytogenetics and Molecular Biology Research Laboratory, Centre of Research for Development (CORD), University of Kashmir, Srinagar 190006, J&K, India
[3] Department of Zoology, Central University of Kashmir, Ganderbal 191201, J&K, India; javeedh3@gmail.com
[4] Department of Livestock Production and Management (LPM), Faculty of Veterinary Sciences and Animal Husbandry, Sher-e-Kashmir University of Agricultural Sciences and Technology (SKUAST-K), Shalimar, Srinagar 190025, J&K, India; irfanvet@gmail.com
[5] Division of Fisheries Resource Management, Faculty of Fisheries, Sher-e-Kashmir University of Agricultural Sciences and Technology (SKUAST-K), Shalimar, Srinagar 190025, J&K, India; arzujaan65@gmail.com
[6] Department of Biology and Environmental Engineering, University of Craiova, 13, A.I. Cuza, 200585 Craiova, Romania; popescu_simona3@yahoo.com
[7] Department of Zoology, College of Science, King Saud University, P.O. Box 2455, Riyadh 11451, Saudi Arabia; bparay@ksu.edu.sa
[8] College of Pharmacy, Woosuk University, Wanju-gun 55338, Korea; draneela@woosuk.ac.kr
* Correspondence: sumayya.nazirhabib@gmail.com (S.N.); mdniamat@hotmail.com (M.N.A.)

Citation: Nazir, S.; Ali, M.N.; Tantray, J.A.; Baba, I.A.; Jan, A.; Popescu, S.M.; Paray, B.A.; Gulnaz, A. Study of Ultrastructural Abnormalities in the Renal Cells of *Cyprinus carpio* Induced by Toxicants. *Toxics* **2022**, *10*, 177. https://doi.org/10.3390/toxics10040177

Academic Editors: François Gagné, Stefano Magni and Valerio Matozzo

Received: 4 February 2022
Accepted: 28 March 2022
Published: 2 April 2022

Publisher's Note: MDPI stays neutral with regard to jurisdictional claims in published maps and institutional affiliations.

Abstract: Transmission Electron Microscopic (TEM) assessments were performed on the renal cells of common carp *Cyprinus carpio* to observe the deleterious effects of two organophosphate insecticides, Phorate and Dimethoate. Pesticides such as Phorate and Dimethoate often pollute aquatic systems where they may negatively impact fish, but so far, the ultrastructural toxicity of these pesticides remains poorly understood. Here, we use Transmission Electron Microscopy (TEM) to determine how acute exposure to sublethal concentrations of these two pesticides may affect the renal cells of common carp *Cyprinus carpio*. For each insecticide, the fish were divided in four experimental conditions: a control and three different exposure concentrations of the pesticide. The Phorate treated fish were exposed to three sublethal concentrations of 0.2 mg/L, 0.4 mg/L, 0.6 mg/L for a duration of 24, 48 & 72 h. The dimethoate treated fish were exposed to three sublethal concentrations of 0.005 mL/L, 0.01 mL/L, 0.015 mL/L for a duration of 24, 48 and 72 h. The two-dimensional transmission electron microscopy revealed ultrastructural abnormalities in the treated fish renal cells when exposed to two toxicants including deformation in the glomerulus, vacuolization of cytoplasm, degenerative nucleus and damaged mitochondria. Furthermore, the ultrastructural abnormalities were more prominent with the increase in the concentrations of both the insecticides and also with their exposure period. Overall, these results provide important baseline data on the ultrastructural toxicity of Phorate and Dimethoate and will allow important follow-up studies to further elucidate the underlying cellular mechanisms of pesticide toxicity in wildlife.

Keywords: Phorate; dimethoate; TEM; renal cells; ultrastructure; toxicants

1. Introduction

Pesticides have been used successfully to improve agricultural productivity and meet the increasing demand for food, but they often end up in the natural environment where they may impact non-target organisms. Pesticides can act on organisms other than pest species. This is of particular concern in developing countries, where increasing intensification of agriculture leads to high pesticide use [1]. Pesticides cover a wide range of substances including insecticides, acaricides, fungicides, molluscicides, herbicides, nematocides and rodenticides. Pesticides are widely used in agriculture to control insects, nematodes, fungi, etc. that affect food and other crops. These are easy to apply, being cost effective and, most importantly, they are readily available practical means of pest control. On the flip side, pesticides may have unwanted effects on the natural environment as they can enter through various routes including direct application, spray drift, atmospheric deposition and surface runoff. Among the different types of pesticides, insecticides are often used. Depending upon the mode of action, insecticides have been classified by the Insecticide Resistance Action committee (IRAC, 2017) into various categories, viz., botanicals (nicotine, rotenone, pyrethrum, etc.), organochlorines (e.g., DDT), organophosphates (Phorate, Dimethoate), carbamates, pyrethroids etc [2]. There are many pathways by which pesticides leave their sites of application and are distributed throughout the aquatic ecosystem. Different concentrations of the pesticides are present in many types of wastewater and several studies have revealed them to be toxic to aquatic organisms including fish [3,4]. Fish are sensitive to polluted water. Hence, fish have long being used to monitor the quality of the aquatic environment and fish histology is increasingly being used as an indicator of environmental stress [5,6].

Phorate is a systemic and broad spectrum organophosphorus (OP) insecticide, commonly used in agriculture to control sucking and chewing insecticides, leaf hoppers and mites. It is also used in pine forests and on root and field crops including corn, cotton, coffee and some ornamental plants and bulbs [7,8]. Phorate is primarily formulated as granules to be applied at planting in a band or directly to the seed furrow. In biota, it inhibits acetylcholinesterase activity by phosphorylating the serine hydroxyl group in the substrate binding domain, which results in the accumulation of acetylcholine and induces neurotoxicity. Though its use has been strictly restricted by the United States Environmental Protection Agency (USEPA), it is still being used in several countries such as India, China, Italy and Egypt [9].

Dimethoate is a systemic organophosphate insecticide used on a large variety of field grown agricultural crops, tree crops, and ornamentals. Dimethoate was first registered in 1962 in the US and later its use for non-agricultural practices (e.g., domestic purposes) was banned as of 2000. It is available as Rogor and like other organophosphates it acts as an acetylcholinesterase inhibitor and works as a nerve poison at synapses of neuromuscular junctions, which is evident by abnormal body movements and jerks [10–12]. Various studies on the toxicity of Phorate on aquatic organisms, especially fish, have been carried out, confirming its toxicity and genotoxic role [13–15]. Histopathological toxicity of brain of *Cyprinus carpio* exposed to Phorate was also reported by Lakshmaiah, (2017) [16]. Lakshmaiah (2016) reported acute toxicity of Phorate on succinate dehydrogenase enzyme activity, which is an important enzyme for Krebs Cycle [17]. Additionally, its genotoxicity potential was reported by Saquib et al., (2012) in male wistar rats and in human amniotic epithelial (WISH) cells [7]. Toxicity of dimethoate in fish fauna has also been reported by many researchers [12,18]. Demet and Canan (2011) reported toxic effects of Dimethoate on hematological, biochemical and behavioural patterns in *Oncorhynchus mykiss* exposed to sublethal concentrations of 0.0735, 0.3675, and 0.7350 mg/L for 5, 15, and 30 days [19]. Ganeshwade (2012) reported biochemical changes in the gills of freshwater fish *Puntius ticto* when exposed to lethal (5.012 ppm) and two sublethal (2.506 and 1.253 ppm) concentrations of Dimethoate for 96 h and 60 days, respectively [20]. Histopathological studies were reported in the kidney of *Cyprinus carpio* after exposure to dimethoate (EC 30%) by Singh (2012) [21]. Binukumari and Vasanthi (2013) observed the toxic effect of the Dimethoate 30% EC on protein metabolism of *Labeo rohita*, when exposed to a concentration of 0.398 ppm for 24, 48 and 72 h, respectively [22]. Singh (2013) exposed common carp, *Cyprinus* to dimethoate 0.40 mg/L for short-term exposure of 96 h and reported toxic effects on its liver [23]. Singh (2014) reported toxic effects of Dimethoate (EC 30%) on gill morphology, oxygen consumption and serum electrolyte levels of common carp, *Cyprinus Carpio.* The fish were exposed to a sub lethal concentration of 0.96 mg/L (60% of 96 h LC50) of dimethoate at a 24, 48 and 96 h exposure duration [24]. Singh (2017) reported testicular toxicity of *Cyprinus carpio* when exposed to dimethoate at 0.96 mg/L and 0.48 mg/L, respectively in a short-term (96 h) and long-term study (36 days) [25].

As is evident, the toxic effects of these two organophosphates have been reported but no work has been reported on their potential role as cytotoxins, i.e., their role to cause ultrastructural abnormalities in fish cells. So, the present study has been designed to investigate the role of two organophosphate insecticides: Phorate and dimethoate, for their potential to damage cells in an in vivo system and contribute to acute toxicity. Hence, an investigation resulting from cellular events leading to cytotoxicity is proposed. Teleost fish have proved to be good models to evaluate the genotoxicity and effects of pollutants such as insecticides on animals, as their biochemical responses are similar to those of mammals and other vertebrates [14,26–28]. The advantage of using fish models includes the facility by which Teleostei, especially the small species, can be maintained and handled inside the laboratory under experimental conditions of toxic exposure [29]. Fish frequently respond to chemical exposure as superior vertebrates, which validate this model to study potential teratogenic and carcinogenic compounds in humans [30]. Common carp *Cyprinus carpio* is a robust fish which can tolerate a wide range of temperatures and is adaptable to varying environmental conditions. It is an economically important fish due to its nutritional value. Thus, it serves as an important vehicle for contamination.

The main objective of the present study was to observe Phorate and Dimethoate induced ultrastructural abnormality in the renal cells of common carp, *Cyprinus carpio* (as the kidney is not only the excretory organ but also functions as osmoregulatory organ of the fish). It will provide baseline data about the ultrastructural toxicity of these insecticides, which are widely used and finally reaches into our precious water bodies, thereby proving hazardous to both aquatic fauna (especially fish), as well as to its consumers. The present study was initiated to understand the acute toxicity of organophosphate insecticide exposure which could make a potential contribution towards the identification of the environmental toxicants. Such an acute toxicity testing would measure the adverse effects that occur within a short period of time. Thus, such information would be fruitful to serve as a basis for hazard classification and will provide information regarding the target tissue/organ toxicity by different doses of the toxicants.

2. Materials and Methods

The fish species *Cyprinus carpio* L. (family: Cyprinidae) were chosen for the present study. Young specimens of *Cyprinus carpio communis* (age: <1 year; weight 30–40 g; length 10–12 cm) were used. The healthy young adult specimens of *Cyprinus carpio communis* were collected from the Dal Lake (34°5′–34°6′ N latitude and 74°8′–74°9′ E longitudes) using a net and hand-picking method, and then these collected fish were transported to a laboratory in a specially-designed container with oxygen supply. After collection, fish were acclimatized for 15 days in 60 L artificially aerated glass aquaria (5 fish each) containing dechlorinated tap water for 12/12 natural photoperiod (pH 7.6–8.4, temperature 25 ± 3). During acclimatization, fish were fed a commercial diet (Feed Royal®, Maa Agro Foods, Visakhapatnam, Andhra Pradesh, India). In order to avoid ammonia accumulation, the water in these aquariums were changed daily with dechlorinated tap water. During the test period, no feed was given to keep the insecticide concentrations constant throughout the test period of 72 h [18,19]. Water quality of the test solution was determined according to standard procedures [31]. The control fish were kept in experimental water without adding these insecticides, keeping all other conditions constant.

2.1. Test Chemicals, Selection and Dosage of the Insecticides

The pesticides used for the present study were two widely used organophosphorus insecticides: Phorate and Dimethoate. Commercial grade formulations of insecticides were used because only commercial preparations are used in agriculture. The commercial grade of Phorate was obtained from agro farmers, Court Road, Srinagar manufactured by Sikkim

Pesticide Industries. Rogor® (30% dimethoate) was obtained from Big Farmers, Jahangir Chowk near Budshah bridge manufactured by Cheminova, A/S Denmark. Dimethoate is fairly stable than phorate in aqueous solutions (Table 1).

Table 1. Phorate and Dimethoate.

S.NO	NAME	FORMULA	CAS	REG. NO	NATURE
1	Phorate	$C_7H_{17}O_2P_{S3}$	O, O-diethyl s-[(ethyl thio)methyl] phosphorodithioate	298-02-02	Highly toxic systemic insecticide used as granules. It has fumigant action.
2	Dimethoate	$C_5H_{12}O_3P_{S2}$	O, O-diethyl s-[2-(methylamino)-2-oxyethyl]phosphorodithioate	60-51-5	It is acutely toxic, has possible links to cancer.

On the basis of the literature data (LC50 values for each insecticide), three sublethal concentrations of these insecticides were selected for the experiment, 0.2 mg/L 0.4 mg/L and 0.6 mg/L for Phorate and 0.005 mL/L, 0.01 mL/L, 0.015 mL/L for Dimethoate.

2.2. Experimental Design

In order to assess the ultrastructural abnormalities induced by the Phorate and Dimethoate in the fish renal cells, the fish for each insecticide were placed in four experimental conditions: a control and three different exposure concentrations. Based on the selected doses of each insecticide and the time of exposure. All groups had an equivalent no. of fishes, i.e., 5 fish per group in a glass aquarium. After completion of the exposure period of 72 h, the fish were euthanized, and the kidneys were removed to assess ultrastructural abnormalities using transmission electron microscopy (TEM). It was briefly washed in 0.09% normal saline soon after biopsy of the animal in order to remove the associated mucus or blood from it. The kidney was cut into small pieces of 1×1 mm. Then, these tissues were fixed in a mixture of 2% paraformaldehyde and 2.5% glutaraldehyde in 0.1 M phosphate buffer for 8–12 h at 4 °C. Then, the tissues were washed in buffer three times, each for one hour duration at 4 °C. The tissues were post fixed in 1% osmium tetroxide for 1 h at 4 °C. The tissues were then dehydrated through an acetone series (30%, 50%, 70%, 90% and 100%) for 15 min each. The dehydrated tissues were then cleared off the acetone by propylene oxide for 30 min and then infiltrated and embedded in a liquid resin epoxy. After embedding, the resin blocks were then thin-sectioned and these ultrathin sections (70 nm) were placed on metal grids and stained with electron dense stains, with uranyl acetate and lead citrate. These samples were then examined and photographed for ultrastructural changes by TEM (TECNAI 200 kV, FEI/Phillips, Hillsboro, OR, USA) at AIIMS, New Delhi for observations and photography.

3. Results

Kidney samples from the control fish showed normal structural features. The nucleus was oval in shape with heterochromatin evenly distributed throughout the nucleoplasm and in a binucleate condition. The heterochromatin was uniformly distributed (Figure 1b). The mitochondria were tubular with distinct cristae (Figure 1a); stacks of rough endoplasmic reticulum were easily seen around the nucleus and mitochondria (Figure 1d) lysosomes were also seen. The basement membrane also had normal nuclei. Additionally, the pedicels which arise from the specialised epithelial cells (podocytes) were easily seen forming narrow infiltration slits in the glomerulus (Figure 1c).

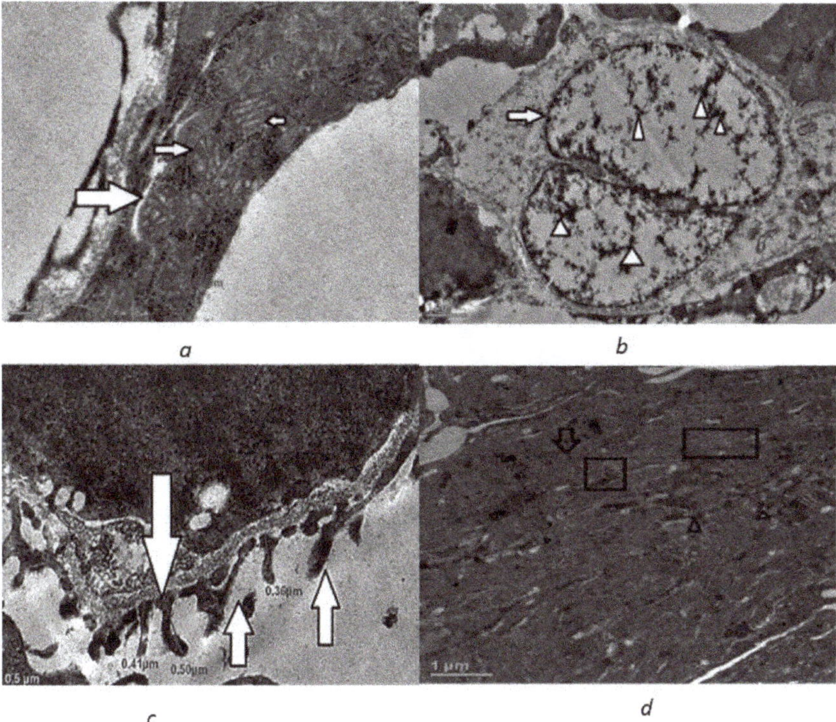

Figure 1. Transmission electron micrographs of control renal cells with intact normal architecture of cellular organelles: (**a**) normal mitochondria with distinct cristae (white filled arrow) (scale 0.2 µm); (**b**) nucleus, though binucleate (white filled arrow) but with uniformly distributed chromatin (white arrow head) (scale 1 µm); (**c**) pedicel with infiltration slits (white filled arrow) (scale 0.5 µm); (**d**) stacks of endoplasmic reticulum (boxed), mitochondria (arrow) and lysosomes (arrow head) (scale 1 µm).

In both the Phorate- and Dimethoate-treated fish, the ultrastructural changes were evident (Figures 2–4). Marked changes were observed in the shape of nuclei in which the nuclear envelope showed shrinkage and, hence, an irregular outline. The mitochondria displayed degenerative changes with condensed and rounded appearance, having disorganized lamellar cristae. The endoplasmic reticulum was also disorganized and fragmented and less in number. Vacuolization was also very much evident in the cytoplasm of the treated fish.

In the glomerulus of Phorate-treated fish, most of the podocytes exhibited swollen, fused and short pedicel with a significant increase in filtration slit width. These degenerative changes became more and more evident with the increase in the concentration of the Phorate for the exposure time period of 72 h. With the increase in dose, the number of mitochondria increased with distinct degenerative changes. The degenerative changes became more pronounced with the increase in the concentration of the genotoxins, as was revealed by severe deformation and nuclear blebbing of the nucleus and other structures. This increase in the dose led to an increase in the apoptotic stimuli, which was confirmed by the apoptotic body near the nucleus (Figure 4a), and nuclear membrane shrinkage was observed in all the treated fish renal cells.

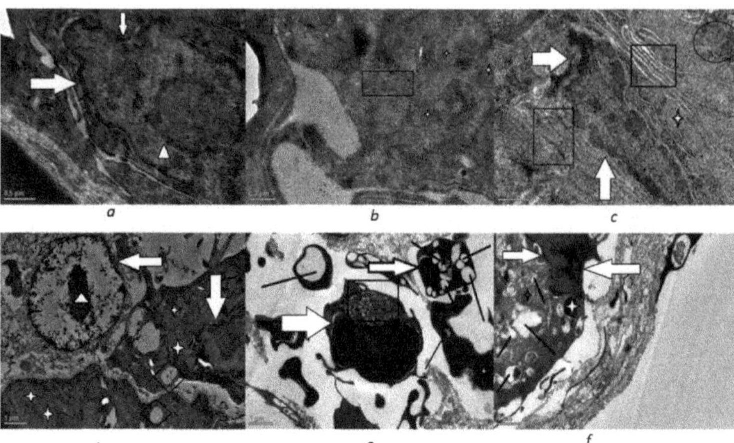

Figure 2. Transmission electron micrographs of 0.2 mg/L Phorate exposed renal cells exposed for 72 h revealing degenerative changes: (**a**) deformed nucleus (white filled arrow) and condensed heterochromatin (arrow head) (scale 0.5 µm; (**b**) degenerated endoplasmic reticulum (□) and condensed degenerated mitochondria(white star) (scale 0.2 µm). (**c**) more degenerative and deformed nucleus (white arrow head), dilated and deformed ER (□), increased lysosomes (circled) and deformed mitochondria (white star) (scale 0.5 µm); (**d**) deformed nucleus (white filled arrow) and electron dense condensed chromatin (white arrow head) and vacuole (boxed ◊) (**e**) showing severe electron dense deposits in the deformed nucleus (white filled arrow) with whorl of degenerated ER (□), severe cytoplasmic vacuolation (thick black line) and deformed mitochondria (star) (scale 1 µm); (**f**) deformed nucleus (white filled arrow), degenerated mitochondria (star), severe vacuolation (thick black line) (scale 1 µm).

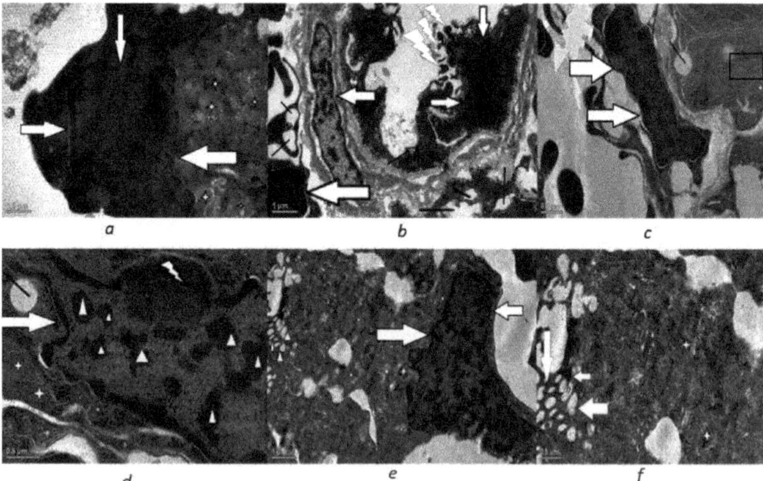

Figure 3. Transmission electron micrographs of 0.6 mg/L Phorate exposed renal cells exposed for 72 h: (**a**) deformed nucleus (white head arrow) and other structures not distinct, mitochondria (star) (scale 1 µm); (**b**) electron dense deposits in deformed nucleus (white filled arrow), deformed fused swollen pedicels (flashed white arrow) and extreme vacuole presence(thick black line) (scale

1 µm). (**c,d**) transmission electron micrographs of 0.005 mL/L Dimethoate exposed renal cells exposed for 72 h: (**c**) degenerated nucleus with electron dense deposits throughout nucleoplasm (white filled arrow); condensed degenerated mitochondria (star), vacuole (thick black line) and curved degenerated ER (□) (scale 0.5 µm); (**d**) deformed nucleus (white filled arrow) with condensed electron dense chromatin (arrow head) and peripheral nucleolus (flashed white arrow), presence of vacuoles(thick black line) (scale 0.5 µm) (**e,f**) transmission electron micrographs of 0.01 mL/L Dimethoate exposed renal cells exposed for 72 h: (**e**) less distinct cytoplasmic structures-deformed nucleus (white filled arrow), degenerated mitochondria (star), severe cytoplasmic vacuoles (white arrow head) (scale 1 µm); (**f**) extreme vacuolation (white filled arrow) and electron dense degenerated mitochondria (star) (scale 0.5 µm).

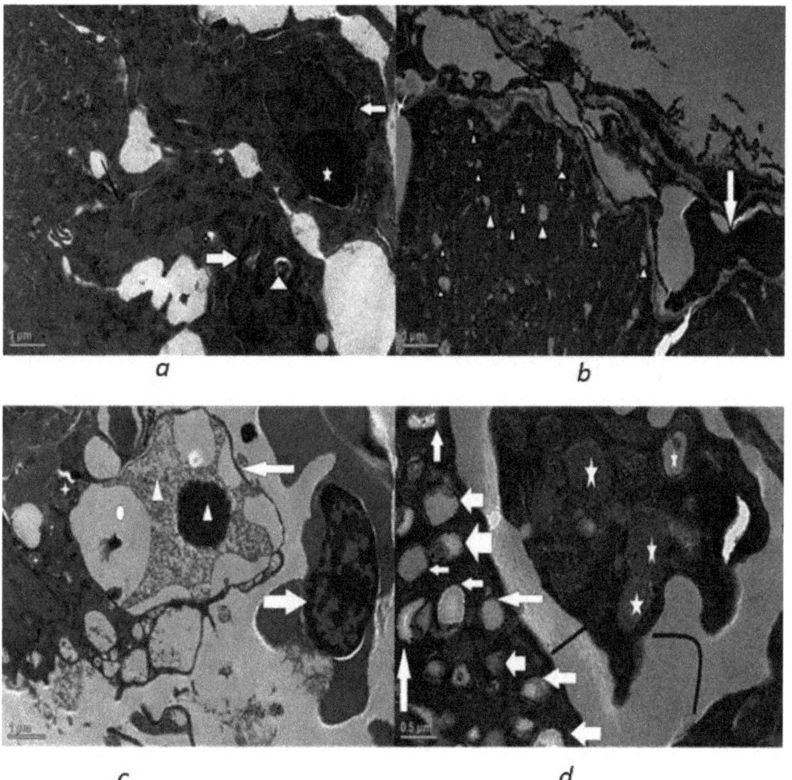

Figure 4. Transmission electron micrographs of 0.015 mL/L Dimethoate exposed renal cells exposed for 72 h: (**a**) highly deformed nucleus (white filled arrow) with apoptotic body (big star), cytoplasmic vacuole (thick black line), vacuolated mitochondria (small star) (scale 1 µm); (**b**) severe cytoplasmic vacuolization (star head), deformed nucleus(white filled arrow) (scale 1 µm); (**c**) deformed nucleus (white filled arrow) with nuclear vacuoles (O) and condensed chromatin (arrow head) and nucleolus (scale 1 µm); (**d**) highly deformed mitochondria (star), increased intercellular space (thick straight and curved line), and extreme cytoplasmic vacuolation (white filled arrow) (scale 0.5 µm).

4. Discussion

Ultra-morphological analyses through electron microscopy are the best tool to detect various cell intoxication symptoms. Transmission electron microscopy (TEM) produces a two-dimensional image of the cell. It provides the minute details of surface alterations which cannot be observed by the simple light microscope such as an increase or decrease in the number of organelles, nuclear degeneration, loss of cytoplasmic integrity, etc. However, one of the main drawbacks of electron microscopy is that this technique cannot be used to quantify the apoptotic cells [32]. Microscopy is therefore used in the interpretation of the various parameters used to measure the extent of toxicity and contributes to the establishment of measures that aim to prevent environmental contamination, a risk not only to the ecosystem but also to human health [33]. It will support follow-up studies to further elucidate the underlying cellular mechanisms of pesticide toxicity in wildlife. It is pertinent to mention here that the number of fish used in the present study was limited, therefore more work needs to be carried out on true replicates so that there is no room of doubt for the confirmation of the results.

The results of the present study on the ultrastructural changes observed in the kidney cells of common carp, *Cyprinus carpio communis*, exposed to two organophosphate insecticides viz. Phorate and Dimethoate have been compared with the previous findings of other researchers, who observed the toxicological effects of other toxicants in different species. Nandita [34] conducted TEM studies on the kidney of *Ctenopharyngodon iddellus* exposed to an organophosphate monocrotophos (36% *w/w* for 15, 30 and 45 days of duration). In the treated fish, several ultrastructural abnormalities were observed in the kidney. The prominent changes included degenerative changes in the nucleus and mitochondria, and cytoplasm in the proximal convoluted tubule showed vacuolization. The number of mitochondria were found to be tremendously increased upon exposure. The degenerative changes were found to be directly linked to the dose concentration of the toxicant as the cellular structures were not clearly visible with a high dosage of the toxicant. Shrinkage of glomerulus, enlargement of Bowman's capsule, dissociation of epithelial lining and vacuolization of the cytoplasm were also observed in the kidney of fish, *Cirrhinus mrigala*, exposed to metal, lead acetate (chronic toxicity study of 30 days with sub lethal concentrations of 14.1 and 28.2 ppm) [35]. Pesticides cause physiological and biochemical changes in the fish species and greatly influence their behavioural activities. The tubular degeneration was also observed in the kidney of zebrafish, *Danio rerio*, when exposed to a combination of various pesticides (sublethal concentration of 8.4 and 4.2 µg/L of Chloropyrifos 50% + Cypermethrin 5% EC for 7, 14, 21 and 28th day) [36]. Kidney of Asian stinging catfish, *Heteropneutes. fossilis* treated with chlorpyrifos (1/50 of LC_{50} and 1/10 of LC_{50} for a duration of 30 days) also revealed tubular degeneration, vacuolization and loss of glomerular [37]. Similar effects were observed in the kidneys of the brown trout, *Salmo trutta m. fario* when exposed to cadmium and zinc poisoning (4.4 mg/kg and 11 mg/kg, respectively, for a 46-week exposure period [38]. Aluminium also showed its toxic effects on the kidney of Tilapia, *Tilapia zillii*, by causing glomerular shrinkage. This freshwater fish was exposed to three levels of aluminium (25, 50, 100 µg/L) for 96 h [39]. TEM of Almix herbicide at sublethal concentrations of 66.67 mg/L for 30 days on freshwater teleost kidney was found to induce necrosis in the nucleus, severe vacuolization, whorl pattern of ER, and also mitochondrial degeneration [40].

Glomerular podocytes, on the other hand, are highly specialized cells and their most unique features are interdigitated foot processes with filtration slits in between. They are bridged by the slit diaphragm, which plays a major role in establishing the selective permeability of the glomerular filtration barrier [41]. Swollen and fused podocytes and increased infiltration slits observed through TEM in the glomerulus is likely to cause glomerular disease, resulting in failure of the filtration barrier [41]. The ultrastructural examination of the kidney of rainbow trout, *Oncorhynchus mykiss,* exposed to synthetic dyes revealed melanomarcophages increased in number and melanin was more abundant in the cells of treated fish [42].

The kidney thus serves as a major route of excretion of metabolites of xenobiotics [40] and receives the largest proportion of postbranchial blood. It is more likely to undergo histopathological alterations under pesticide stress [14] and, hence, deserves to be examined at cellular level, which is possible only through transmission electron microscopy. The cellular degeneration which occurred in the kidney due to the Phorate and Dimethoate toxicity might have also affected the osmoregulatory function of the fish as the kidney is not only the excretory organ, but also functions as osmoregulatory organ of the fish. The teleost kidney is one of the first organs to be affected by pollutants in water [41]. Thus, the overall ultrastructural changes observed in the kidney of common carp in the present study proved that these organophosphorus insecticides viz. Phorate and Dimethoate are potent genotoxicants which showed their deleterious effects on the osmoregulatory organ of the fish. Previous findings have also proved Phorate to be genotoxic on various other parameters, including the activities of carbohydrate metabolic enzyme, succinate dehydrogenase in common carp [14] and also altered protein content in various tissues of common carp [8]. Phorate also induced pronounced pathological toxicity in the brain of common carp, *Cyprinus carpio* and the extent of damage and degeneration were progressive with the exposure time period and the concentration of Phorate [16]. Phorate also has deleterious effects on humans as it causes cellular damage, as well as changes in DNA structure [7].

The degree of damage to the gill morphology of the Dimethoate-treated fish showed direct correlation to the exposure period, (24, 48 and 96 h when exposed to sub lethal concentration of 0.96 mg/L), while the oxygen consumption and ventilation rate showed a significant decrease during the exposure period to *Cyprinus carpio* [24]. Studies have also shown that Dimethoate can adversely affect the reproductive fitness of male common carps, even at sublethal concentration, both in short-term and long-term exposure, 96 h and 36 days, respectively, when exposed to 0.96 mg/L and 0.48 mg/L [25]. The carp fingerlings exhibit a number of abnormalities in their behaviour when exposed to Dimethoate such as gulping of surface water, uncoordinated movements, tremors and convulsions, excessive mucus secretion, and imbalanced swimming, ending in a collapse to the bottom of the aquarium with wide mouth open [10–12]. Histopathological examination of liver of carp exposed to a sublethal concentration of Dimethoate revealed disorganization of hepatocytes, rupture of blood vessels, vacuolization of cytoplasm, pyknotic nuclei and necrosis of the hepatic tissue [23]. *Oncorhynchus mykiss* fish showed remarkable behavioural abnormalities such as loss of balance, erratic swimming, and convulsion and a decrease in protein concentration with the increase in exposure to Dimethoate [19]. Shrinkage of glomerulus, dilation of lumen, vacuolization and tubular degeneration were histopathological changes observed in the kidney of common carp exposed to Dimethoate [21]. Dimethoate-exposed Labeo showed a decrease in protein content [22]. Nortan [27] studied ultrastructural effects of diethylnitrosamine (DENA) and trichloro ethylene (TCE) on the kidney of Japanese medaka, *Oryzias latipes*. Tissues of control and DENA and TCE exposed medaka for 32 weeks were observed by TEM. The kidney of the treated group showed cytotoxic effects along the length of proximal tubule, which were more pronounced in region II, distented proximal tubule, cellular necrosis, membrane fragments, pyknotic nuclei, nuclear membrane disintegration. Marlasca et al., [42] reported degranulation and dilation of endoplasmic reticulum of hepatocytes and renal cells with packed melanosomes in all the developmental stages of rainbow trout exposed to two synthetic dyes (for 30 days) by TEM. Spazier et al., [43] reported severe cellular disintegration of spleen of Anguilla fish exposed to contaminated waters. The cells revealed necrotic changes marked by swelling of the mitochondria. Godar et al., [44] observed apoptosis by dissociation of ribosomes, ER dilation, degeneration of nucleus and chromatin condensation through TEM on inducing the photo toxicity of ultraviolet radiations (UV) of different wave band regions on L5178Y-R murine lymphoma (LYR) cells. Anglade et al., [45] observed apoptotic changes such as loss of contact with surrounding tissues, shrinkage and chromatin condensation in myelinated neurons of substantia nigra in four normal aged humans. Hagar et al., [46] reported ultrastructural anomalies in β-cells of rat pancreas on exposure to dimethoate, as there was a tremendous increase in the number of vacuoles and dilated ERs. The treated group received dimethoate orally via gavage (21 mg/kg) daily for two months. Brajuskovic et al., [47] reported ultrastructural changes such as a reduction in the size of mitochondria of B-chronic lymphocytic leukaemia (B-CLL). The proliferation of smooth endoplasmic reticulum and the presence of several cytosolic electron dense granules were reported in the liver hepatocytes of two the fish species exposed to contaminated waters [48]. The ultrathin sections of the kidney of a malathion-exposed (dose of 27 mg/kg b.wt/day for one month) rat revealed positive signs of degeneration including narrowing of capillary space and deposition of electron dense material in the podocytes and vacuolation of podocyte cytoplasm with fusion of its foot processes [49]. In a nutshell, these examples clearly reveal that our findings are in line with previous studies that suggest that, like other toxicants, these two insecticides, i.e., Phorate and dimethoate, also induce abnormal degenerative changes in cells.

5. Conclusions

The results of the present study thus revealed that even the sublethal concentrations of Phorate and Dimethoate induced ultrastructural abnormalities in the kidney of the fish, *Cyprinus carpio communis* and TEM serves as a useful and effective tool for detecting the toxicity of any substance. From the discussion, it is quite evident that the changes or the ultrastructural abnormalities observed during the present study clearly revealed classical morphological changes which occur during apoptosis. Additionally, it was clear that these two widely used insecticides are highly toxic, as was supported by the present TEM study, as well as other previous reports on these two insecticides. Hence, the current study is a proof-of-principle that TEM can be used to examine the cellular impact of pesticide exposure. Since both Phorate and Dimethoate have deleterious effects on fish fauna, their indiscriminate use should be strictly prohibited through legislation.

Author Contributions: Conceptualization, S.N. and M.N.A.; methodology, S.N.; software, validation, S.N., J.A.T., B.A.P., A.J., A.G. and M.N.A.; formal analysis, investigation, data curation, writing—original draft preparation, S.N., J.A.T., I.A.B., S.M.P., B.A.P., A.J., A.G. and M.N.A.; writing—review and editing, S.N., J.A.T. and M.N.A.; visualization, supervision, S.N., J.A.T., S.M.P. and M.N.A. All authors have read and agreed to the published version of the manuscript.

Funding: The research received no external funding.

Institutional Review Board Statement: Not applicable.

Informed Consent Statement: Not applicable.

Data Availability Statement: Not applicable.

Acknowledgments: The authors are highly thankful to TC Nag, Department of Anatomy, AIIMS, New Delhi and Sheikh Mansoor Advanced Centre for Human Genetics SKIMS Srinagar for valuable suggestions and guidance. The authors are also highly thankful to P.G. Department of Zoology, University of Kashmir and Centre of Research for Development, University of Kashmir for providing the laboratory facilities. The authors would like to extend their sincere appreciation to the Researchers Supporting Project Number (RSP-2021/144), King Saud University, Riyadh, Saudi Arabia.

Conflicts of Interest: The authors declare no conflict of interest.

References

1. Kafula, Y.A.; Philippe, C.; Pinceel, T.; Munishi, L.K.; Moyo, F.; Vanschoenwinkel, B.; Brendonk, L.; Thore, E.S.J. Pesticide sensitivity of *Nothobranchius neumanni*, a temporary pond predator with a non generic life history. *Chemosphere* **2022**, *291*, 132823. [CrossRef] [PubMed]
2. Insecticide Resistance Action Committee. 2017. Available online: http://www.irac-online.org/documents/moa-classification/ (accessed on 12 May 2019).
3. Banae, M.; Mirvaghefei, A.R.; Rafei, G.R.; Majazi, A.B. Effect of sub lethal Diazinon concentrations on blood plasma biochemistry. *Int. J. Environ. Res.* **2008**, *2*, 189–198.
4. Nwani, C.D.; Lakra, W.S.; Nagpureb, N.S.; Kumar, R.; Kushwaha, B.; Srivastava, S.K. Mutagenic and genotoxic effects of carbosulfan in freshwater fish *Channa punctatus* (Bloch) using micronucleus assay and alkaline single-cell gel electrophoresis. *Food Chem. Toxicol.* **2010**, *48*, 202–208. [CrossRef] [PubMed]
5. Stentiford, G.D.; Longshaw, M.; Lyons, B.P.; Jones, G.; Green, M.; Feist, S.W. Histopathological biomarkers in estuarine fish species for the assessment of biological effects of contaminants. *Mar. Environ. Res.* **2003**, *55*, 137–159. [CrossRef]

6. Dar, S.A.; Yousuf, A.R.; Balkhi, M.H. An introduction about genotoxicology methods as tools for monitoring aquatic ecosystem: Present status and future perspectives. *Fish Aquac. J.* **2016**, *7*, 158.

7. Saquib, Q.; Attia, S.M.; Siddiqui, M.A.; Aboul-Soud, M.A.M.; Al-Khedhairy, A.A.; Giesy, J.P.; Musarrat, J. Phorate-induced oxidative stress, DNA damage and transcriptional activation of p53and caspase genes in male Wistar rats. *Toxicol. Appl. Pharmacol.* **2012**, *259*, 54–65. [CrossRef]

8. Lakshmaiah, G. Toxic impact of lethal and sublethal concentrations of Phorate on the total protein levels in different tissues of common carp *Cyprinus carpio*. *Int. J. Pharma Bio Sci.* **2014**, *5*, 422–428.

9. Abhilash, P.C.; Singh, N. Pesticide use and application: An Indian scenario. *J. Hazard. Mater.* **2009**, *165*, 1–12. [CrossRef]

10. Singh, R.N.; Pandey, R.K.; Singh, N.N.; Das, V.K. Acute toxicity and behavioral responses of common carp *Cyprinus carpio* (linn.) to an organophosphate (Dimethoate). *World J. Zool.* **2009**, *4*, 70–75.

11. Singh, R.N.; Yadava, C.L.; Singh, K. Dimethoate induced alterations in carbohydrate metabolism of common carp, *Cyprinus carpio* (Linn.). *Indian J. Sci. Res.* **2017**, *15*, 175–180.

12. Qayoom, I.; Shah, F.A.; Mukhtar, M.; Balkhi, M.H.; Bhat, F.A.; Bhat, B.A. Dimethoate induced behavioural changes in juveniles of *Cyprinus carpio* var. *communis* under temperate conditions of Kashmir, India. *Sci. World J.* **2016**, *2016*, 4726126. [CrossRef] [PubMed]

13. Lakshmaiah, G.; Indira, P. Investigation of acute toxicity of phorate (an organophosphate) on fresh water fish *Cyprinus carpio*. *Int. J. Pharma Bio Sci.* **2014**, *5*, 104–110.

14. Lakshmaiah, G. Effect of phorate on histopathological changes in the kidney of common carp *Cyprinus carpio* exposed to lethal concentrations. *Int. J. Zool. Stud.* **2016**, *1*, 25–28.

15. Patel, S.K.M.; Indulkar, S.T.; Shaikh, A.L.A.H.; Pai, R. Lethal and sublethal toxicity of an organophosphate pesticide, Phorate 10G on fingerlings of *Tilapia* sp. *Int. J. Pure App. Biosci.* **2017**, *5*, 1153–1157.

16. Lakshmaiah, G. Brain histopathology of the fish *Cyprinus carpio* exposed to lethal concentrations of an organophosphate insecticide phorate. *Int. J. Adv. Res. Dev.* **2017**, *2*, 668–672.

17. Lakshmaiah, G. Effect of phorate lethal concentrations on the histological aspects of liver in common carp *Cyprinus carpio* (Linnaeus, 1758). *Int. J. Chem. Stud.* **2016**, *4*, 6–9.

18. Pandey, R.K.; Singh, R.N.; Singh, S.; Singh, N.N.; Das, V.K. Acute toxicity bioassay of dimethoate on freshwater air breathing catfish, *Heteropneustes fossilis* (Bloch). *J. Environ. Biol.* **2009**, *30*, 437–440.

19. Dogan, D.; Can, C. Hematological, biochemical, and behavioral responses of *Oncorhynchus mykiss* to dimethoate. *Fish Physiol. Biochem.* **2011**, *37*, 951–958. [CrossRef]

20. Ganeshwade, R.M. Biochemical changes induced by dimethoate (Rogor 30% EC) in the gills of fresh water fish (*Puntius ticto* (Hamilton). *J. Ecol. Nat. Environ.* **2012**, *4*, 181–185. [CrossRef]

21. Singh, R.N. Histopathological alterations in the kidney of *Cyprinus carpio* after exposure to dimethoate (EC 30%). *Indian J. Sci. Res.* **2012**, *3*, 127–131.

22. Binukumari, S.; Vasanthi, J. The toxic effect of pesticide dimethoate 30% EC on the protein metabolism of the fresh water fish, *Labeo rohita*. *Int. J. Curr. Microbiol. App. Sci.* **2013**, *2*, 79–82.

23. Singh, R.N. Effects of dimethoate (30%EC) an organophosphate pesticide on liver of common carp *Cyprinus carpio*. *J. Environ. Biol.* **2013**, *34*, 657–661. [PubMed]

24. Singh, R.N. Effects of dimethoate (EC 30%) on gill morphology, oxygen consumption and serum electrolyte levels of common carp, *Cyprinus carpio* (Linn). *Int. J. Sci. Res. Environ. Sci.* **2014**, *2*, 192–198. [CrossRef]

25. Singh, R.N. Testicular toxicity of dimethoate in common carp *Cyprinus carpio* (linn). *Indian J. Sci. Res.* **2017**, *15*, 152–156.

26. Thore, E.S.J.; Philippe, C.; Brendonck, L.; Pinceel, T. Towards improved fish tests in ecotoxicology—Efficient chronic and multi-generational testing with the killifish *Nothobranchius furzeri*. *Chemosphere* **2021**, *273*, 129697. [CrossRef] [PubMed]

27. Norton, W.N. *Acute Exposure of Medaka to Carcinogens: An Ultrastructural, Cytochemical and Morphometric Analysis of Liver and Kidney*; Southeastern Louisiana University Hammond: Hammond, LA, USA, 1992.

28. Anvarifar, H.; Amirkolaie, A.K.; Jalali, A.M.; Miandare, H.K.; Sayed, A.H.; Ucuncu, S.I.; Ouraji, H.; Ceci, M.; Romano, N. Environmental pollution and toxic substances: Cellular apoptosis as a key parameter in a sensible model like fish. *Aquat. Toxicol.* **2018**, *204*, 144–159. [CrossRef]

29. Rocha, C.A.M.D.; Santos, R.A.D.; Bahia, M.D.O.; Cunha, L.A.D.A.; Ribeiro, H.F.; Burbano, R.M.R. The micronucleus assay in fish species as an important tool for xenobiotic exposure risk assessment—A brief review and an example using Neotropical fish exposed to methylmercury. *Rev. Fish. Sci.* **2009**, *17*, 478–484. [CrossRef]

30. Al-Sabti, K.; Metcalfe, C.D. Fish micronuclei for assessing genotoxicity in water. *Mutat. Res.* **1995**, *343*, 121–135. [CrossRef]

31. APHA; AWWA; WPCF. *Standard Method for Examination of Water and Wastewater*, 20th ed.; APHA: Washington, DC, USA, 1998.

32. Philpott, N.J.; Turner, A.J.C.; Scopes, J.; Westby, M.; Marsh, J.C.W.; Gordon-Smith, E.C.; Dalgleish, A.G.; Gibson, F.M. The use of 7-Amino actinomycin D in identifying apoptosis: Simplicity of use and broad spectrum of application compared with other techniques. *Blood* **1996**, *87*, 2244–2251. [CrossRef]

33. Fontanetti, C.S.; Christofoletti, C.A.; Pinheiro, T.G.; Souza, T.S.; Pedro-Escher, J. Microscopy as a tool in toxicological evaluations. *Microsc. Sci. Technol. Appl. Educ.* **2010**, *2*, 1001–1007.

34. Singla, N. An attempt to study ultrastructural changes in kidney of *Ctenopharyngodon idellus* (Cuvier and Valenciennes) induced by monocrotophos using TEM. *Scitech J.* **2015**, *2*, 28–33.

35. Chavan, V.R. Lead induced morphological and biochemical fluctuations in kidney of *Cirrhinus mrigala*. *Res. J. Life Sci. Bioinform. Pharm. Chem. Sci.* **2016**, *2*, 1–11.
36. Rajini, A.; Revathy, K.; Selvam, G. Histopathological changes in tissues of *Danio rerio* exposed to sub lethal concentration of combination pesticide. *Indian J. Sci. Technol.* **2015**, *8*, 1–12. [CrossRef]
37. Khatun, N.; Rahman, T.; Mahanta, R. Histopathological studies of chlorpyrifos toxicity in Catfish. *Glob. J. Med. Res.* **2016**, *6*, 48–54.
38. Besirovic, H.; Alic, A.; Prasovic, S.; Drommer, W. Histopathological effects of chronic exposure to cadmium and zinc on kidneys and gills of brown trout (*Salmo trutta m. fario*). *Turk. J. Fish. Aquat. Sci.* **2010**, *10*, 255–262. [CrossRef]
39. Hadi, A.A.; Alwan, S.F. Histopathological changes in gills, liver and kidney of fresh water fish, *Tilapia zillii*, exposed to aluminum. *Int. J. Pharm. Life Sci.* **2012**, *3*, 2071–2081.
40. Samanta, P.; Pa, S.; Mukherjee, A.K.; Senapati, T.; Ghosh, A.R. Cytopathological effects of Almix herbicide on gill, liver and kidney of *Oreochromis niloticus* under field and laboratory conditions. *Toxicol Open Access* **2016**, *2*, 1–7. [CrossRef]
41. Massar, B.; Dey, S.; Dutta, K. Structural changes in kidneys of common carp (*Cyprinus carpio* L.) inhabiting a polluted reservoir, Umiam in Meghalaya, India. *J. Adv. Microsc. Res.* **2014**, *9*, 105–109. [CrossRef]
42. Marlasca, M.J.; Valles, B.; Riva, M.C.; Crespo, S. Sublethal effects of synthetic dyes on rainbow trout *Oncorhynchus mykiss*: A light and electron microscope study. *Dis. Aquat. Org.* **1992**, *12*, 103–112. [CrossRef]
43. Spazier, E.; Storch, V.; Braunbeck, T. Cytopathology of spleen in eel *Anguilla anguilla* exposed to a chemical spill in the Rhine River. *Dis. Aquat. Org.* **1992**, *14*, 1–22. [CrossRef]
44. Godar, D.E.; Miller, S.A.; Thomas, D.P. Immediate and delayed apoptotic cell death mechanisms: UVA versus UVB and UVC radiation. *Cell Death Differ.* **1994**, *1*, 59–66. [PubMed]
45. Anglade, P.; Vyas, S.; Hirsch, E.C.; Agid, Y. Apoptosis in dopaminergic neurons of the human substantia nigra during normal aging. *Histol. Histopathol.* **1997**, *12*, 603–610. [PubMed]
46. Hagar, H.H.; Azza, H. A biochemical, histochemical, and ultrastructural evaluation of the effect of dimethoate intoxication on rat pancreas. *Toxicol. Lett.* **2002**, *133*, 161–170. [CrossRef]
47. Bajruskovic, G.R.; Anoelija, B.S.; Slobodan, A.M.; Snezana, J.C.; Slavica, F.K. The ultrastructural investigation of mitochondria in B-CLL cells during apoptosis. *Arch. Oncol.* **2004**, *12*, 139–141. [CrossRef]
48. Domingos, F.X.V.; Assis, H.C.S.; Silva, M.D.; Damian, R.C.; Almeida, A.I.M.; Cestari, M.M.; Randi, M.A.F.; Ribeiro, C.A.O. Anthropic impact evaluation of two Brazilian estuaries trough biomarkers in fish. *J. Braz. Soc. Ecotoxicol.* **2009**, *4*, 21–30. [CrossRef]
49. Mossalam, H.H.; Abd-El Aty, O.A.; Morgan, E.N.; Youssaf, S.M.S.; Mackawy, A.M.H. Biochemical and ultra structural studies of the antioxidant effect of aqueous extract of *Hibiscus sabdariffa* on the nephrotoxicity induced by organophosphorous pesticide (malathion) on the adult albino rats. *Life Sci. J.* **2011**, *8*, 561–574.

Journal of
Xenobiotics

Article

An Ecological Approach to Control Pathogens of *Lycopersicon esculentum* L. by Slow Release of Mancozeb from Biopolymeric Conjugated Nanoparticles

Ravinder Kumar [1,*], Vikash Nain [2] and Joginder Singh Duhan [1,*]

[1] Department of Biotechnology, Chaudhary Devi Lal University, Sirsa 125055, India
[2] Department of Food Science and Technology, Chaudhary Devi Lal University, Sirsa 125055, India
* Correspondence: rsulakh@gmail.com (R.K.); duhanjs68@gmail.com (J.S.D.); Tel.: +91-9416072588 (R.K.); +91-9416725009 (J.S.D.)

Abstract: To control insects, weeds, and infections in crops, old-fashioned pesticide formulations (with massive quantities of heavy metals and a variety of chemicals) are used. By biological amplification via the food chain, many of these established pesticide formulations have accumulated in living systems and caused environmental pollution. To form a nanoparticulate matrix with a diameter ranging from 322.2 ± 0.9 to 403.7 ± 0.7 nm, mancozeb was embedded in chitosan–gum acacia (CSGA) biopolymers and loadings were confirmed via TEM and FTIR. Differential scanning calorimetry analyses were carried out as part of the investigation. Inhibition of *Alternaria alternata* by nanoparticles (NPs) with 1.0 mg/mL mancozeb (CSGA-1.0) was 85.2 ± 0.7 % at 0.5 ppm, whereas for *Stemphylium lycopersici* it was 62.1 ± 0.7% in the mycelium inhibition method. NPs demonstrated antimicrobial action in pot house environments. After ten hours, the mancozeb was liberated from the nanoformulations due to polymer matrix diffusion and relaxation, compared to 2 h for commercial mancozeb. Even while drug-loaded conjugated nanoparticles have equivalent antifungal activities, they have a lower release rate and, hence, reduced toxicology compared to commercial mancozeb. Therefore, this method can be employed to implement sustainable farming techniques in the future.

Keywords: chitosan–gum acacia nanocomposite; tomato pathogens; slow-release of mancozeb; eco-friendly; biopolymers

Citation: Kumar, R.; Nain, V.; Duhan, J.S. An Ecological Approach to Control Pathogens of *Lycopersicon esculentum* L. by Slow Release of Mancozeb from Biopolymeric Conjugated Nanoparticles. *J. Xenobiot.* **2022**, *12*, 329–343. https://doi.org/10.3390/jox12040023

Academic Editors: François Gagné, Stefano Magni and Valerio Matozzo

Received: 4 October 2022
Accepted: 5 November 2022
Published: 9 November 2022

Publisher's Note: MDPI stays neutral with regard to jurisdictional claims in published maps and institutional affiliations.

1. Introduction

Old-fashioned pesticide preparations involve enormous amounts of heavy metals and diverse chemicals that are used to control agricultural insects, weeds, and pathogens. Conversely, many of these established pesticide formulations have mounted up in living systems through biological magnification via the food chain. Thus, they also contaminate soil and water atmospheres, harm living entities, and cause disturbances in the equilibrium of the ecological unit [1]. Approximately 90% of applied agrochemicals are lost as run-off during the application, affecting both the environment and farmers' application costs [2]. Pest resistance upsurges due to indiscriminate pesticide use, which reduces soil biodiversity, kills beneficial soil bacteria, causes biomagnification pollinator decreases, and eliminates the bird's natural habitat [3]. Tomato (*Lycopersicon esculentum*) is a major worldwide crop that contributed over IND 218 billion to the Indian economy in 2018. The tomato crop is susceptible to various fungal infections throughout pre-harvest and post-harvest periods (https://www.statista.com/statistics/1080566/india-economiccontribution-of-tomatoes/, accessed on 7 September 2022). Primary pathogenic fungi that impact tomato plant growth and development are mainly *A. alternata* and *S. lycopersici*. Mancozeb, a broad-spectrum contact fungicide, has shown strong fungicidal activity in various horticultural crops and accounted for 20% of the global fungicide market (Fungicides Market, 2017–2025 (https://www.prnewswire.com, accessed on 7 September

2022). The fungicide disrupts lipid metabolism and respiration by acting on the sulfhydryl groups of amino acids and enzymes in fungal cells. It is easily water-soluble and can pollute water bodies; its stability is affected by conditions, including light, temperature, humidity, and pH [4]. The maximum residue limit for mancozeb (dithiocarbamates) in living matter is 0.01–25 ppm [5], yet residue levels higher than this have been found in tomatoes [6].

These flaws can be mitigated by encasing the chemical fungicide in a polymer casing, such as chitosan or gum acacia, for long-term release. They are important for encapsulating active ingredients because of their biocompatibility and capacity to encapsulate molecules with diverse physicochemical properties [7]. Chitosan is a natural biodegradable polymer made from the deacetylation of chitin (found naturally in crustaceans, insects, and mollusks) and has a wide range of antifungal, antibacterial, and medicinal uses. It has been recognized as a plant growth stimulant [8]. Chitosan and its nanoparticles (NPs) can also control infections and prevent crop pathogens in cereals and horticulture crops by activating defense-related enzymes [9–11]. Gum acacia is a natural gum made from the hardened sap of acacia species. It is a complex mixture of glycoproteins and polysaccharides, with arabinose and galactose as the primary sugars. Because of its viscous nature, it is a non-ionic, non-toxic, and biocompatible bio-polymer utilized in drug delivery [11].

Nanomaterials are synthesized using various entities, manipulated, and utilized to treat various plant ailments [12–14]. According to researchers, microorganisms have been identified as potential eco-friendly nano-factories for managing plant diseases [15–18]. Nanotechnology balances minimal concentrations, maximum pest control, and safe concentrations, resulting in lowering pest management costs [19]. NPs have large surface area to volume ratios and are easily absorbed. Compared to parent formulations, nano-encapsulation technology has been used for commercial pesticides, promising increased potency in precise discharges, targeted deliveries, and environmental and physical stability [20,21]. The generation of reactive oxygen species (ROS), which causes oxidative stress and cell death, is thought to be linked to the antimicrobial effects of NPs [22].

This study described an eco-friendly alternative to using harmful chemicals (commercial mancozeb fungicide) to control phytopathogenic tomato fungi and simultaneously help control environmental pollution and soil health. This may be due to the sustained-release behavior of fungicides from biopolymeric chitosan–gum acacia nanocomposites and presents advanced knowledge of the existing methods and techniques of pesticide applications with reduced leaching.

2. Materials and Methods

2.1. Reagents and Materials Used

Sigma-Aldrich (St. Louis, MO, USA) provided the chitosan (deacetylation 75%) and sodium tripolyphosphate (TPP). Qualigen, India, supplied gum acacia. Fungicide mancozeb was acquired locally. Hi-Media, Ltd., Mumbai, provided the dialysis tubing. Tomato seeds (Arun Hisar) were bought from the state agricultural university's vegetable section. A polyhouse was used to conduct an in vivo pot experiment. The Vero cell line was kept at the National Research Centre on Equines, Hisar.

ITCC, New Delhi, provided the pathogenic fungi (*Alternaria alternata* and *Stemphylium lycopersici*). The cultures were resurrected according to the manufacturer's instructions.

2.2. Synthesis of Blank and Mancozeb Loaded Chitosan–Gum Acacia Conjugated Nanoparticles

Ionic gelation and polyelectrolyte complexation were used to make conjugated nanoparticles. Liquidizing 1.0 mg/mL of chitosan in 1.0% glacial acetic acid (*v/v*) and stirring overnight to dissolve entirely yielded a stock of chitosan at a native pH. Different amounts of solid mancozeb were introduced in three conical flasks of 100 mL capacities, along with a 20 mL stock of chitosan in each flask. Adding a stock solution of aqueous gum acacia (1.0 mg/mL, native pH) to each of the above flasks, dropwise, resulted in final concentrations of mancozeb of 0.5, 1.0, and 1.5 mg/mL, respectively. Magnetic stirring continued for another ten minutes. After that, dropwise, we added 2.0 mL of 1.0% TPP to the solution and stirred for 45 min. Each flask received 10 µL of Tween-20; we stirred the flasks for another 30 min. The suspension was centrifuged at 12,000 rpm for 20 min; the pellet was rinsed with 10 mL of double-distilled water (DDW), and the pellet was kept at 4 °C in a 1.5 mL centrifuge vial for future investigation.

Except for the inclusion of mancozeb, all of the previous processes were followed to synthesize the blank NPs.

2.3. Optimization of Experimental Design

Design-Expert Software (Version 8.0.4, Stat-Ease, Inc., MN, USA) was used to optimize the experiment. Mancozeb-loaded chitosan–gum acacia NPs were optimized using a standard protocol via a central composite design. Three factors, i.e., concentrations of chitosan, gum acacia, and mancozeb, varied, and TPP concentration was constantly retained. The particle size was chosen as the response variable.

2.4. Characterization

2.4.1. Size, Polydispersity Index, and Zeta Potential

The synthesized conjugated NPs were characterized by a particle size analyzer (PSA) using Zetasizer Nano ZS90 (Malvern Instrumentations, Malvern, UK). Moreover, 50 µL NPs were dispensed in a cuvette and disseminated in a 950 µL DDW to measure the percentage intensity at 25 °C.

2.4.2. Fourier Transform Infrared (FTIR) Spectroscopy

At 80 °C under 0.4 bar, the particles were freeze-dried using a freeze-dryer (Christ, Germany). As a cryoprotectant, mannitol (10%) was utilized. With a fine powder of 5.0 mg, Fourier transform infrared (FTIR) spectroscopy was accomplished. At an average temperature between 4000 and 400 cm^{-1}, FTIR spectra were documented using potassium bromide (KBr) in a 1:10 ratio with an AVATAR 370 FTIR (Therma Nicolet, San Jose, CA, USA). A graph pattern was employed to determine the ionic interaction between samples to measure mancozeb loading. Spectroscopic Tools, 2019 (Thomas St. (http://www.science-and-fun.de/tools/, accessed on 7 September 2022) was used to evaluate the data.

2.4.3. Transmission Electron Microscopy (TEM)

The Tecnai^{TM} (FEI, Hillsboro, OR, USA) TEM was used to confirm the mancozeb loading. Before analysis, samples were placed in a water bath sonicator. A drop of diluted conjugated nanoparticles was placed on a carbon-coated copper grid, followed by room-temperature air-drying. The photos were acquired using a 200 kV operational voltage using a facility from SAIF, AIIMS, New Delhi.

2.4.4. Differential Scanning Calorimetry (DSC)

At LPU, Jalandhar's CIF facility, a 3.0 mg sample was utilized to make thermographs using a DSC 4000 System (Perkin Elmer, Waltham, MA, USA) with a heating and cooling rate of 10 °C/min. Samples were heated from 30 to 445 °C. Pure nitrogen gas (99.99%) was pumped into the system at a rate of 20 mL/min.

2.5. In Vitro Study

2.5.1. Encapsulation Efficiency (EE) and Loading Capacity (LC)

The supernatant was collected in a sterile tube after the nanocomposites were centrifuged at 15,000 rpm for 35 min. NanoDrop2000c (Thermo Scientific, Wilmington, DE, USA) was used to acquire mancozeb in the supernatant at 290 nm using the supernatant of their equal blank-attached nanoparticles as simple adjustments. The following equation was used to compute the encapsulation efficiency (EE%):

Total mancozeb added-mancozeb in supernatant/total mancozeb added $\times 100 =$ Encapsulation efficiency (percentage).

The following equation was used to compute the loading capacity (LC%):

$$\text{(Mass of mancozeb in CSGA conjugated NPs)/(Weight of CSGA conjugated NPs recovered)} \times 100 \quad (1)$$

2.5.2. Slow-Release Profile of Conjugated CSGA NPs

Mancozeb's in vitro release from CSGA-conjugated nanoparticles was derived via dialysis tubing (Hi-Media Ltd., Mumbai). A total of 10 mg of the sonicated nanocomposite was added to 1.0 mL of sterile phosphate-buffered saline, PBS (pH 7.2). This was dished in a dialysis membrane with closed clips at one end and immersed in 10 mL of the same PBS in separate tiny beakers. At 37 °C, it was incubated in a shaker at 160 rpm; 1.0 mL of phosphate-buffered saline was pipetted out of each beaker after predetermined intervals of time (30 min), and the same volume of buffer was supplied to each beaker. The absorbance of the resultant solution was measured at 290 nm to estimate the quantity of mancozeb in the buffer using the standard curve.

2.5.3. Antimicrobial Activity

The impact of NPs on the mycelial growth of a specific pathogen was established using the mycelium inhibition technique on potato dextrose agar (PDA, 2%) with certain modifications [23]. To make the final concentration of conjugated nanoparticles as 0.5, 1.0, and 1.5 ppm, autoclaved potato dextrose agar medium (Hi-Media, Mumbai, India; temp. 40 °C) was placed onto separate sterile Petri plates (90 mm × 15 mm) and allowed to harden. A 5.0 mm mycelial disc was removed from a seven-day-old test pathogen culture and placed in the center of the test Petri dish, where it was incubated at 28 ± 1 °C under constant monitoring. Growth was measured after four days for three replications, and the treated plates were compared to the control plates (without nanocomposites) to calculate the percent suppression of mycelia by NPs using an earlier approach [24].

$$dc - dt/dc\ 100 = \text{percent inhibition} \quad (2)$$

where dc is the control mycelial diameter and dt is the mycelial treatment diameter.

2.6. In Vivo Study

2.6.1. Bioefficacy in Pot House Conditions

To determine the bioefficacy of polymeric conjugated nanoparticles in controlling early blight and leaf spots in tomatoes (*Lycopersicon esculentum* L.), the study was conducted in pots filled with sandy soil in a glasshouse in natural light and temperature.

2.6.2. Treatment of Seeds and Disease Detection

Seeds were properly rinsed and treated with 4% sodium hypochlorite for 10 min before being thoroughly cleaned. The seeds were dipped in CMC (carboxymethyl cellulose, 5.0 g in 100 mL DDW) for 10 min and air-dried. The seeds were then air-dried after being treated with conjugated nanoparticles (10 ppm) for 2 h and 30 min. Five tomato seeds per pot were planted in pots filled with soil (pH 7.7 at 20 °C) infected with pathogenic fungus [13]. The forty-day-old plants were sprayed with 15 mL of aqueous conidial solution (3.1×10^7 CFU/mL) of specific pathogens and covered with clear plastic bags to maintain the humidity essential for disease outbreaks. Following the disease outbreak, a foliar spray of CSGA NPs (10 ppm and 15 mL/pot) was used to test the bioefficacy. As a positive control, commercial mancozeb was employed.

Disease severity (DS) was recorded randomly in the standard grade of 0–5 before the polymeric NP spray.

$$\begin{aligned} \text{Disease severity (\%DS)} &= \\ \text{(Sum of all disease ratings)/(Total plants assessed} &\times \text{ maximum rating scle)} \times 100 \end{aligned} \quad (3)$$

The disease control efficacy (% DCE) was calculated after the polymeric NP spray using the formula in [25].

For the overall health and vitality of the test plant, bioefficacy was measured using plant development characteristics, such as plant height, root–shoot ratio, and dry biomass. The dry weight of each tomato per plant was determined by placing the entire plant with roots in a brown envelope (3 plants per envelope) and drying it for seven days at 40 °C in a hot air oven.

2.7. Statistical Treatment of Data

All experiments were executed in triplicate, and the results are presented as mean $\pm$ standard deviation (SD). Statistical variances among sets were determined using one-way ANOVA. The statistical significance was accepted at a level of p-value ≤ 0.05 by a t-test. To handle statistical data, Microsoft Office Excel 2013 was utilized (Microsoft Corporation, Albuquerque, NM, USA).

3. Results and Discussion

3.1. Nanoparticle Size Optimization, Stability, and Physicochemical Characterization

The quantity of chitosan and gum acacia used in the experiment affected the particle sizes of the NPs in the initial trials. The optimization graph shows that the particle size increases with increasing chitosan and gum acacia concentrations (Figure 1a). In the case of chitosan, however, the impact is more pronounced.

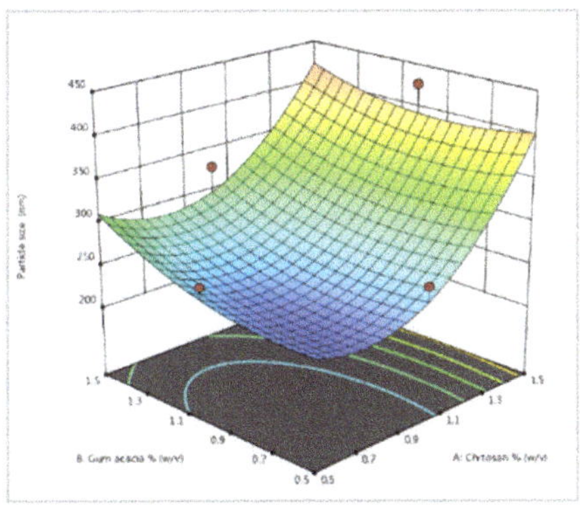

(a)

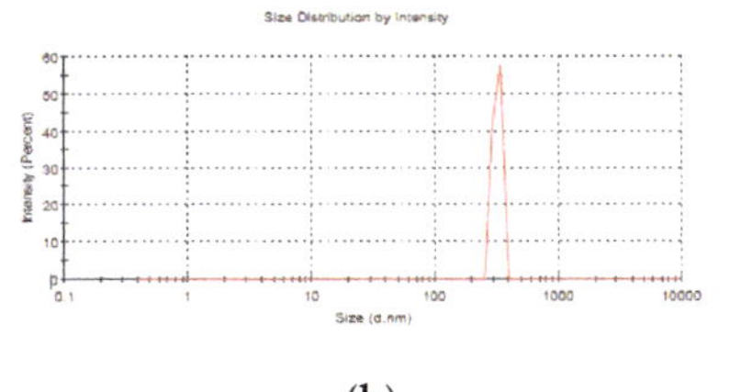

(b)

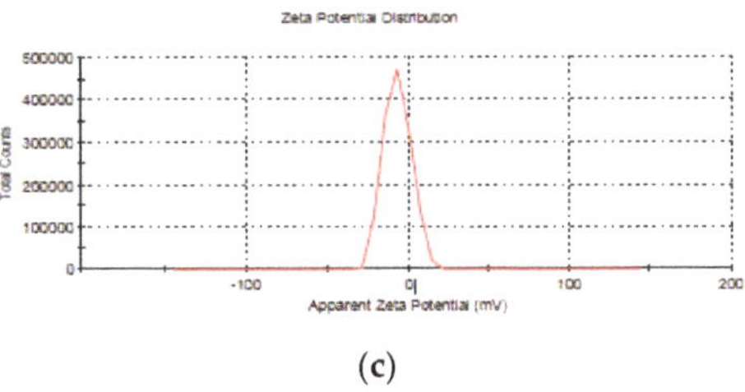

(c)

Figure 1. (**a**) Optimization of the concentration of gum acacia and chitosan for the particle size by response surface methodology (RSM); (**b**) mancozeb-loaded NP size; (**c**) mancozeb-loaded NP zeta potential.

The average diameter of blank NPs was 322.2 $\pm$ 0.9 nm, a 1.00 $\pm$ 0.1 PDI, and a zeta potential of -23.2 ± 0.08 mV. The size of NPs grew in response to increasing mancozeb concentrations (Table 1).

Table 1. CSGA NP size, PDI, and zeta potential at native pH; freshly prepared, and their storage stability after 20 days in DDW at 4 °C.

CSGA NPs	Size (nm)	PDI	Zeta Potential (mV)
Freshly prepared nanoparticles			
CSGA Blank	322.2 ± 0.9	1.00 ± 0.1	−23.2 ± 0.08
CSGA-1.0	403.7 ± 0.7	0.789 ± 0.1	−6.99 ± 0.5
Storage stability of conjugated nanoparticles at 4 °C			
CSGA Blank	343.9 ± 0.5	0.255 ± 0.3	21.3 ± 0.4
CSGA-1.0	413.1 ± 0.8	0.301 ± 0.7	25.0 ± 0.3

Mean ± standard deviation in replication of three.

The molecular weight, degree of deacetylation of the chitosan employed, the stirring speed, and time determine the size of the NPs [26]. A single strong peak at 403.7 ± 0.7 nm was detected for CSGA-1.0 (NPs containing 1.0 mg/mL mancozeb) with a zeta potential of −6.99 ± 0.5 mV (Figure 1b,c). Zeta potential values show the stability of NPs up to ±30 mV [27]. The nanocomposites formed were noticed in the synthesis mixture as a white, foggy haziness that settled at the bottom of the flask.

3.2. Storage Stability Determination

By preserving NPs in double-distilled water (DDW) at 4.0 °C for twenty days, the storage stability of the particles was established. The CSGA-1.0 size changed slightly from 403.7 ± 0.7 to 413.1 ± 0.8 nm (Table 1), which might be related to NP clumping since size differences in aggregated systems are common for various causes [28]. The zeta potential values obtained in this investigation ranged from +15.4 to +25.4 mV, indicating their stability.

The CSGA-1.0 formulation will be chosen for future research in vivo and in vitro because of its smaller size, higher loading capacity, and better storage stability.

3.3. Ionic Group Interaction Study Using Fourier Transform Infrared Spectroscopy (FTIR)

Firm peaks indicate asymmetrical NH2 stretching vibration in CSGA-conjugated NPs at 3455.94 cm^{-1}, whereas O-H stretching vibration is indicated by peaks at 3297.16 cm^{-1} (Figure 2a,b). In earlier research [29] with CSCRG NPs, it was found that a high peak at 3444.75 cm^{-1} in raw chitosan polymer suggested an asymmetrical NH_2 stretching vibration [30]. Peaks in gum acacia powder indicated O-H stretching vibration at 3455.94 cm^{-1}, C-H stretching vibration at 2929.66 cm^{-1}, and o-amino- or o-hydroxyaryl ketones at 1635.25 cm^{-1} [31]. The peak at 1071.13 cm^{-1}, on the other hand, indicated CH_3 rocking vibration. Due to H-bonding between the COOH group of gum acacia and the NH_2 of chitosan, the peak at 3400–3500 cm^{-1} seems to be wide. The fungicide loading within the nanocomposite is shown as a peak at 1635.17 cm^{-1} in mancozeb-loaded CSGA NPs. This is also supported by a previous study in which a researcher validated the encapsulation of acetamiprid in a nanoform by seeing a peak at 1633 cm^{-1}, which corresponds to the stretching vibration of C=N/C-N [27].

3.4. Transmission Electron Microscopy (TEM)

TEM results (Figure 2c,d) confirmed the formation of discrete, small, round conjugated nanoparticles with fungicide seen inside the NPs as a dark gray spot for loaded NPs [32]. TEM micrographs verified crystalline nanocomposite formation and better understood the biopolymeric nanoparticle's shape and size. The TEM pictures in this study show that the formed NPs were well dispersed, designating a monodispersed nature of synthesized nanoparticles and following previous research [27].

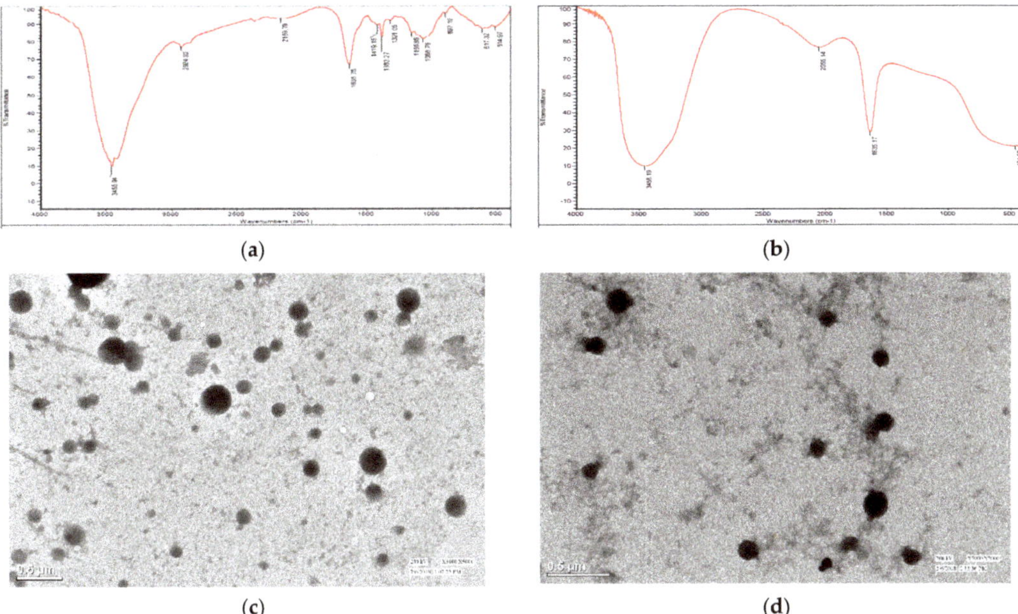

Figure 2. Fourier transform infrared spectra and TEM micrographs of conjugated nanoparticles; FTIR spectra of (**a**) gum acacia, (**b**) mancozeb-loaded NPs, (**c**) TEM micrograph (at 200 kV, ×5000 magnification, 0.5 μm scale bar) of blank NPs, (**d**) TEM of mancozeb-loaded NPs (at 200 kV, X7000 magnification, 0.5 μm scale).

3.5. Differential Scanning Calorimetry (DSC)

The down peak in the graph represents an endothermic reaction, and the up peak represents an exothermic reaction (Figure 3). In mancozeb-loaded chitosan–gum acacia NPs, a sharp endothermic peak is observed at 190.95 °C with an enthalpy change (ΔH) of 99.406 J/g, which may be attributed to the melting of mancozeb [33]. Commercial mancozeb around 180 °C has a ΔH of 159.8806 J/g (Figure 3e). In contrast, blank and drug-loaded polymeric NPs have lower ΔH of 33.3051 J/g and 99.406 J/g, respectively (Figure 3c,d), which may be due to the glass transition of CSGA nanocomposites [34]. The lower ΔH, as compared to commercial mancozeb, makes both nanoformulations (blank and mancozeb-loaded NPs) more thermally stable than commercial mancozeb [29].

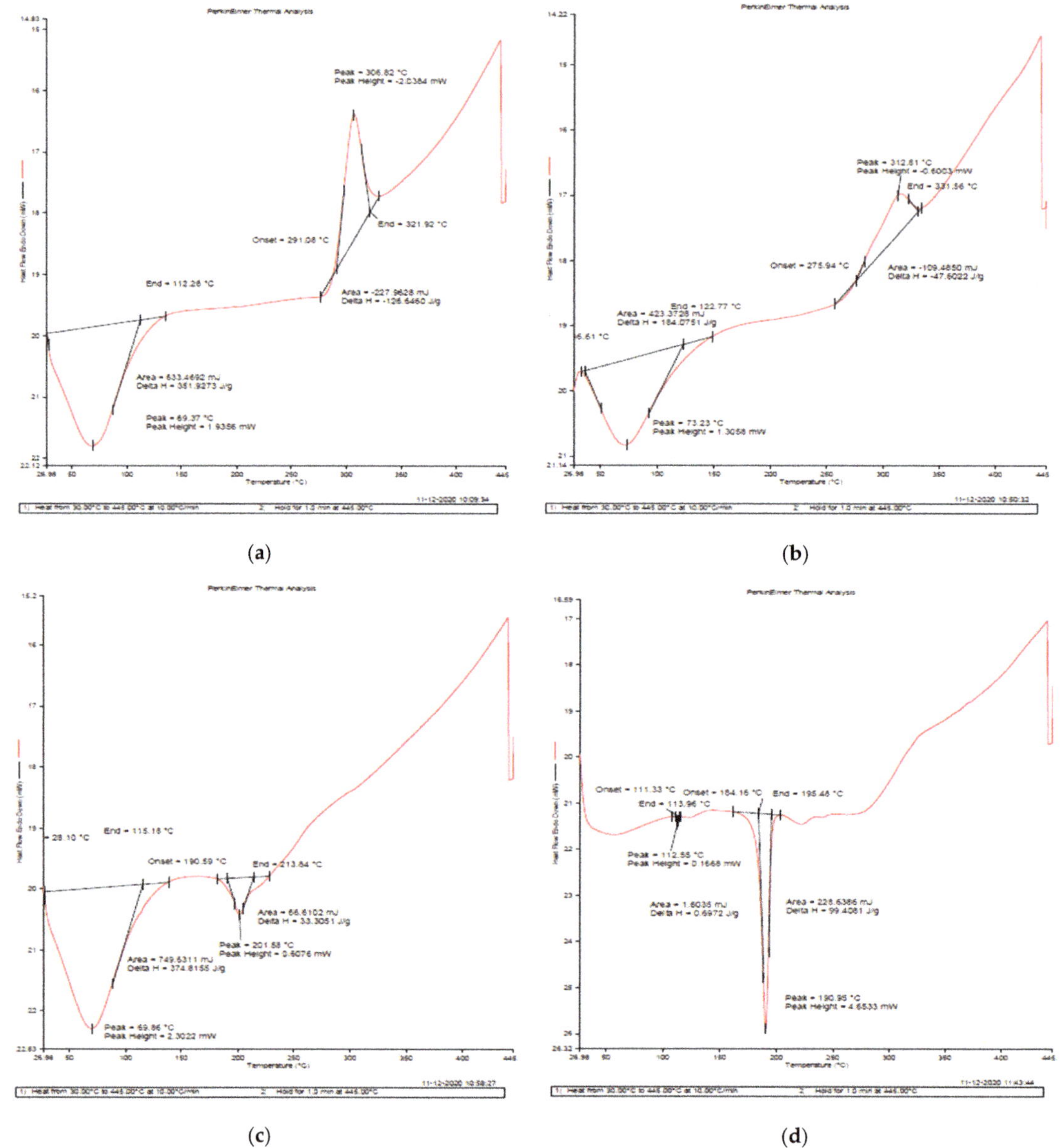

Figure 3. *Cont.*

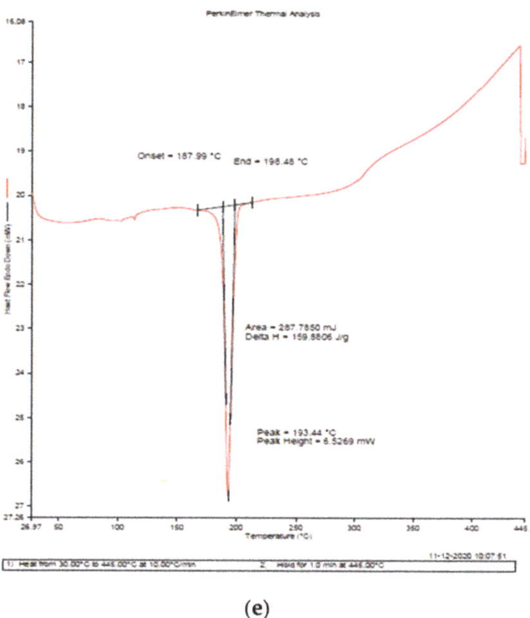

(e)

Figure 3. DSC thermograms of raw biopolymeric materials, (**a**) chitosan, (**b**) gum acacia and test samples, (**c**) CSGA blank, NPs (**d**) CSGA-1.0 NPs, (**e**) mancozeb.

3.6. Encapsulation Efficiency and Loading Capacity

The encapsulation efficiency (EE) percentage and the loading capacity (LC) percentage of all three loaded nanoforms are shown in Table 2. A concentration-dependent pattern in the EE (%) and non-concentration-dependent increase in the percentage LC of mancozeb-loaded CSGA-conjugated nanoparticles was observed. Minimum and maximum EEs recorded were 15.95 ± 0.25 and 57.13 ± 0.29 for the formulations CSGA-0.5 and CSGA-1.5, respectively, with LC of $76.25 \pm 0.26\%$, which might be due to excess mancozeb in the CSGA-1.5 sample [35]. Maximum LC (81.10 ± 0.18) was found for NPs containing 1.0 mg/mL mancozeb sample with optimum fungicide available for loading; results of the present study followed an earlier study [23,36].

Table 2. Mancozeb encapsulation and loading capacity of CSGA NPs.

Formulation	Encapsulation Efficiency (%)	Loading Capacity (%)
CSGA-0.5	15.95 ± 0.25	47.57 ± 0.39
CSGA-1.0	36.62 ± 0.31	81.10 ± 0.18
CSGA-1.5	57.13 ± 0.29	76.25 ± 0.26

Mean $\pm$ standard deviation in replication of three.

3.7. Controlled Release Behavior

The drug release from the conjugated nanoparticles was time-dependent and sustained-release owing to diffusion (Figure 4). The commercial fungicide released 100% within two hours, while NPs containing 0.5 mg/mL mancozeb (CSGA-0.5) had the best release mechanisms within 8 h, releasing 64% of total mancozeb content, while CSGA-1.5 formulation released 81% of total mancozeb content for the same time [37]. Because the active component of mancozeb is confined in the center of the polymer core, the release rate of nanoform is slower than that of commercial mancozeb [38,39]. Compared to commercial fungicides, this in vitro gradual release of NPs boosts plant absorption, reduces soil leaching, and reduces soil and water pollution. In a comparative investigation, the cumulative release percentage for carbendazim-loaded NPs was 50.4 ± 0.13 percent, compared to $67.7 \pm 0.1\%$ for pure carbendazim at pH 7.4 [40]. Another researcher [26] used an ionic gelation approach to collect marketable hexaconazole in chitosan NPs and obtained a sustained release of 99.91% over 86 h. There may be some chemical bonding between chitosan, gum, and fungicide as the polymers are oppositely charged, and fungicides also have both positive and negative charges, enhancing the release time. This enhanced interaction was also seen as a biocidal additive in paints and coatings [41].

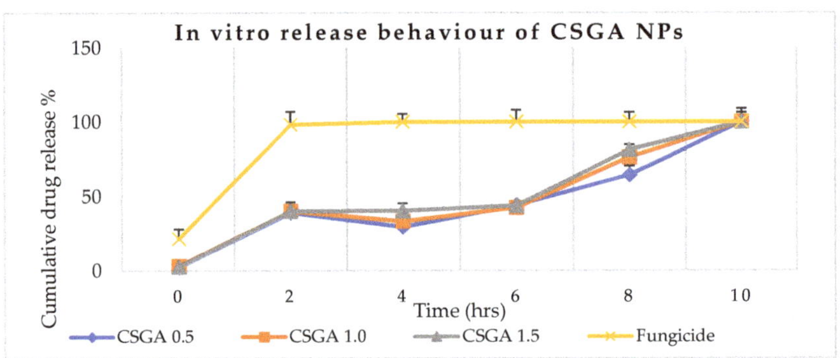

Figure 4. In vitro sustain release of mancozeb from chitosan–gum acacia-conjugated nanoparticles (CSGA). The yellow line represents commercial fungicide, while blue, saffron, and gray represent CSGA-0.5, 1.0, and 1.5, respectively.

3.8. In Vitro Antifungal Activity

Maximum inhibition ($85.2 \pm 0.7\%$) was seen in the case of *Alternaria alternata* in a mancozeb-loaded formulation at 0.5 ppm and is almost the same as commercial mancozeb at 1.5 ppm (Figure 5a), which might be attributed to the formulation's increased zeta potential. Mancozeb-loaded CSGA NPs had the best antimicrobial efficacy against *S. lycopersici* at 1.0 and 1.5 ppm (100% inhibition at both concentrations). These findings matched prior research in which carbendazim-loaded polymeric nanoparticles were evaluated against *F. oxysporum* and *A. parasiticus* and showed 100% fungal suppression at 0.5 and 1.0 ppm [23]. In contrast, it exhibited $62.1 \pm 0.7\%$ inhibition at 0.5 ppm, equivalent to commercial mancozeb ($56.1 \pm 0.7\%$). Table 3 and Figure 5a,b) indicate the percentage inhibition shown by NPs.

Table 3. In vitro antifungal efficacy of mancozeb-loaded (1.0 mg/mL) CSGA-1.0 nanocomposites.

Fungi	Nanoformulation with Mancozeb (ppm)	CSGA NPs Fungi Diameter (mm)	NPs % Inhibition = dc − dt/dc × 100	Mancozeb (ppm)	Mancozeb Fungi Diameter (mm)	Mancozeb % Inhibition = dc − dt/dc × 100
	Blank NPs, N 1.0	18.5 ± 0.45	76.1 ± 0.22 b		-	-
A. alternata	Loaded NPs, NF 0.5	11.5 ± 0.23	85.2 ± 0.61 b	F 0.5	12.0 ± 1.4	84.5 ± 1.4 b
(ITCC6343)	Loaded NPs, NF 1.0	15.5 ± 0.71	80.0 ± 0.42 b	F 1.0	11.5 ± 0.7	85.2 ± 0.7 b
	Loaded NPs, NF1.5	12.5 ± 0.96	83.9 ± 0.54 b	F 1.5	10.5 ± 0.7	86.5 ± 0.7 b
	Blank NPs, N 1.0	15.5 ± 0.7	53.0 ± 0.7		-	-
S. lycopersici	Loaded NPs, NF 0.5	12.5 ± 0.7	62.1 ± 0.7	F 0.5	14.5 ± 0.7	56.1 ± 0.7 c
(ITCC5431)	Loaded NPs, NF 1.0	0	100 ± 0 a	F 1.0	0	100 ± 0 a
	Loaded NPs, NF 1.5	0	100 ± 0 a	F 1.5	0	100 ± 0 a

Each value is a mean of a triplicate. Mean $\pm$ standard deviation followed by the same letter in the treatment column indicate that values are not significantly different at $p \leq 0.05$, as determined by a *t*-test.

Nanofungicides have better efficiency than traditional fungicides owing to the formation of reactive oxygen species (ROS), membrane permeability due to their tiny size, and enzyme-related defense responses in plants against fungi [22,42]. Another study obtained more substantial repellents (around 80%) against two key agricultural pests (whitefly (*Bemisia tabaci*) and spider mites) while working with a hydrogel containing botanical repellents encapsulated in zein nanoparticles [43]. In previous research, *Rhizoctonia* sp. and *Alternaria* sp. strains were shown to be the most susceptible to chitosan–zinc nanocomposites, followed by chitosan–copper nanocomposites [24].

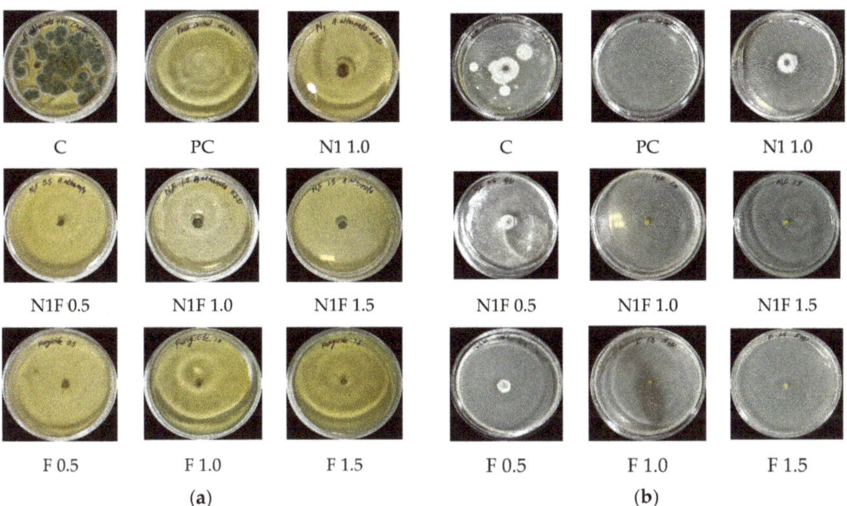

Figure 5. In vitro antifungal efficacy of blank- and fungicide- (1.0 mg/mL) loaded CSGA NPs at three concentrations (0.5, 1.0, and 1.5 ppm) using the mycelium inhibition method against tomato pathogens (**a**) *A. alternata*, (**b**) *S. lycopersici*; C—control, PC—pure control (Petri plates with PDA alone), N1—blank CSGA NPs, N1F—fungicide, F—fungicide.

3.9. In Vivo Antifungal Efficacy

The control plants exhibited a disease severity (DS) of 42.9 ± 3.3% for early blight and 40.9 ± 0.8% for leaf spot disease after seven days of conidial spraying of a pathogenic fungus, as shown in Table 4. Disease control efficiency (DCE, in percent) in pathogen-treated plants for loaded NPs was found to be 66.2 ± 5.0% and 70.7 ± 1.6%, respectively, against early blight and leaf spot diseases, which is comparable to sick pathogen plants treated with commercial fungicide (66.0 ± 3.5% and 68.5 ± 1.1%) (Table 4). Similar to metolachlor alone, polymeric (PEG-PLGA) nanoparticles loaded with the weed killer metolachlor showed strong herbicidal action against *Digitaria sanguinalis*, *Oryza sativa*, and *Arabidopsis thaliana* in the research [44].

Table 4. Treatment effects of CSGA NPs on percentage disease severity (DS) and disease control efficacy (DCE) in tomato plants under pot house conditions against early blight and leaf spots.

Treatment	A. alternata		S. lycopersici	
	% DS	% DCE	% DS	% DCE
Pure control, C	16.1 ± 1.4	00.0 ± 0.0	12.7 ± 1.5	00.0 ± 0.0
Control + Pathogen, CP	42.9 ± 3.3	00.0 ± 0.0	40.9 ± 0.8	00.0 ± 0.0
Fungicide, F	10.1 ± 1.9	76.5 ± 5.8 a	10.2 ± 1.8	75.1 ± 1.8 a
Fungicide + Pathogen, FP	14.6 ± 3.4	66.0 ± 3.5 b	12.9 ± 2.3	68.5 ± 1.1 b
Blank NPs, N1	12.9 ± 0.5	69.9 ± 3.7 c	16.0 ± 1.7	60.9 ± 1.9 c
Blank NPs + Pathogen, N1P	15.6 ± 3.4	63.6 ± 1.5	15.0 ± 2.8	63.3 ± 4.6
Loaded NPs, N1F	10.0 ± 1.2	76.7 ± 3.4 a	10.5 ± 1.2	77.3 ± 1.6 a
Loaded NPs + Pathogen, N1FP	14.5 ± 1.4	66.2 ± 5.0 a	12.0 ± 2.9	70.7 ± 1.6 a

Each value is a mean of a triplicate. Mean ± standard deviation followed by the same letter in the treatment column indicate that values are not significantly different at $p \leq 0.05$, as determined by a *t*-test.

In the pot settings, chitosan nanoparticles, and chitosan–silver nanocomposites were as efficient as traditional fungicides, such as copper oxychloride 525]. Another study used silver nanoparticles produced by rhizospheric chickpea microorganisms to treat chickpea wilt illness in vivo (*Cicer arietinum*). The current investigation of disease inhibition is in line with earlier research [15]. Earlier, applying chitosan hexaconazole–dazomet to treat *G. boninense*-caused basal stem rot disease in palm trees reduced the disease by 74.5% [45]. These findings corroborate the validity of the current research.

3.10. Plant Growth Parameters Study

For this study, three parameters, namely germination percentage, dry mass per plant, and root-shoot ratio, were chosen for the pot house conditions.

3.10.1. Effect of Treatment on Germination Percentage

The tomato is an important crop sown globally. No comprehensive research data are available on the effects of chitosan–gum acacia nanoparticles on plant germination indices in pot or field conditions. The lowest germination (60%, in both) was observed in plants treated with fungicide alone (F) and plants treated with blank NPs, which were made sick with the pathogen *A. alternata* (N1P1). The highest germination (78%) was recorded in plants treated with *S. lycopersici* and loaded NPs (N1FP2). It was pretty high as compared to plants treated with commercial mancozeb alone (F-60%), plants treated with pathogen *S. lycopersici* alone (CP2-72%), and fungicide-treated plants, which were made sick with *S. lycopersici* (FP2-62%). From the above data, it can be conferred that the loaded NPs (N1FP2) treated with *S. lycopersici* showed good germination (78%) compared to commercial mancozeb (60%). The same pattern of germination in the present study was found in earlier research, where the germination of the seeds treated with nanoformulation was 96%, while pure carbendazim showed decreased (60%) germination [23].

3.10.2. Tomato Dry Mass per Plant

The highest mass of 1328.3 mg was obtained for plants treated with fungicide-loaded NPs, and made sick with *S. lycopersici* (N1FP2), followed by 1102.5 mg for fungicide-treated plants made sick with *S. lycopersici* (FP2). Plants treated with fungicide alone (F) had a mass of 630 mg, which is relatively lower than all the NP-treated plants. Hence, all treatments, except N1P1 (plants treated with blank NPs, and made sick with *A. alternata*), had a positive effect on the plant biomass or weight per plant compared to the fungicide alone at a concentration of 10 ppm of foliar spray of NPs with 1.0 mg/mL of mancozeb, i.e., CSGA-1.0. In an earlier study, high concentrations of ZnO and silver NPs harshly affected the growth of tomato and wheat plants, respectively [46,47]. Earlier research also found that root and shoot growths in tomatoes were more affected by silver usage than seed germination [48].

3.10.3. Root–Shoot Ratio of Plants

The most significant shoot lengths of 19.5 and 19 cm were observed in plants treated with mancozeb-loaded nanocomposites infected with *A. alternata* and *S. lycopersici*, respectively. There may be some interactions between mancozeb-loaded NPs and pathogens, which caused an increase in shoot length. In a study, CeO_2 NPs were found in soybean roots and root nodules after growth in soil treated with NPs [49]. A similar transfer in corn plants was found, and CeO_2 NPs around vascular vessels supported that the particles found their way to the transportation system and moved through the xylem compelled by transpiration [50]. The most miniature shoots (11 cm) were seen in control plants, commercial fungicide-treated plants, followed by plants treated with blank and loaded nanocomposites (11.5 cm). The most significant root length (15.3 cm) was found in plants treated with commercial mancozeb and infected with *A. alternata* while the smallest root length of 8.0 cm was observed in control plants. So it can be concluded that nanocomposites positively affected the shoot and root lengths of plants compared to non-treated plants.

4. Conclusions

Conventional mancozeb has metals, such as zinc and manganese, which interfere with the living entity and cause significant environmental, soil, and health problems because of leaching via soil and water. Considering the hazards of commercial mancozeb, the biopolymeric nanocomposites in the current study showed a good EE and LC by sandwiching mancozeb in oppositely-charged biopolymers. Dialysis tubing showed slow and sustained release of mancozeb from encapsulated nanocomposite, which may be the reason for higher or comparable DCE percentages against *A. alternata* and *S. lycopersici* compared to commercial mancozeb, and can be used for environmental protection. Thus, these nanoformulations may be explored for the site-directed delivery of mancozeb for disease control in plants with minimalistic fungicide loaded to minimize environmental, soil, and human health problems.

Author Contributions: Conceptualization, J.S.D. and R.K.; methodology, R.K.; software, R.K.; validation, J.S.D. and R.K.; formal analysis, R.K.; investigation, R.K.; data curation, R.K.; writing—original draft preparation, R.K.; writing—review and editing, V.N.; visualization, V.N.; supervision, J.S.D.; project administration, J.S.D. All authors have read and agreed to the published version of the manuscript.

Funding: This research received no external funding.

Institutional Review Board Statement: Not applicable.

Informed Consent Statement: Not applicable.

Data Availability Statement: Data will be made available upon request.

Acknowledgments: The authors thank the Director, SAIF, AIIMS, New Delhi; STIC, Kochi, Department of Pharmaceutical Sciences, GJUS&T, Hisar, LPU, Jalandhar for the characterization of the nanoformulations. The authors appreciate Manju Barnela and Munish Ahuja (GJUS&T, Hisar) for her help with the optimization and release behaviour studies. The authors are thankful to the Chairperson, the Department of Biotechnology, Chaudhary Devi Lal University, Sirsa, for providing the lab and other infrastructure facility to carry out this research.

Conflicts of Interest: The authors declare no conflict of interest.

References

1. Carvalho, F.P. Pesticides, environment and food safety. *Food Energy Secur.* **2017**, *6*, 48–60. [CrossRef]
2. Stephenson, G.R. Pesticide use and world food production: Risks and benefits. In *Environmental Fate and Effects of Pesticides*; Coats, J.R., Yamamoto, H., Eds.; American Chemical Society: Washington, DC, USA, 2003; pp. 261–270.
3. Tilman, D.; Cassman, K.G.; Matson, P.A.; Naylor, R.; Polasky, S. Agricultural sustainability and intensive production practices. *Nature* **2002**, *418*, 671–677. [CrossRef] [PubMed]
4. Tudi, M.; Daniel, R.H.; Wang, L.; Lyu, J.; Sadler, R.; Connell, D.; Chu, C.; Phung, D.T. Agriculture development, pesticide application and its impact on the environment. *Int. J. Environ. Res. Public Health* **2021**, *18*, 1112. [CrossRef] [PubMed]
5. Sharma, D.; Choudhury, P.P. *Pesticide Use and Residue Management in Vegetables*; Technical Bulletin No 74; ICAR-IIHR: Bangalore, India, 2018; 40p.
6. Atuhaire, A.; Kaye, E.; Mutambuze, I.L.; Matthews, G.; Friedrich, T.; Jørs, E. Assessment of dithiocarbamate residues on tomatoes conventionally grown in Uganda and the effect of simple washing to reduce exposure risk to consumers. *Environ. Health Insights* **2017**, *11*, 1–8. [CrossRef]
7. Raza, Z.A.; Khalil, S.; Ayub, A.; Banat, I.M. Recent developments in chitosan encapsulation of various active ingredients for multifunctional applications. *Carbohydr. Res.* **2020**, *492*, 108004. [CrossRef]
8. Acemi, A.; Bayrak, B.; Çakır, M.; Demiryürek, E.; Gün, E.; El Gueddari, N.E.; Özen, F. Comparative analysis of the effects of chitosan and common plant growth regulators on in vitro propagation of *Ipomoea purpurea* (L.) Roth from nodal explants. *Vitro Cell Dev. Biol.-Plants* **2018**, *54*, 537–544. [CrossRef]
9. Francesconi, S.; Steiner, B.; Buerstmayr, H.; Lemmens, M.; Sulyok, M.; Balestra, G.M. Chitosan hydrochloride decreases *Fusarium graminearum* growth and virulence and boosts growth, development and systemic acquired resistance in two durum wheat genotypes. *Molecules* **2020**, *25*, 4752. [CrossRef]
10. Karthikeyan, C.; Tharmalingam, N.; Varaprasad, K.; Mylonakis, E.; Yallapu, M.M. Biocidal and biocompatible hybrid nanomaterials from biomolecule chitosan, alginate and ZnO. *Carbohydr. Polym.* **2021**, *274*, 118646. [CrossRef]

11. Veiga, M.D.; Ruiz-Caro, R.; Martín-Illana, A.; Notario-Pérez, F.; Cazorla-Luna, R. Polymer gels in vaginal drug delivery systems. In *Polymer Gels*; Springer: Singapore, 2018; pp. 197–246.

12. Bansal, P.; Kaur, P.; Duhan, J.S. Biogenesis of silver nanoparticles using *Fusarium pallidoroseum* and its potential against human pathogens. *Ann. Biol.* **2017**, *33*, 180–185.

13. Bansal, P.; Kaur, P.; Kumar, A.; Duhan, J.S. Microwave assisted quick synthesis method of silver nanoparticles using citrus hybrid "Kinnow" and its potential against early blight of tomato. *Res. Crop.* **2017**, *18*, 650–655. [CrossRef]

14. Mohamed, Y.M.A.; Elshahawy, I.E. Antifungal activity of photo-biosynthesized silver nanoparticles (AgNPs) from organic constituents in orange peel extract against phytopathogenic *Macrophomina phaseolina*. *Eur. J. Plant Pathol.* **2022**, *162*, 725–738. [CrossRef]

15. Kaur, P.; Thakur, R.; Duhan, J.S.; Chaudhury, A. Management of wilt disease of chickpea in vivo by silver nanoparticles biosynthesized by rhizospheric microflora of chickpea (*Cicer arietinum*). *J. Chem. Technol. Biotechnol.* **2018**, *93*, 3233–3243. [CrossRef]

16. Gupta, M.; Seema, K. Living Nano-factories: An eco-friendly approach towards medicine and environment. In *Bio-Manufactured Nanomaterials*; Springer: Cham, Switzerland, 2021; pp. 95–124.

17. Paul, A.; Roychoudhury, A. Go green to protect plants: Repurposing the antimicrobial activity of biosynthesized silver nanoparticles to combat phytopathogens. *Nanotechnol. Environ. Eng.* **2021**, *6*, 10. [CrossRef]

18. Al-Khattaf, F.S. Gold and silver nanoparticles: Green synthesis, microbes, mechanism, factors, plant disease management and environmental risks. *Saudi J. Biol. Sci.* **2021**, *28*, 3624–3631. [CrossRef] [PubMed]

19. Prabha, A.S.; Thangakani, J.A.; Devi, N.R.; Dorothy, R.; Nguyen, T.A.; Kumaran, S.S.; Rajendran, S. Nanotechnology and sustainable agriculture. In *Nanosensors for Smart Agriculture*; Elsevier: Amsterdam, The Netherlands, 2022; pp. 25–39.

20. Duhan, J.S.; Kumar, R.; Kumar, N.; Kaur, P.; Nehra, K.; Duhan, S. Nanotechnology: The new perspective in precision agriculture. *Biotechnol. Rep.* **2017**, *15*, 11–23. [CrossRef] [PubMed]

21. Silva, A.O.; Cunha, R.; Hotza, D.; Machado, R. Chitosan as a matrix of nanocomposites: A review on nanostructures, processes, properties, and applications. *Carbohydr. Polym.* **2021**, *272*, 118472. [CrossRef] [PubMed]

22. Agarwal, H.; Menon, S.; Venkat Kumar, S.; Rajeshkumar, S. Mechanistic study on antibacterial action of zinc oxide nanoparticles synthesized using green route. *Chem. Biol. Interact.* **2018**, *286*, 60–70. [CrossRef]

23. Kumar, S.; Kumar, D.; Dilbaghi, N. Preparation, characterization, and bio-efficacy evaluation of controlled release carbendazim-loaded polymeric nanoparticles. *Environ. Sci. Pollut. Res.* **2017**, *24*, 926–937.

24. Kaur, P.; Thakur, R.; Barnela, M.; Chopra, M.; Manuja, A.; Chaudhury, A. Synthesis, characterization, and in vitro evaluation of cytotoxicity, and antimicrobial activity of chitosan–metal nanocomposites. *J. Chem. Technol. Biotechnol.* **2015**, *90*, 867–873. [CrossRef]

25. Kaur, P.; Duhan, J.S.; Thakur, R. Comparative pot studies of chitosan, and chitosan-metal nanocomposites as nano-agrochemicals against fusarium wilt of chickpea (*Cicer arietinum* L.). *Biocatal. Agric. Biotechnol.* **2018**, *14*, 466–471. [CrossRef]

26. Maluin, F.N.; Hussein, M.Z.; Yusof, N.A.; Fakurazi, S.; Idris, A.S.; Hilmi, Z.; Jeffery Daim, L.D. Preparation of chitosan–hexaconazole nanoparticles as fungicide nanodelivery system for combating *Ganoderma* disease in oil palm. *Molecules* **2019**, *24*, 2498. [CrossRef] [PubMed]

27. Kumar, S.; Chauhan, N.; Gopal, M.; Kumar, R.; Dilbaghi, N. Development, and evaluation of alginate-chitosan nanocapsules for controlled release of acetamiprid. *Int. J. Biol. Macromol.* **2015**, *81*, 631–637. [CrossRef] [PubMed]

28. Zucker, R.M.; Ortenzio, J.; Degn, L.L.; Lerner, J.M.; Boyes, W.K. Biophysical comparison of four silver nanoparticles coatings using microscopy, hyperspectral imaging and flow cytometry. *PLoS ONE* **2019**, *14*, e0219078. [CrossRef] [PubMed]

29. Kumar, R.; Najda, A.; Duhan, J.S.; Kumar, B.; Chawla, P.; Klepacka, J.; Malawski, S.; Sadh, P.K.; Poonia, A.K. Assessment of antifungal efficacy and release behavior of fungicide-loaded chitosan-carrageenan nanoparticles against phytopathogenic fungi. *Polymers* **2022**, *14*, 41. [CrossRef]

30. Piran, F.; Khoshkhoo, Z.; Hosseini, S.E.; Azizi, M.H. Controlling the antioxidant activity of green tea extract through encapsulation in chitosan-citrate nanogel. *J. Food Qual.* **2020**, *2020*, 7935420. [CrossRef]

31. Burhan, A.M.; Abdel-Hamid, S.M.; Soliman, M.E.; Sammour, O.A. Optimisation of the microencapsulation of lavender oil by spray drying. *J. Microencapsul.* **2019**, *36*, 250–266. [CrossRef]

32. Kumar, R.; Duhan, J.S.; Manuja, A.; Kaur, P.; Kumar, B.; Sadh, P.K. Toxicity assessment and control of early blight and stem rot of *Solanum tuberosum* L. by mancozeb-loaded chitosan–gum acacia nanocomposites. *J. Xenobiot.* **2022**, *12*, 74–90. [CrossRef]

33. Ali, H.; Kilic, G.; Vincent, K.; Motamedi, M.; Rytting, E. Nanomedicine for uterine leiomyoma therapy. *Ther. Deliv.* **2013**, *4*, 161–175. [CrossRef]

34. Noor, F.; Zhang, H.; Korakianitis, T.; Wen, D. Oxidation and ignition of aluminum nanomaterials. *J. Chem. Phys.* **2013**, *15*, 20176–20188. [CrossRef]

35. Hamadou, A.H.; Huang, W.C.; Xue, C.; Mao, X. Formulation of vitamin C encapsulation in marine phospholipids nanoliposomes: Characterization and stability evaluation during long term storage. *LWT* **2020**, *127*, 109439. [CrossRef]

36. Kaur, I.; Agnihotri, S.; Goyal, D. Fabrication of chitosan–alginate nanospheres for controlled release of cartap hydrochloride. *Nanotechnology* **2021**, *33*, 025701. [CrossRef] [PubMed]

37. Wang, Q.; Shen, M.; Li, W.; Li, W.; Zhang, F. Controlled-release of fluazinam from biodegradable PLGA-based microspheres. *J. Environ. Res. Public Health Part B* **2019**, *54*, 810–816. [CrossRef] [PubMed]

38. Ashitha, A.; Mathew, J. Characteristics and types of slow/controlled release of pesticides. In *Controlled Release of Pesticides for Sustainable Agriculture*; Springer: Cham, Switzerland, 2020; pp. 141–153.
39. Yusoff, S.N.M.; Kamari, A.; Ishak, S.; Halim, A.L.A. N-hexanoyl-O-glycol chitosan as a carrier agent for water-insoluble herbicide. *J. Phys. Conf. Ser.* **2018**, *1097*, 012053. [CrossRef]
40. Kumar, S.; Bhanjana, G.; Sharma, A.; Dilbaghi, N.; Sidhu, M.C.; Kim, K.H. Development of nanoformulation approaches for the control of weeds. *Sci. Total Environ.* **2017**, *586*, 1272–1278. [CrossRef] [PubMed]
41. Edge, M.; Seal, K.; Allen, N.S.; Turner, D.; Robinson, J. Enhanced Performance of Biocidal Additives in Paints and Coatings. In *Industrial Biocides: Selection and Application*; Karsa, D.R., Ashworth, D., Eds.; RSC: Cambridge, UK, 2002; pp. 84–94.
42. Thirugnanasambandan, T. Advances of Engineered Nanofertilizers for Modern Agriculture. In *Plant-Microbes-Engineered Nano-Particles (PM-ENPs) Nexus in Agro-Ecosystems*; Springer: Cham, Switzerland, 2021; pp. 131–152. [CrossRef]
43. De Oliveira, J.L.; Campos, E.V.; Camara, M.C.; Della Vechia, J.F.; de Matos, S.T.; de Andrade, D.J.; Goncalves, K.C.; Nascimento, J.D.; Polanczyk, R.A.; de Araujo, D.R.; et al. Hydrogels containing botanical repellents encapsulated in zein nanoparticles for crop protection. *ACS Appl. Nano Mater.* **2019**, *3*, 207–217. [CrossRef]
44. Tong, Y.; Wu, Y.; Zhao, C.; Xu, Y.; Lu, J.; Xiang, S.; Zong, F.; Wu, X. Polymeric nanoparticles as a metolachlor carrier: Water-based formulation for hydrophobic pesticides and absorption by plants. *J. Agric. Food Chem.* **2017**, *65*, 7371–7378. [CrossRef]
45. Maluin, F.N.; Hussein, M.Z.; Azah Yusof, N.; Fakurazi, S.; Idris, A.S.; Zainol Hilmi, N.H.; Jeffery Daim, L.D. Chitosan-based agronanofungicides as a sustainable alternative in the basal stem rot disease management. *J. Agric. Food Chem.* **2020**, *68*, 4305–4314. [CrossRef]
46. Raliya, R.; Nair, R.; Chavalmane, S.; Wang, W.N.; Biswas, P. Mechanistic evaluation of translocation and physiological impact of titanium dioxide and zinc oxide nanoparticles on the tomato (*Solanum lycopersicum* L.) plant. *Metallomics* **2015**, *7*, 1584–1594. [CrossRef]
47. Yasmeen, F.; Razzaq, A.; Iqbal, M.N.; Jhanzab, H.M. Effect of silver, copper and iron nanoparticles on wheat germination. *Int. J. Biosci.* **2015**, *6*, 112–117.
48. Mehrian, S.K.; Heidari, R.; Rahmani, F.; Najafi, S. Effect of chemical synthesis silver nanoparticles on germination indices and seedlings growth in seven varieties of *Lycopersicon esculentum* Mill (tomato) plants. *J. Clust. Sci.* **2016**, *27*, 327–340. [CrossRef]
49. Priester, J.H.; Ge, Y.; Mielke, R.E.; Horst, A.M.; Moritz, S.C.; Espinosa, K.; Holden, P.A. Soybean susceptibility to manufactured nanomaterials with evidence for food quality and soil fertility interruption. *Proc. Natl. Acad. Sci. USA* **2012**, *109*, E2451–E2456. [CrossRef] [PubMed]
50. Zhao, L.; Sun, Y.; Hernandez-Viezcas, J.A.; Hong, J.; Majumdar, S.; Niu, G.; Gardea-Torresdey, J.L. Monitoring the environmental effects of CeO$_2$ and ZnO nanoparticles through the life cycle of corn (*Zea mays*) plants and in situ μ-XRF mapping of nutrients in kernels. *Environ. Sci. Technol.* **2015**, *49*, 2921–2928. [CrossRef] [PubMed]

MDPI

St. Alban-Anlage 66

4052 Basel

Switzerland

Tel. +41 61 683 77 34

Fax +41 61 302 89 18

www.mdpi.com

MDPI Books Editorial Office

E-mail: books@mdpi.com

www.mdpi.com/books